Highly Efficient Thermal Renewable Energy Systems

The text comprehensively highlights the latest methodologies, models, techniques, and applications along with a description of modeling, optimization, and experimental works in the energy sector. It further explains key concepts such as finite element analysis tools, hybrid energy systems, mechanical components design, and optimization, solar coupled systems, and vertical heat exchanger.

This book

- Discusses the role and integration of solar, geothermal, and hydrogen-based thermal energy storage (TES) technologies in different sectors for space heating and cooling applications.
- Covers mechanical modeling and optimization of hybrid energy storage systems for performance improvement and focuses on hydrogen production, storage, and safety measures.
- Explores the integration of IoT and global energy interaction technologies, highlighting their potential benefits in driving the transition toward a sustainable and resilient global energy system.
- Explains different aspects of clean technologies such as batteries, fuel cells, ground energy storage, solar thermal system, and the role of green hydrogen in decarbonizing sectors like transportation and energy.
- Showcases a clear idea of sustainable development using renewable energy, focusing on policymaking, challenges in transition from conventional to renewable energy, and future directions in energy sector.

It is primarily written for senior undergraduates and graduate students, and academic researchers in the fields of mechanical engineering, production engineering, industrial engineering, and environmental engineering.

Advances in Manufacturing, Design and Computational Intelligence Techniques

Series Editor-Ashwani Kumar- Senior Lecturer, Mechanical Engineering, at Technical Education Department, Uttar Pradesh, Kanpur, India

The book series editor is inviting edited, reference and textbook proposal submission in the book series. The main objective of this book series is to provide researchers a platform to present state-of-the-art innovations, research related to advanced material applications, cutting-edge manufacturing techniques, innovative design, and computational intelligence methods used for solving nonlinear problems of engineering. This book series includes a comprehensive range of topics and their application in engineering areas such as additive manufacturing, nanomanufacturing, micromachining, biodegradable composites, material synthesis and processing, energy materials, polymers and soft matter, nonlinear dynamics, dynamics of complex systems, MEMS, green and sustainable technologies, vibration control, AI in power station, analog-digital hybrid modulation, advancement in inverter technology, adaptive piezoelectric energy harvesting circuit, contactless energy transfer system, energy efficient motors, bioinformatics, computer-aided inspection planning, hybrid electrical vehicle, autonomous vehicle, object identification, machine intelligence, deep learning, control-robotics-automation, knowledge-based simulation, biomedical imaging, image processing, and visualization. This book series compiled all aspects of manufacturing, design, and computational intelligence techniques from fundamental principles to current advanced concepts.

Hybrid Metal Additive Manufacturing: Technology and Applications
Edited by Parnika Shrivastava, Anil Dhanola, and Kishor Kumar Gajrani

Thermal Energy Systems: Design, Computational Techniques, and Applications
Edited by Ashwani Kumar, Varun Pratap Singh,
Chandan Swaroop Meena, and Nitesh Dutt

Highly Efficient Thermal Renewable Energy Systems: Design, Optimization and Applications
Edited by Vikas Verma, Sivasakthivel Thangavel, Nitesh Dutt,
Ashwani Kumar, and Rohitha Weerasinghe

https://www.routledge.com/Advances-in-Manufacturing-Design-and-Computational-Intelligence-Techniques/book-series/CRCAIMDCIT?publishedFilter=alltitles&pd=published,forthcoming&pg=1&pp=12&so=pub&view=list?publishedFilter=alltitles&pd=published,forthcoming&pg=1&pp=12&so=pub&view=list

Highly Efficient Thermal Renewable Energy Systems

Design, Optimization and Applications

Edited by
Vikas Verma, Sivasakthivel
Thangavel, Nitesh Dutt, Ashwani Kumar
and Rohitha Weerasinghe

CRC Press
Taylor & Francis Group
Boca Raton London New York

CRC Press is an imprint of the
Taylor & Francis Group, an **informa** business

Designed cover image: Shutterstock

First edition published 2024
by CRC Press
2385 NW Executive Center Drive, Suite 320, Boca Raton FL 33431

and by CRC Press
4 Park Square, Milton Park, Abingdon, Oxon, OX14 4RN

CRC Press is an imprint of Taylor & Francis Group, LLC

ISBN: 9781032595641 (hbk)
ISBN: 9781032753379 (pbk)
ISBN: 9781003472629 (ebk)

DOI: 10.1201/9781003472629

Typeset in Sabon
by codeMantra

Contents

12 Comparative assessment of hydrogen production methods from renewable energy: a review 194

ABDUL QUAIYOOM, NIKHIL DEV, BHUPENDER SINGH,
AND ASHOK KUMAR YADAV

16 Green hydrogen productions: methods, designs and smart
    applications   261

MIRA CHITT, SIVASAKTHIVEL THANGAVEL, VIKAS VERMA,
AND ASHWANI KUMAR

20 Atmospheric water generation with different grades of molecular sieve   330

MANOJ KUMAR, AVADHESH YADAV, RAVI KANT RAVI,
VED PRAKASH SHARMA, AND VENKATESWARA R. KODE

Aim and Scope

Energy is the key input for the development process of any country. The utilization of renewable energies is not at all new; in the history of mankind, renewable energies have been used as the primary possibility of generating energy. This only changed with the industrial revolution from hard coal to crude oil to natural gases. All these energy sources' availabilities are limited and not uniformly distributed throughout the world. The utilization of fossil energy carriers involves a series of undesirable side effects, which are less and less tolerated by industrialized societies increasingly sensitized to possible environmental and climate effects at the beginning of the 21st century. It is also important to note that climate change is primarily due to the extensive use of fossil fuel-based energy sources. Energy accounts for two-thirds of total greenhouse gas emission. This is why the search for environmental, climate-friendly, and socially acceptable alternatives suitable to cover the energy demand has become increasingly important. Energy access is crucial for poverty alleviation, economic growth, and improved living standards of people. To fill the above gap, the proposed book will work like a bridge to cover policy making and application for three key renewable energy resources.

The book *Highly Efficient Thermal Renewable Energy Systems: Design, Optimization and Applications* focuses on three key renewable energies, i.e., solar, geothermal, and hydrogen energy and its storage techniques, design and development of devices, and its optimization. In the book design, development, storage, processing, and applications are addressed by the authors in 20 chapters. These chapters cover everything from the potential of energy generation systems to IoT and global energy interaction technologies. In these chapters, the authors have presented the balanced discussion of literature review, solution, methodology, experimental setup, results validation, and future scope with targeted field as materials, manufacturing, mechanical, energy sector, and energy-efficient buildings with hybrid energy applications. Also, this book includes the role and integration of solar, geothermal, and hydrogen energy storage-based hybrid technologies in different sectors for space heating and cooling applications. In addition, mechanical modeling and optimization of hybrid energy storage systems for improving performance are also covered.

The book *Highly Efficient Thermal Renewable Energy Systems: Design, Optimization and Applications* is practical-oriented along with high-lighting important renewable thermal energy storage concepts often not included in other academic textbooks. This book is essentially intended for use by undergraduate, postgraduate, Ph.D. students, and researchers in multidisciplinary fields. Also focusing on providing scientific knowhow and orientation for the benefit of readers and society in the emerging technologies areas for the utilization of solar energy, geothermal energy, and hydrogen energy resources for clean energy storage to meet the energy demand. Detailed analysis of renewable energy storage and technologies is undertaken throughout this book, providing new methodologies, models, techniques, and applications along with description of modeling, optimization, and experimental works. The editors strongly believe that this book will be very helpful to students, engineers, energy experts, and young scientist professionals to upgrade their knowledge and research work for those who wish to keen about to learn more about solar, geothermal, and hydrogen energy resources for clean energy production.

The main objective of this book is to enable researchers and students with intensive educational material on the advancements of solar, geothermal, and hydrogen energy-based highly efficient thermal energy storage devices and their applications for sustainable development. The high-quality research content will work as a reference book for postgraduate students, Ph.D. researchers, materials scientists, mechanical engineers, budding researchers, energy scientists, manufacturing engineers, design engineers, mechatronic students, and energy study professionals. The applicability of this book covers wide range of industries such as production sector, design engineer, research & development engineer, sustainable building, civil engineers, researchers in energy sectors, and policy makers for renewable energy etc. and it will help to audience conducting research in these industries with renewable energy for sustainability.

Editors

Dr. Vikas Verma

Dr. Sivasakthivel Thangavel

Dr. Nitesh Dutt

Dr. Ashwani Kumar

Dr. Rohitha Weerasinghe

Preface

The book *Highly Efficient Thermal Renewable Energy Systems: Design, Optimization and Applications* has 20 chapters covering from potential of energy generation systems to IoT and global energy interaction technologies. Also, this book includes role and integration of solar, geothermal and hydrogen energy technologies for different industrial applications. In addition, the book covers modeling and optimization of hybrid renewable energy systems for improving the energy performance. Further, it also described the theoretical analysis, Exergy analysis, implementation of renewable energy-based hybrid system, energy storage systems (ESS), and their applications.

Chapter 1 highlights the solar, geothermal, and hydrogen energy systems. It provides a brief introduction, recent advancements, and future directions in energy sector. **Chapter 2** deals with the technological advancements in sustainable and renewable solar energy systems. Solar energy's potential lies in its ubiquity and environmental friendliness. The Earth receives varying amounts of solar radiation due to atmospheric conditions and geographical factors, which can be harnessed to meet energy demands. Solar panels can be installed on sea surfaces, beach rooftops, and commercial centers, offering multifaceted benefits including reduced humidity and support for water purification. **Chapter 3** provides an insightful exploration into the realm of thermal energy storage (TES). It delves into the diverse array of TES techniques, shedding light on the materials used along with their unique applications that contribute to enhanced energy efficiency. Through the adoption of TES, our energy future can be more sustainable and resilient.

In continuation, **Chapter 4** examines solar energy adoption and implementation in underdeveloped nations and presents both potential and obstacles. Solar energy is a sustainable and ecologically benign form of energy that may help these regions with energy access, economic growth, and climate change mitigation. This chapter examines the multiple prospects offered by solar energy, such as increased energy availability, cost savings, job development, energy independence, and environmental advantages. **Chapter 5** explores the integration of IoT and global energy interaction technologies, highlighting their potential benefits in driving the transition towards a sustainable and resilient global energy system. Chapter 6 covers various aspects of designing and modeling renewable energy systems including solar and geothermal and hybrid systems. The major topics discussed include the use of FEA tools, optimization of thermal energy storage systems, solar and radiant systems, and geothermal and hybrid energy systems design. This chapter explores the application of different techniques of the design and modeling of solar, geothermal, and hybrid energy systems. Chapter 7 provides a concise overview of optimization techniques applied to solar thermal and hybrid energy systems. It encompasses a range of techniques, including mathematical modeling, simulation, genetic algorithms, and machine learning, all tailored to enhance the performance and efficiency of these systems. Chapter 8 attempts to improve the performance prediction accuracy delivered by the fuzzy logic expert system. For this, a novel FLES (N-FLES) integrated with the concepts of the Apriori algorithm is proposed. Chapter 9 aims to comprehensively explore the convergence of solar and geothermal energy with Thermal Energy Storage (TES) for enhancing the energy efficiency and resilience of building applications. In geothermal energy. Chapter 10 presents an overview of geothermal heat pump technology, discusses its applicability in Oman's unique geological and climatic conditions, and presents a comprehensive analysis of the benefits, challenges, and feasibility of adopting low-level geothermal heat pumps in the country.

Chapters 11–14 deal with green hydrogen energy. Chapter 11 underscores the role of green hydrogen in decarbonizing sectors like transportation and energy. Aiming to inform researchers, engineers, and policymakers, this chapter contributes to advancing the understanding of green hydrogen's pivotal role in the transition to a sustainable energy future. Chapter 12 conducts a meticulous examination, comparison, and assessment of the multifaceted impacts on the environment, society, and economics that stem from various technologies utilized in hydrogen generation. The study undertakes a systematic comparison of the environmental ramifications intrinsic to these eight distinct technological approaches. Chapter 13 delves into hydrogen manufacturing methods, explores green hydrogen production, and introduces cutting-edge techniques in the field. The versatility of hydrogen extends to its use in fuel cells and as an energy carrier in storage devices, offering a host of benefits but also presenting challenges. Chapter 14 analyzes the necessity of the solar and hydrogen industries as well as their essential ideas, the global energy situation, highlights of research projects undertaken to enhance the sector, prospective applications, and obstacles to the development of better sectors in the future.

Chapter 15 highlights recent developments in solar energy research and development that have made solar energy systems more inexpensive for commercial use. There are direct and indirect methods for using solar energy to produce electricity. In the direct method, utilizing PV modules, solar radiation is transformed into energy. In the indirect method, concentrated solar power (CSP) systems with linear Fresnel collectors and parabolic trough collectors are used to capture thermal energy.

In **Chapter 16**, an overview of hydrogen production through electrolysis is discussed, highlighting various methods and designs, as well as exploring the wide-ranging applications of green hydrogen. The electrolysis process involves splitting water molecules into hydrogen and oxygen using electrical energy. The primary objective of **Chapter 17** is to evaluate the heat transfer and pressure loss characteristics of counter flow SAH (CFSAH) by adopting novel discrete multi-V-shaped and staggered rib geometry. The experimentations are performed for various flows with Reynolds number (Re) ranging from 3000 to 21,000. **Chapter 18** presents important aspects of a practical methodology applied to design a gear pump using the ANSYS Gambit software parametric abilities followed by a FVM analysis of its assembly to identify the most stressed components. A model for the prediction of the dynamic behavior of a solar-operated gear pump is presented.

Chapter 19 expounds on various solar desalination techniques, ranging from solar distillation to advanced methods such as solar multistage flash distillation (MSF), solar reverse osmosis (RO), and solar nano-filtration. The intricacies of these techniques are examined, highlighting their mechanisms and the vital role of solar energy in powering them. Moreover, this chapter illuminates the importance of energy efficiency and technological advances, emphasizing the need to optimize energy conversion and utilize innovative collector and membrane technologies for improved performance. **Chapter 20** presents an experimental investigation that has been performed to get the amount of water generated from a thin layer of atmospheric air by using molecular sieves (desiccant material). Three grades of molecular sieves 4A, 5A, and 13 X have been used. The experiments were conducted in Northern Indian climatic conditions at 29° 58′ (latitude) North and 76° 53′ (longitude) East.

The above summary of each chapter shows the book's multidisciplinary nature, which makes it ideal for usage in interdisciplinary undergraduate and postgraduate curricula, and makes it useful for teaching and research purposes. In addition, the organization of the book facilitates self-study by independent pupils. It is also beneficial for engineers, economists, and decision-makers since it helps them comprehend the background, current, and prospective future of the subject under discussion.

Editors

Dr. Vikas Verma

Dr. Sivasakthivel Thangavel

Dr. Nitesh Dutt

Dr. Ashwani Kumar

Dr. Rohitha Weerasinghe

Acknowledgment

We express our gratitude to **CRC Press (Taylor & Francis Group)** and the editorial team for their suggestions and support during the completion of this book. We are grateful to all authors for submitting their quality work and reviewers for their illuminating views on each book chapter presented in the book *Highly Efficient Thermal Renewable Energy Systems: Design, Optimization and Applications.*

This book is dedicated to all engineers, researchers and academicians…

Editors

Dr. Vikas Verma

Dr. Sivasakthivel Thangavel

Dr. Nitesh Dutt

Dr. Ashwani Kumar

Dr. Rohitha Weerasinghe

Contributing Authors

Mohammad Ahmadizadeh
Persian Gulf University
Bushehr, Iran

Kiumars Khani Aminjan
Faculty of Aerospace
Malek Ashtar University of
    Technology
Tehran, Iran

Rajesh Attri
Department of Mechanical
    Engineering
J.C. Bose University of Science and
    Technology, YMCA
Faridabad, India

Sunil Baloda
Department of Mechanical
    Engineering
The Technological Institute of
    Textile and Sciences
Bhiwani, India

Aarti Rana Chauhan
Department of Electrical
    Engineering
G H Raisoni Institute of
    Engineering & Business
    Management
Jalgaon, Maharasthra

Madhur Chauhan
Department of Electrical
    Engineering
G H Raisoni Institute of
    Engineering & Business
    Management
Jalgaon, Maharashtra

Mira Chitt
Department of Mechanical
    Engineering
Global College of Engineering and
    Technology (GCET)
Muscat, Sultanate of Oman

Nikhil Dev
Department of Mechanical
    Engineering
J.C. Bose University of Science and
    Technology, YMCA
Faridabad, India

Nitesh Dutt
Department of Mechanical
    Engineering
COER University
Roorkee, India

Priyanka Gajare
Department of Electrical
    Engineering
G H Raisoni Institute of Engineering
    & Business Management
Jalgaon, India

Apurva Goyal
School of Commerce & Management
IIMT University
Meerut, India

Arvind Gupta
Department of Mechanical
    Engineering
J.C. Bose University of Science and
    Technology, YMCA
Faridabad, India

Milad Heidari
Department of Mechanical
    Engineering
Global College of Engineering and
    Technology (GCET)
Muscat, Sultanate of Oman

Siddharth Jain
School of Advanced Engineering
University of Petroleum and Energy
    Studies
Dehradun, India

Nazrul Islam Khan
Department of Mechanical
    Engineering
Netaji Subhas University of
    Technology
New Delhi, India

Morteza Khashehchi
Department of Mechanical
    Engineering
Global College of Engineering and
    Technology (GCET)
Muscat, Sultanate of Oman

Venkateswara R. Kode
Chemical Engineering Department
Christian Brothers University
Memphis, Tennessee

Ashwani Kumar
Department of Mechanical
    Engineering
Technical Education Department
    Uttar Pradesh (under
    Government of Uttar Pradesh)
Kanpur, India

Manoj Kumar
Department of Mechanical
    Engineering
National Institute of Technology
    Srinagar
Srinagar, India

Mukesh Kumar
Department of Mechanical
    Engineering
IIEST Shibpur
Howrah, India

Puneet Kumar
School of Commerce &
    Management
IIMT University
Meerut, India

Rajesh Kumar
Department of Mechanical
    Engineering
Dayalbagh Educational Institute
Agra, India

Sridharan M
Department of Mechanical
    Engineering
SRM Institute of Science and
    Technology
Tiruchirappalli, India

Manish D Mahale
Department of Electrical
   Engineering
G H Raisoni College of Engineering
   and Business Management
Jalgaon, India

Sushma M Mahale
Department of Civil Engineering
NMKC College of Engineering
Jalgaon, India

Taleb Moazzeni
Department of Electrical and
   Electronics Engineering
Global College of Engineering and
   Technology (GCET)
Muscat, Sultanate of Oman

Eman Al Naamani
Global College of Engineering and
   Technology (GCET)
Muscat, Sultanate of Oman

Bhaskar Nagar
Department of Mechanical
   Engineering
J.C. Bose University of Science and
   Technology, YMCA
Faridabad, India

Mohit Nayal
School of Advanced Engineering
University of Petroleum and Energy
   Studies
Dehradun, India

Bipasa B Patra
Department of Electrical
   Engineering
Padmashri Dr. V. B. Kolte College
   of Engineering
Malkapur, India

Abdul Quaiyoom
Department of Mechanical
   Engineering
J.C. Bose University of
   Science and
   Technology, YMCA
Faridabad, India

Pooyan Rahmanivahid
Department of Mechanical
   Engineering
Global College of Engineering and
   Technology (GCET)
Muscat, Sultanate of Oman

Ravi Kant Ravi
Department of Mechanical
   Engineering
Dr. B.R. Ambedkar National
   Institute of Technology
Jalandhar, India

Senthil Kumar S
Department of Mechanical
   Engineering
SRM Institute of Science and
   Technology
Tiruchirappalli, India

Abhishek Kr. Sharma
School of Advanced
   Engineering
University of Petroleum and
   Energy Studies
Dehradun, India

Naveen Sharma
Department of Mechanical
   Engineering
Netaji Subhas University of
   Technology
New Delhi, India

Ved Prakash Sharma
Department of Mechanical
   Engineering
National Institute of Technology
   Srinagar
Srinagar, India

Bhupender Singh
Department of Mechanical
   Engineering
J.C. Bose University of Science and
   Technology, YMCA
Faridabad, India

Chandraveer Singh
Department of Mechanical
   Engineering
G.B. Pant Institute of Engineering
   and Technology
Pauri Garhwal, India

Ranjana Singh
School of Commerce &
   Management
IIMT University
Meerut, India

Varun Pratap Singh
School of Advanced Engineering
University of Petroleum and Energy
   Studies
Dehradun, India

Sivasakthivel Thangavel
Department of Mechanical
   Engineering
Global College of Engineering and
   Technology (GCET)
Muscat, Sultanate of Oman

Vikas Verma
Department of Energy
Tezpur University Assam (Central
   University, Government of India)
Tezpur, India

Rohitha Weerasinghe
Mechanical Engineering
University of Bedfordshire
Luton, United Kingdom

Ashok Kumar Yadav
Department of Mechanical
   Engineering
Raj Kumar Goel Institute of
   Technology
Ghaziabad, India

Avadhesh Yadav
National Institute of Solar Energy
Gurugram, India

About the Editors

Dr. Vikas Verma has a doctorate from the Indian Institute of Technology, Roorkee, India. He is an Assistant Professor in the Department of Energy at Tezpur University, Assam, India from since April 2017. He has contributed research works in reputed journals in the fields of renewable energy, hydrogen energy, nanofluid, and HVAC system. He has reviewed international publications and delivered invited talks and oral presentations in international platforms. He was nominated and represented India in Durban, South Africa during the Third BRICS Young Scientist Conclave in 2018. His research work on Solar Assisted Geothermal Heat Pump for Space Heating Application was covered by media such as *India Today*, *Times of India*, and *The Education Times* in 2018.

Dr. Sivasakthivel Thangavel has been an associate professor in the Department of Mechanical Engineering at the Global College of Engineering and Technology, Muscat, Sultanate of Oman since May 2018. He earned his Ph.D. in Mechanical Engineering from IIT Roorkee, India. Throughout his career, Dr. Siva worked as a JSPS Postdoctoral Fellow and researcher in China, Japan, and France, publishing more than 30 papers in refereed journals and conference proceedings. He is a certified higher education teacher and earned a postgraduate certificate in academic practice in 2020 from the University of the West of England. Moreover, Dr. Siva has over eight years of experience teaching undergraduate and postgraduate students at various higher educational institutes in India, China, Japan, and Oman. He has taught diverse modules across the field of mechanical engineering. His Orcid id is 0000-0001-6802-4015 and Scopus Id is 55962921000.

Dr. Nitesh Dutt earned his Ph.D. in Mechanical Engineering from IIT Roorkee in 2018, where his research, supported by a Board of Research in Nuclear Sciences (BRNS) project, delved into "Heat up and cooling studies of debris of IPHWR." Remarkably, he gained experience in establishing a dedicated Nuclear lab under the guidance of his supervisor and BARC scientists within the Mechanical and Industrial Engineering Department at IIT Roorkee. With an M. Tech in Renewable Energy from the same institution, Dr. Nitesh's expertise spans diverse domains. As a GATE-qualified professional, he has presented his work internationally and contributed research articles to SCI journals, peer-reviewed platforms, book chapters, and conferences. He has been a contributing author in three edited books published by Taylor and Francis Group (CRC Press), holds four patents, and serves as the Editor-in-Chief of the *International Journal of Energy Resources Applications* (IJERA). Specializing in nuclear reactor safety, Heat and Mass Transfer, Thermodynamics, Fluid Mechanics, Solar Energy Applications, and Computational Fluid Dynamics (CFD), Dr. Nitesh is an Associate Professor in the Department of Mechanical Engineering at COER University, Uttarakhand, India, since 2018, with a previous stint at Graphic Era University Dehradun, India from July 2011 to July 2012. Garnering over nine years of rich academic and research experience, he was honored with research awards in 2022 by the College of Engineering Roorkee (COER), Roorkee, India. Dr. Nitesh also led the TEQIP-III Sponsored project and is actively involved in research under the COER-IIT Roorkee MOU partnership. He has participated in numerous workshops, conferences, and has convened Short Term Courses (STC), further showcasing his commitment to advancing knowledge and academic excellence.

Dr. Ashwani Kumar earned a Ph.D. (Mechanical Engineering) in the area of Mechanical Vibration and Design. He is currently a Senior Lecturer, Mechanical Engineering (Gazetted Officer Group B) at Technical Education Department Uttar Pradesh (under the Government of Uttar Pradesh) Kanpur, India since December 2013. He was an assistant professor in the Department of Mechanical Engineering, Graphic Era University Dehradun India (NIRF Ranking 55) from July 2010 to November 2013. He has more than 13 years of research, academic and administrative experience including Coordinator for AICTE-Extension of Approval, Nodal officer for

PMKVY-TI Scheme (Government of India), internal coordinator-CDTP scheme (Government of Uttar Pradesh), Industry Academia relation officer, Assistant Centre Superintendent (ACS)-Institute Examination Cell, Zonal Officer to conduct Joint Entrance Examination (JEE-Diploma), Sector Magistrate for Lok Sabha- Vidhan Sabha Election, etc. Being an academician and researcher, he is Series Editor of four book series Advances in Manufacturing, Design and Computational Intelligence Techniques[1]/Renewable and Sustainable Energy Developments[2]/Smart Innovations and Technological Advancements in Mechanical and Materials Engineering[3]/Computational Intelligence and Biomedical Engineering[4] published by CRC Press (Taylor & Francis USA)[1-3] and Wiley Scrivener Publishing[4] USA. He is a guest editor of special issues titled as Sustainable Buildings, Resilient Cities and Infrastructure Systems (Buildings *ISSN: 2075-5309*, I.F. 3.8). He is Editor-in-Chief for *International Journal of Materials, Manufacturing and Sustainable Technologies* (IJMMST, *ISSN: 2583-6625*), and Editor of *International Journal of Energy Resources Applications* (IJERA, *ISSN: 2583-6617*). He is a guest editor and editorial board member of 8 international journals and acts as a review board member of 20 prestigious (Indexed in SCI/SCIE/Scopus) international journals with high impact factor i.e., Applied Acoustics, Measurement, JESTEC, AJSE, SV-JME, and LAJSS. In addition, he has published **100+ research articles** in journals, book chapters, and conferences. He has authored/co-authored cum edited **30+ books** of Mechanical, Materials and Energy Engineering. He has published **three patents**. He is associated with International Conferences as an Invited Speaker/Session Chair/Advisory Board/Review Board member/Program Committee Member. He has delivered many invited talks in webinar, FDP, and Workshops. He has been awarded as **Best Teacher** for excellence in academic and research. He has successfully guided 15 B. Tech., M.Tech, and Ph.D. thesis. He is an external doctoral committee member of SRM. University, New Delhi. He is currently involved in the research area of AI & ML in Mechanical Engineering, Smart Materials & Manufacturing Techniques, Thermal Energy Storage, Building Efficiency, Renewable Energy Harvesting, Sustainable Transportation, and Heavy Vehicle Dynamics. His Orcid is **0000-0003-4099-935X**, Google Scholar web link is **https://scholar.google.com/citations?hl=en&user=KOILpE kAAAAJ** and research gate web link is **https://www.researchgate.net/ profile/Ashwani-Kumar-45**.

Dr Rohitha Weerasinghe is a senior lecturer at the University of Bedfordshire England in Mechanical Engineering. He has worked at the University of the West of England and University of Moratuwa, Sri Lanka previously, as a senior lecturer in Mechanical Engineering. He was a senior member of the energy group of University of Moratuwa. He has worked extensively in the renewable energy and automotive engineering sectors in Sri Lanka and in 2003 won the National Transport Award for designing and constructing a solar-assisted electric car. He worked as the Deputy Dean (Academic Affairs) at Global College of Engineering, Oman in 2018 and 2019. Rohitha has also worked as an EPSRC Research fellow at the University of Sussex between 2005 and 2007. He published in numerous refereed publications and journals and has a patent in automotive thermal heat recovery. He has been a co-chair of international conferences such as ICAEER 2018, ICAEER 2019, ARFEE 2019, ICAEER 2020, and ICAEER 2021. He has also published and refereed in numerous conferences. He is the awardee of a number of grants both in Sri Lanka and in the UK including an INNOVATE UK grant. Dr Rohitha Weerasinghe also has significant industry experience as an Energy Specialist, especially in renewable energy and energy strategies. He has also been an international consultant in energy for the UNDP in the past and represented Sri Lanka as an expert in regional energy policy making. Rohitha received his Ph.D. from Imperial College London after reading in gas turbine combustion. He is also a Fellow of the Institution of Mechanical Engineers and a Senior Fellow of the Higher Education Academy.

Chapter 1

Introduction to thermal energy storage

Solar, geothermal and hydrogen energy

Vikas Verma

Tezpur University Assam (Central
University, Government of India)

Sivasakthivel Thangavel

Global College of Engineering and Technology, Muscat Oman

Nitesh Dutt

COER University, Roorkee

Ashwani Kumar

Technical Education Department Uttar Pradesh, Kanpur
(Under Government of Uttar Pradesh)

Rohitha Weerasinghe

University of Bedfordshire, Luton

1.1 INTRODUCTION

The abundance of natural resources bestowed upon our planet is indeed plentiful, and their optimal utilization is paramount in achieving sustainable development and environmental justice while safeguarding the well-being of both present and future generations. Renewable energy sources, constantly replenished, stem from diverse origins such as sunlight, wind, water, biomass, and geothermal heat. These renewables serve as a reliable source of both heat and electricity, playing a pivotal role in reducing greenhouse gas emissions, lessening our dependence on fossil fuels, promoting energy security, and stimulating economic growth. Notably, in 2020, we witnessed a 3% growth in the use of renewable energy, even as demand for other fuel sources declined [1].

DOI: 10.1201/9781003472629-1

Advancements in capture and storage technologies, coupled with the recent global drive toward adopting a Net Zero strategy, have led to a significant expansion in renewable and green energy production. These developments span from the installation of small-scale solar panels on homes to the establishment of large-scale facilities like offshore wind farms. They encompass the utilization of geothermal energy for power generation and heating as well as the generation of electricity using hydrogen as a fuel source. In addition, there has been notable progress in the utilization of biomass to generate biogas for heating, electricity, and transportation.

Solar energy offers a versatile array of applications, including cooling, heating, lighting, power generation, water desalination, and transportation. Solar energy stands as a clean, renewable, and abundant energy source. Solar thermal energy, a form of renewable energy, enables the harnessing of heat energy received on the Earth's surface to produce electricity, heating, or cooling. This process involves directing sunlight onto a receiver through solar collectors, such as mirrors or lenses, which then transfers the heat to a fluid that can be used directly or converted into steam to power turbines or refrigeration systems.

Geothermal energy, emanating from the Earth's core, proves to be a reliable and renewable source of energy. It finds applications in power generation, manufacturing processes, agriculture, and building heating and cooling. Enhanced geothermal systems (EGS), capable of tapping into deeper and hotter resources, present a crucial opportunity to expand the role of geothermal energy in the clean energy transition.

Hydrogen energy involves the effective utilization of hydrogen as an energy carrier, derived from various sources such as petroleum, biomass, and water electrolysis. Hydrogen serves a multitude of applications, ranging from powering electric vehicles and electricity generation to industrial processes and heating and cooling solutions. It also plays a vital role in energy storage, being capable of storing excess renewable electricity and releasing it when needed. Hydrogen, as a clean fuel, produces only water when combusted or used in fuel cells.

However, these energy sources do face challenges, including intermittent land and water use, technical regulations, social barriers, and implementation costs. Therefore, it is crucial to develop and implement strategies that support the deployment and integration of these energy sources into our energy systems and societies. Through these efforts, solar, geothermal, and hydrogen energy can significantly contribute to the global energy transition and the achievement of sustainable development goals.

1.2 POTENTIAL OF SOLAR THERMAL ENERGY

Solar energy stands as a dynamic and readily available source of renewable energy, playing a pivotal role in combating climate change while delivering clean and cost-effective energy for both electricity generation and thermal applications in various sectors. Solar thermal energy, in particular, holds significant promise as it harnesses the maximum daily solar radiation, offering optimal energy output. Solar technologies leverage the Sun's radiative energy in diverse ways, producing power, heat, cooling, and even fuel. It's essential to view solar thermal heating and cooling (SHC), concentrated solar power (CSP), and photovoltaic energy (PV) as complementary technologies, rather than isolating them individually [2].

In regions with suitable climatic conditions, solar energy holds great potential, especially in developing nations. Industrialized countries have an obligation to support these efforts by aiding in the development, testing, demonstration, and market introduction of solar technologies.

Solar thermal energy has the capability to fulfill the entire heating and cooling requirements of residential areas while significantly impacting the energy supply of commercial and industrial sectors. Presently, large buildings with central air conditioning systems predominantly utilize solar-assisted cooling. However, the growing demand for air-conditioned homes and small office buildings is creating new opportunities for this technology. Solar thermal technologies also offer substantial potential for residential hot water and space heating systems, effectively reducing the need for traditional heating fuel.

Moreover, the pressing need for freshwater has spurred the demand for solar thermal seawater desalination systems. The relatively low operating temperatures of desalination systems make them ideal candidates for solar thermal collectors, but further research and development (R&D) is essential to refine these systems.

Solar thermal plants find common use in industries such as food and beverage, chemicals, and textiles as well as in tasks such as car washes. Flat-plate collectors, noted for their efficiency at lower temperatures, are commonly employed for processes requiring lower heat. In addition, solar heat plays a role in heating production facilities, further enhancing its practical applications [3].

1.2.1 Global scenario of solar thermal energy

Starting from 1980, a multitude of collaborative initiatives have been initiated with the support of governments from OECD member states, along with influential organizations such as the World Bank and UNIDO [3].

In 2018, the international solar thermal market's capacity was measured at 496.15 GW, and it is forecast to expand at an annual rate of 5.6%, aiming to achieve 767.73 GW by 2026 [4]. The year 2021 alone witnessed the operation of 115 million solar thermal systems, with an 85% dominance of new installations, primarily in solar hot water systems catering to single- and multi-family residential properties, hotels, and public buildings [5]. In 2020, approximately 250 million households globally utilized solar thermal systems for water heating. To align with the NZE scenario (net zero emissions by 2050), which calls for equipping 400 million homes with solar thermal systems by 2030, around 290 million new systems must be installed within this decade. Projections anticipate the global demand for solar thermal energy to amount to $36 billion by 2026. Furthermore, the International Energy Agency (IEA) predicts that by 2050, solar thermal, alongside geothermal energy, will meet 75% of the world's total heat demand [6].

According to estimates from the International Renewable Energy Agency (IRENA), by 2050, solar thermal power could satisfy 16% of the world's energy demand and mitigate 6.1 gigatons of CO_2 emissions annually [7]. As of 2022, the global installed capacity for CSP had expanded by 200 MW, reaching a total of 6.3 GW. During the period from 2010 to 2021, the cost of electricity generated by CSP plants decreased by a substantial 68% [8].

1.3 SOLAR THERMAL ENERGY STORAGE TECHNIQUES

One way to store solar heat for later use is by means of solar thermal energy storage. Various methods exist for storing solar thermal energy, depending on the material and approach employed. The principles of solar thermal energy storage are similar to the concepts employed for thermal energy storage.

- The most common technique is sensible heat storage (SHS). It simply indicates a medium's temperature that has either increased or decreased. In general, the materials and techniques employed in SHS are safe and cost-effective to a great extent. A water reservoir is one of the most affordable and often used solutions. However, there are certain materials, like molten salts or metals, that can be heated to a certain higher temperature and have a greater capacity for storage. Also, energy can be stored underground in the form of an underground tank or a heat-transfer fluid that circulates through a network of pipes. SHS's dependence on the properties of the storage medium is a disadvantage. The specific heat capacity of the storage material restricts the amount of storage that can be held, so the system must be designed carefully to ensure energy extraction at a constant temperature [9].
- Materials that change phases, also called phase change materials (PCM), such as from solid to liquid or liquid to gas, when they absorb or release heat are used in latent heat storage (LHS). They are able to store a lot of energy at constant temperatures in a small volume, thanks to this. When it comes to LHS, cooling materials

like ice or water are the most popular examples. Additional substances that can be utilized for LHS are metal alloys, chemical compounds, and paraffin wax. Emphasis should be placed on the proper utilization of different PCM, taking into account certain important characteristics such as melting point, energy storage capacity, and thermal conductivity [10].

- Unfortunately, PCM faces the consequences of low thermal conductivity. To improve the thermal conductivity of PCM, a variety of carbon-based nanomaterials, such as carbon nanotubes, grapheme, etc., are utilized while creating the PCM. Through the incorporation of these materials, significant improvements in thermal conductivity have been observed. Similarly, loading PCM with metal and metal oxide nanoparticles as well as with other carbon-based materials has yielded promising results [11].

- The thermochemical form of heat storage is a mechanism of energy storage that involves breaking down or combining molecules using heat or light. The molecules undergo three stages of reaction: endothermic dissociation, storage of reaction products, and exothermic reaction of the dissociated products. The molecules formed create certain chemical bonds that store energy, which can be released later by reversing the reaction. This type of energy storage can store heat for longer periods and at higher temperatures, which is better than solid or latent energy storage. In fact, they have a higher energy storage density than sensible and LHS systems. Long-term storage applications are strong for thermochemical storage as its losses are low [11]. To ensure energy stored optimally, appropriate materials or their combinations should have the ability to store energy with minimal heat loss and release it efficiently when necessary [9]. But it's still in the early stages of R&D and faces challenges such as the availability of required materials, keeping the reaction stable and reversible, and cost of implementation.

A combined storage method has overall improved the efficiency of thermal energy storage, utilizing both sensible and LHS. In SHS, the temperature of the storage medium is increased without any change in phase, while in LHS it occurs due to the transfer of heat into the medium, causing the phase change. By combining the materials involved in both, combined thermal energy storage offers improved efficiency, cost-effectiveness, and a steady temperature output [9].

The technology involved in thermal storage is advancing toward its growth phase; significant investment costs, policy implementation and issues related to its legislation, and limited expertise among all the stakeholders emerge as major obstacles (Figure 1.1).

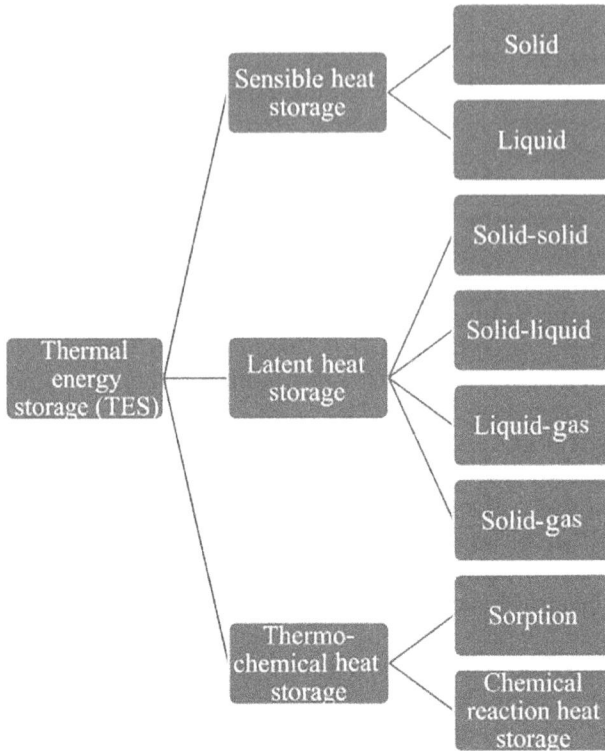

Figure 1.1 Classifications of TES methods [12].

Some examples of solar thermal energy storage systems are:

- The Solana Solar Power Station in the USA utilizes molten salt to capture heat generated by concentrated solar energy plants and generate electricity for 6 hours after the Sun sets [12].
- Drake Landing Solar Community, located in the Canadian province of Alberta, utilizes borehole thermal storage to capture solar energy from the ground and supply space heating to 52 dwellings annually [13].
- The MOST project in Sweden [14], where photo-switchable molecules are used to store solar energy as chemical energy and release it when needed (Figure 1.2).

1.4 APPLICATIONS OF SOLAR THERMAL ENERGY STORAGE

Solar thermal energy storage (STES) has important applications aiding the fields utilizing solar thermal energy such as power generation, heating and cooling, water desalination, and industrial processes [16].

Figure 1.2 SOLANA solar power station in the USA [15].

- One of the main uses of STES is power generation. STES can be used to support CSP plants. STES can increase the reliability and flexibility of CSP systems by storing excess heat during the day and releasing it again at night or when the sky is cloudy. This allows the CSP system to provide distributable power and match power demand and supply even when the Sun is not available. STES can also reduce the cost of CSP plants because it can reduce the need for backup fossil fuel generators and grid connections.
- Another use for STES is to provide heating and cooling for buildings, industrial processes, agriculture, and water supplies. STES can store excess heat during the day and release it at night or during cold weather, reducing fossil fuel and electricity consumption for thermal applications. This improves the energy efficiency and performance of solar thermal systems and reduces greenhouse gas emissions and air pollutants. There are various STES methods for heating and cooling systems such as borehole thermal storage (BTES), aquifer thermal storage (ATES), and pit thermal storage (PTES). BTES uses vertical tubes filled with a liquid that exchanges heat with the surrounding soil and rock. ATES uses natural or man-made groundwater reservoirs that store heat or cold air. PTES uses shallowly dug pits filled with gravel and water as storage media.
- The third application of STES is in desalination, providing fresh water for drinking, irrigation, or industrial use from seawater or brackish water. The process involves the storage of excess heat during the day and releasing it at night or when sunlight is low. This can reduce the operating costs and environmental impact of solar desalination plants, increasing their availability and reliability (Figures 1.3 and 1.4).

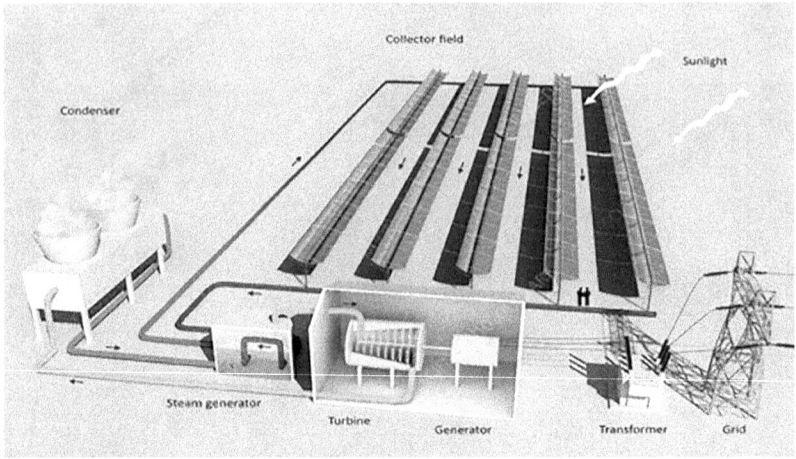

Figure 1.3 Solar thermal power generation [17].

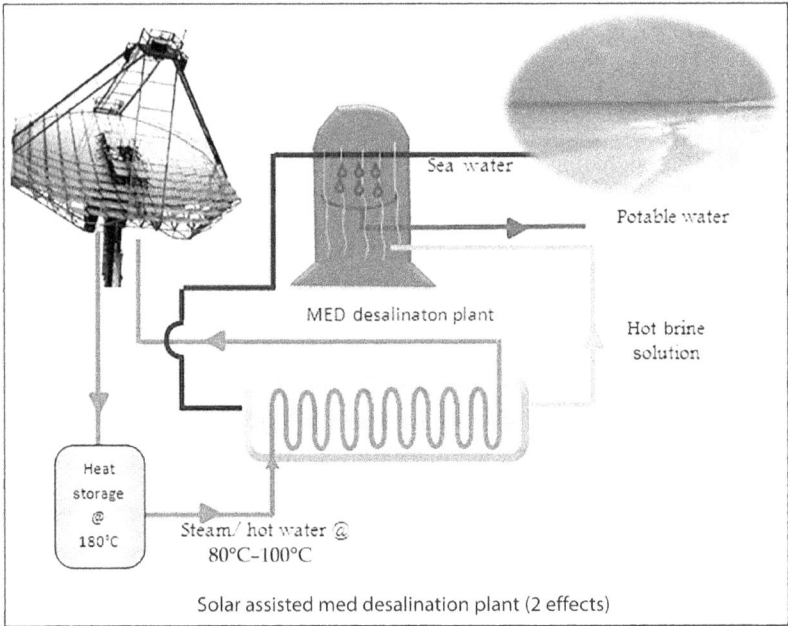

Figure 1.4 Solar-assisted desalination plant with energy storage [18].

1.5 POTENTIAL OF GEOTHERMAL ENERGY

Geothermal energy is heat energy obtained from the ground. It is obtained by utilizing the energy radiated by the radioactive decay at the core of the Earth as well as the solar radiation received on the Earth's crust. Also, there are reservoirs of hot water beneath the surface, where the heat is utilized for their energy output. Geothermal energy can be used for many different purposes such as electricity generation, hot water requirement, and housing temperature regulation. The increased use of geothermal energy can reduce fossil fuel use and global warming emissions.

Geothermal energy has been used to generate heat and/or electricity for thousands of years. For example, geothermal hot spring water has been used since Paleolithic times, and space heating has been used since Roman times.

Geothermal energy has been prominently used in geothermal power plants to generate electricity. Most of the systems use hot water from reservoirs to generate steam to turn a turbine to generate electricity. Since water heat is an integral part of a geothermal power plant, these active systems are usually located where there are thermal reservoirs.

There are different types of geothermal resources that vary in temperature, depth, and quality.

- **Hydrothermal Resources:** These are reservoirs of hot water or steam that have been trapped in pores or rock fractures. It can be brought about by digging a well or hole and pumping the water. Hydrothermal components can be divided into low temperature (<90°C), intermediate temperature (90°C–150°C), and high temperature (>150°C) [19].
- **Enhanced Geothermal Systems (EGS):** These are artificial reservoirs created by injecting water into hot rocks or existing reservoirs with limited hydrothermal activities. The water breaks down rocks and creates cracks, allowing water to circulate and absorb heat. The steam is then pumped back up through the reaction well. EGS can extend the availability of geothermal energy to areas without conventional hydropower resources [20] (Figure 1.5).
- **Geopressured Resources:** These are reservoirs of hot water and natural gas that form under high pressure in deep reservoirs. It is accessed by drilling wells and releasing pressure, allowing water and gas to flow upwards. Geopressurized materials can provide thermal energy and utilize it to provide mechanical motion and recoverable methane [20].

Direct deep use (DDU) applications use low-temperature resources to facilitate large-scale use in residential, commercial, and industrial environments. In particular, low-temperature DDU is useful in applications that include residential heating and cooling, district heating systems, industrial processes, manufacturing, and agricultural operations. Temperature control is typically done with ground-source heat pumps that rely on geothermal energy found only 1.5–5 m below the surface, known as passive geothermal, rather than extracting the heat from the deeper hot water reservoirs used in other applications [22].

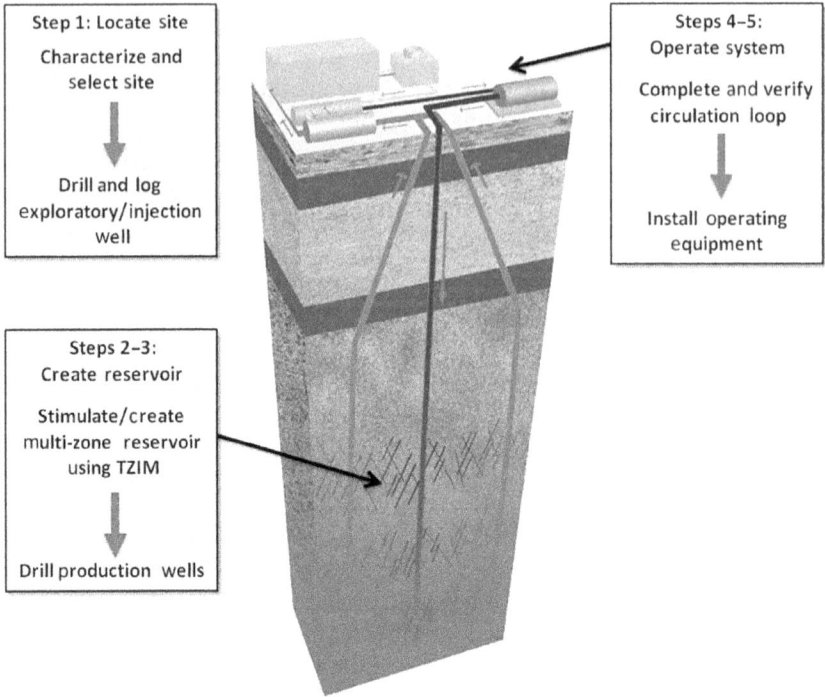

Figure 1.5 Enhanced geothermal systems [21].

Some of the benefits of geothermal energy include [23]:

- Geothermal energy is a clean, renewable energy source that does not produce greenhouse gases or air pollutants that contribute to global warming and health problems.
- Geothermal energy is abundant and available in most parts of the world. It can provide essential power and flexibility to the grid, as well as heating and cooling for buildings and industry.
- Geothermal energy is cost-effective and competitive with traditional electricity and heat sources. This can reduce dependence on fossil fuels and increase the security and diversity of energy supply.
- Geothermal energy can create jobs and stimulate economic growth. It can also promote innovation and entrepreneurship in various fields.

1.5.1 Global scenario of geothermal energy

It was estimated that in 2020, 15.4 GW of geothermal power plants were installed worldwide, and it is estimated that the potential for up to 200 GW of geothermal power generation from hydrothermal systems around the world is possible. The economic potential of generating electricity using hydrothermal systems with existing technologies is estimated at 70 GW by 2050. By comparison, geothermal heating and cooling systems, including heating and cooling of buildings, bathing, swimming, industrial, and agricultural applications, have grown to 107 GWt. Geothermal technological potential is estimated at 5000 GWt worldwide [24].

In 2010, 26 countries had geothermal power generation. In terms of installed capacity, as of 2019, the USA has an installed capacity of 23.86% or 3.68 GW, which has the highest share of all the countries' total power-generating installations [25].

In 2021, global geothermal power generation was 16 GW. Only a handful of countries have surpassed the 1 GW mark [26].

1.5.2 Geothermal energy extraction techniques

There are various methods of storing the geothermal energy emanating from the Earth. Most of the energy is already available in zones, and extraction is the only requirement. However, geothermal energy is a complex form of energy. These resources can vary greatly depending on location, depth, temperature, and fluid characteristics, and the availability of the required heat energy to maintain a steady temperature poses a challenge [27].

That's where geothermal energy storage (GES) comes in. By storing the required heat in the ground, GES technology allows geothermal systems to regulate the balance between energy supply and demand, ultimately boosting efficiency and performance.

There are different methods of GES that vary in their design, operation, and application. Some of the main methods are [17] as follows:

- **Borehole Thermal Energy Storage (BTES):** BTES, also known as Borehole Thermal Energy Storage, involves utilizing vertical pipes filled with fluid to transfer heat to and from the surrounding soil or rock. This versatile method can be employed for various purposes, such as providing heat or extracting heat for an extended period or immediate usage, including space temperature control, household hot water, and industrial processes. Moreover, BTES is also beneficial for EGS as it allows for higher rates of heat extraction and injection.
- **Aquifer Thermal Energy Storage (ATES):** ATES utilizes underground water reservoirs, natural or man-made, to store heat or cold. This storage method is versatile, with the ability to store heat or cold for either short-term or seasonal applications, such as space heating or cooling, domestic hot water, and industrial processes. In addition, ATES can enhance the efficiency and lower the costs of GSHPs and solar thermal collectors when integrated with them. Moreover, it can boost the heat extraction and injection rates of hydrothermal geothermal systems (Figure 1.6).

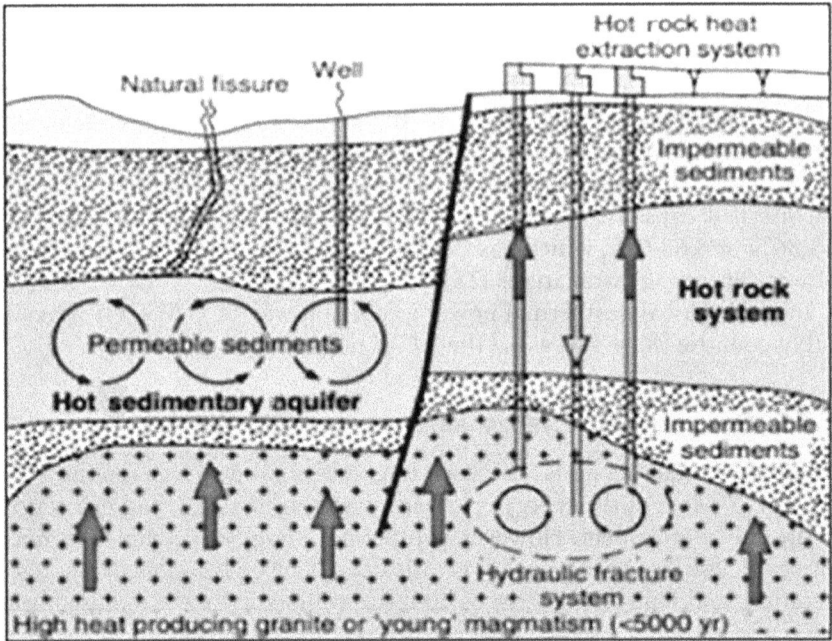

Figure 1.6 Hot rock and hot sedimentary aquifer system [29].

- One method of thermal energy storage is known as Pit Thermal Energy Storage (PTES). It involves utilizing shallow dug pits, filled with gravel and water, as the storage medium. This versatile technique can store heat or cold for a variety of purposes, such as seasonal or short-term applications such as space heating or cooling, domestic hot water, and industrial processes. By incorporating PTES with solar thermal collectors, it can greatly enhance their effectiveness and lower costs. Another benefit of PTES is its potential in low-temperature geothermal systems, where it can significantly increase heat extraction and injection rates.

GES serves as a vital component in optimizing the functioning of geothermal systems by effectively balancing thermal energy supply and demand. Through this, the efficiency and performance of the systems utilizing geothermal heat can be significantly improved.

1.5.3 Applications of geothermal energy and storage

Geothermal energy offers a diverse array of uses that cater to a multitude of needs and requirements across various industries and areas. A few examples of the current and potential applications include:

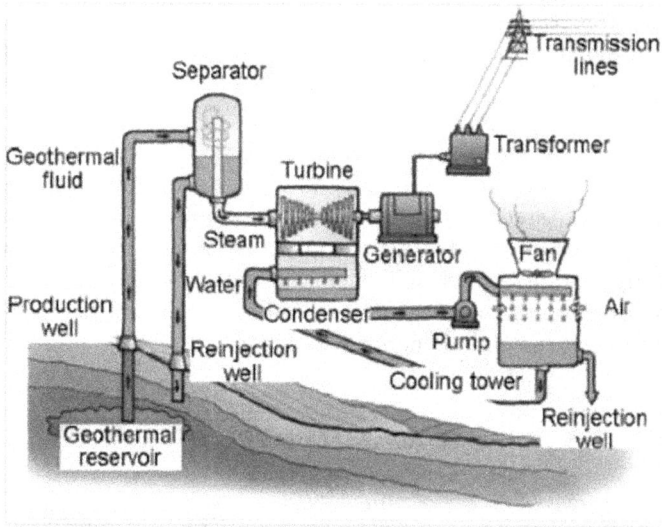

Figure 1.7 Geothermal power plant schematic [28].

- **Power Generation:** Geothermal energy is a viable power source for meeting the electricity demands of households, businesses, industries, utilities, and communities. Dry steam plants utilize steam directly from geothermal reservoirs or geothermal storage systems to power turbines. In flash steam plants, high-pressure hot water is reduced to produce steam. Binary cycle plants use lower temperature hot water to heat a secondary fluid with a lower boiling point, which in turn powers a turbine. Combined cycle plants combine both flash and binary systems to maximize the efficiency and output of power generation (Figure 1.7).
- **Heating and Cooling:** Geothermal energy is a valuable resource for both heating and cooling needs in various applications such as buildings, industry, agriculture, and water supply. With direct use systems, hot water from geothermal sources or wells is directly utilized to provide heat for a variety of purposes, from space heating to industrial processes. Heat pump systems, on the other hand, utilize the Earth's natural heat source or sink to provide heating or cooling for buildings. Ground-source heat pumps utilize buried pipes to transfer heat with the ground, while water-source heat pumps use submerged pipes in a water body. Both systems effectively regulate the temperature for buildings [17].
- **Desalination:** Through the power of geothermal energy, it is possible to transform brackish water into a source of drinking water, irrigation, or even for industrial purposes. By utilizing the heat from the

Earth, geothermal desalination systems employ distillation or membrane processes to effectively extract salt from water. This system not only provides a sustainable solution for water production and supply but also decreases our reliance on fossil fuels and electricity [29].

Geothermal energy and GES offer numerous advantages for the environment, society, and economy. However, they also encounter various obstacles that must be overcome in order to fully maximize their potential within the energy system and society. It is crucial to devise and implement effective strategies and solutions that will facilitate the widespread adoption and integration of geothermal energy and GES. Only then can they emerge as a dominant force in the global energy economy.

1.6 POTENTIAL OF HYDROGEN ENERGY STORAGE

Hydrogen energy is a source of renewable energy that harnesses the potential of hydrogen to create electricity, heat, and other forms of energy. Hydrogen, which is abundant in the universe, can be extracted from diverse sources such as water, natural gas, biomass, or renewable electricity. The possibilities for hydrogen energy storage are limitless, with potential use in various fields and areas such as transportation, power generation, industry, and heating.

The possibilities for utilizing hydrogen energy storage span across various industries and regions such as transportation, power generation, and heating. Hydrogen energy storage holds vast potential, tapping into abundant energy reserves that can be effectively stored and transported over considerable distances. Not only does it provide a solution for managing the inconsistency of conventional renewable sources, but it also offers flexibility and dependability for the grid and district heating systems. Many experts expect hydrogen to be a viable alternative to fossil fuels. This is largely due to hydrogen's capabilities, such as powering fuel cells in electric vehicles with zero emissions, its potential for domestic production, and the fuel cell's high efficiency [30].

When hydrogen fuel cell is paired with an electric motor, it can generate two to three times more energy than a gasoline-fueled internal combustion engine. In addition, hydrogen has the potential to fuel internal combustion engines. In fact, the energy contained in just 1 kg of hydrogen gas is equivalent to that of 2.8 kg of gasoline. The increasing need to incorporate hydrogen as a means of on-board energy storage in emission-free vehicles is driving the innovation of storage techniques better suited for this purpose. The main obstacle is the extremely low boiling point of hydrogen, which is at approximately 20.268 K. Maintaining such a low temperature requires a significant amount of energy [31].

Hydrogen has the potential to serve as a versatile storage solution for renewable energy, providing both short-term load balancing and seasonal energy storage. Hydrogen can be generated through electrolysis using ample summer sunlight and used to power the grid during the colder, cloudier winter months. Given its carbon-neutral properties, hydrogen is the most viable option for seasonal storage [32].

1.6.1 Global status of hydrogen storage

In 2018, the IEA reported that the world's demand for hydrogen was 70 million tons. This demand was primarily driven by the need for refining and ammonia production. Unfortunately, the majority of hydrogen was generated from non-renewable sources like natural gas and coal, resulting in significant carbon emissions. In contrast, a mere 4% of hydrogen was produced through water electrolysis, utilizing electricity either from the grid or from sustainable sources [33].

According to IRENA, the worldwide capacity for hydrogen energy storage stood at approximately 1.6 GW in 2019. The prominent regions for such projects were Europe and North America, primarily for power-to-gas implementations. This involves the conversion of electricity into hydrogen through water electrolysis, which is then either introduced into the natural gas network or utilized for other objectives [34].

According to the IEA, hydrogen has the potential to fulfill 18% of the world's energy needs by 2050, primarily through the use of renewable energy sources [35].

India's Status in Hydrogen Production:

- As of now, India's utilization of hydrogen amounts to 6 million tons [36].
- The refinery industry alone requires almost 3 million tons of hydrogen, making up a significant 46% of the nation's total hydrogen demand. Furthermore, the production of ammonia plays a major role, accounting for 48% of the current demand for hydrogen [37].
- The integration of hydrogen in IC engine technology has made strides with 50 hythane buses (consisting of 18% hydrogen and 82% compressed natural gas) currently operating on the busy streets of New Delhi [32].

The largest hydrogen energy storage is located in an underground cavern of Beaumont, Texas. Managed by Air Liquide, the French industrial gas company, this facility has a storage capacity of approximately 600 million standard cubic feet (17 million m^3) of hydrogen. Since its inception in 2017, it has supplied for numerous clients in the petrochemical, refining, and steel sectors [38–41].

1.6.2 Methods of hydrogen energy storage

There are different methods of hydrogen energy storage that vary in their design, operation, and application. Some of the main methods are:

- **Compressed Hydrogen:** In this method, hydrogen gas is contained in tanks or cylinders at high pressures. This method has practical applications in both transportation and power generation, as it offers the advantages of high energy density and quick refueling times. Nevertheless, the process of compressing hydrogen requires significant energy input and the construction of specialized tanks and infrastructure, resulting in substantial costs.
- **Liquefied Hydrogen:** Cryogenic tanks are used to store hydrogen as a liquid, a commonly utilized method known as liquefied hydrogen. It has several practical applications such as transportation and power generation, thanks to its ability to provide higher energy density compared to compressed hydrogen method. In addition, it offers the advantage of reduced costs for both tanks and infrastructure. Nevertheless, it's worth noting that the process of liquefaction requires a significant energy input and can result in high losses due to boil-off [42–44].
- **Chemical Storage:** One way to store large amounts of hydrogen is through the use of chemical compounds that release this gas when needed. Not only can this method be applied to various industries, including transportation and power generation, but it also offers the benefits of high energy density and cost-effectiveness. However, this approach does come with its limitations, such as the need for extreme temperatures or catalysts for hydrogen release and lower efficiency for regeneration. Different types of chemical storage materials that have been utilized include hydrocarbons, boron hydrides, ammonia, and metal hydrides, among others.
- **Electrochemical Storage:** Electrochemical storage is a highly efficient and flexible method that involves using electrochemical devices to convert electricity into hydrogen through water electrolysis and vice versa through fuel cells. With this technology, electricity can be generated and consumed in a more practical and versatile manner, making it beneficial for tasks such as power generation and grid balancing. However, the cost of electrolyzers and fuel cells is a significant consideration as well as the limited availability and quality of water sources (Figure 1.8).

However, there are certain shortcomings in hydrogen storage such as

Low Energy Density: A major drawback of hydrogen as a fuel is its low energy density per unit volume. Most vehicles using hydrogen as a fuel need a large fuel tank to cover the same distance compared to other fuels.

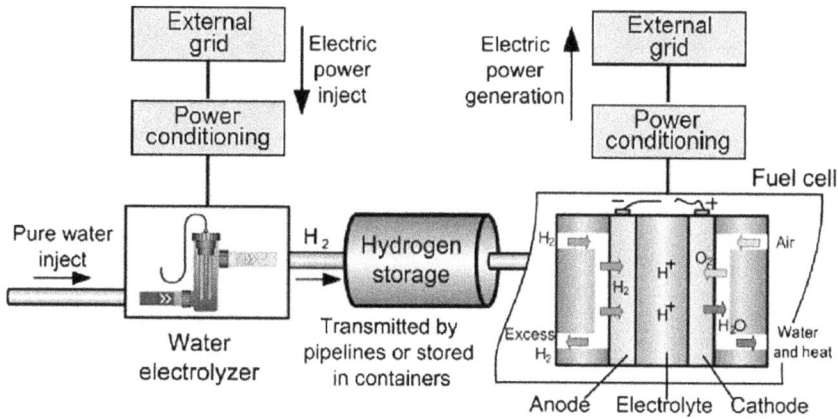

Figure 1.8 Topology of hydrogen storage [39].

Leakage: Owing to its low density and small molecular size, hydrogen is highly prone to leakage and contamination from its container.

Liquefaction and Pressurization Requirements: For storing hydrogen in liquid or mixed-phase form, a significant amount of energy and special storage arrangements are required for the processes of liquefaction and pressurization.

1.6.3 Applications of hydrogen energy storage

There are numerous applications for utilizing hydrogen energy storage, such as:

- **Transportation:** With the use of hydrogen energy storage, vehicles can now have access to clean fuel and utilize the electricity generated using fuel cells or direct use in internal combustion engines. By choosing hydrogen-powered vehicles, several benefits can be experienced over traditional vehicles, including increased efficiency, reduced emissions, extended driving range, faster refueling, and quieter operation [23] (Figure 1.9).

- **Power Generation:** Hydrogen energy storage is a promising solution for power generation. It offers the flexibility to provide reliable power for both on-grid and off-grid applications. The process begins by generating hydrogen from renewable sources like solar or wind energy through methods such as electrolysis. The produced hydrogen can then be stored in tanks or underground reservoirs, ready for use at a later time. When needed, hydrogen can be converted back to electricity using fuel cells or turbines [20].

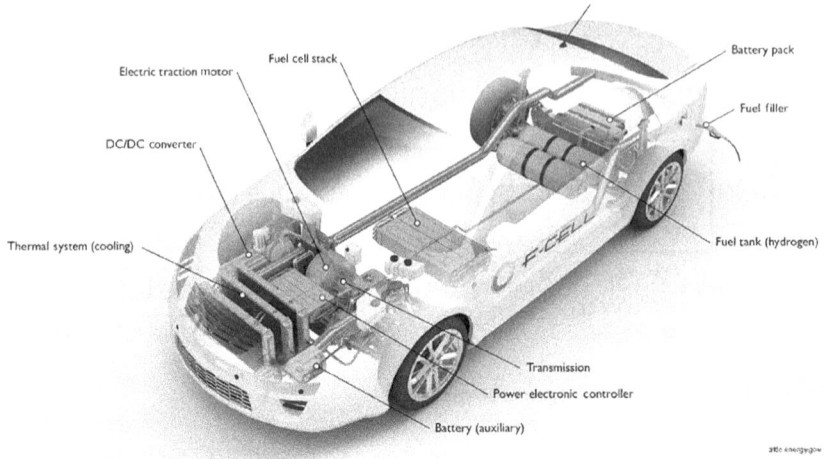

Figure 1.9 Fuel cell vehicle [40].

- **Industries:** Utilizing hydrogen energy storage has the potential to greatly reduce carbon emissions in industrial processes that rely on fossil fuels for heating. This adaptable energy source can be easily incorporated into existing natural gas pipelines or district heating systems to provide heat to various industries. In addition, hydrogen can serve as a valuable feedstock or fuel for important industrial operations such as refining, ammonia and methanol production, and steelmaking. There are significant financial investments, limitations, and variations in hydrogen sources [24].

Research is ongoing to find solutions to enhance hydrogen storage capabilities while simultaneously improving production efficiency. Current research suggests that by wrapping polypyrrole around polydivinyl benzene nanotubes, a photothermal efficiency of 85.2% and energy storage capacity of 145.7 J/g can be obtained. Another promising research involves using a composite of polyethylene glycol and carboxylate dye, which serves as an energy absorber and converter for converting solar energy into thermal energy [41].

A concept of hydrogen economy has been created that makes hydrogen as a prime source of fuel for transportation and generating electricity. It is a highly efficient and environmentally friendly fuel, which makes it a promising substitute for traditional fossil fuels.

Even though the hydrogen economy has certain obstacles to overcome, like expenses involved in production, storage, and transportation. Also infrastructure is lacking for hydrogen refueling stations. Furthermore, the flammability of hydrogen adds to the complexity and careful measures needed for its handling and storage. The practical usage of the materials used for hydrogen storage relies greatly on their ability to efficiently store and release hydrogen in a reversible manner.

1.7 CONCLUSION

The exploration of thermal energy storage within the realms of solar, geothermal, and hydrogen energy represents a promising frontier in the pursuit of sustainable and clean energy solutions [45–47]. These renewable sources offer a pathway to reduce greenhouse gas emissions, enhance energy security, and foster economic growth while addressing the urgent need for environmental conservation. Solar energy, with its versatile applications and innovative solar thermal technology, is emblematic of harnessing the Sun's power to meet various energy needs. Geothermal energy, derived from the Earth's core, serves as a dependable source of power, heating, and cooling, and it holds the potential for even greater expansion through EGS. Meanwhile, hydrogen energy, as an energy carrier derived from various sources, demonstrates exceptional adaptability in powering a wide range of applications and providing energy storage. Despite the challenges faced by these energy sources, including intermittency, land and water use, technical complexities, social barriers, and implementation costs, it is imperative to recognize their immense potential. The development and implementation of strategies to integrate these renewable sources into our energy systems and societies are not just beneficial but also essential for a sustainable and greener future. As we look forward, the integration of solar, geothermal, and hydrogen energy into our energy landscape represents a crucial step toward the global energy transition and the realization of sustainable development goals. By continuing to invest in research, innovation, and infrastructure, we can harness the power of thermal energy storage and create a more sustainable and environmentally just world for generations to come.

REFERENCES

[1] IEA. (2021). Global Energy Review 2021. Retrieved from International Energy Agency: https://iea.blob.core.windows.net/assets/d0031107-401d-4a2f-a48b-9eed19457335/GlobalEnergyReview2021.pdf

[2] IEA. (2019). Solar Energy: Mapping the Road Ahead. Retrieved from International Energy Agency.: https://iea.blob.core.windows.net/assets/7de8652c-47b0-474e-8642-cbf20245b1f6/Solar_Energy_Mapping_the_road_ahead.pdf

[3] Faninger, G. (2009). *The Potential of Solar Heat in the Future Energy System*. Klagenfur: University of Klagenfur

[4] Fortune Business Insights. (2019). Solar Thermal Market Size. Retrieved from Renewables/Solar Thermal Market: https://www.fortunebusinessin-sights.com/industry-reports/solar-thermal-market-101920.

[5] IEA. (2022). Solar Heat Worldwide. Retrieved from IEA-Solar Heating and Cooling Program: https://www.iea-shc.org/solar-heat-worldwide.

[6] EIT. (2022). Solar Thermal Energy: An Industry with Untapped Potential. Retrieved from European Union: https://www.climate-kic.org/innovation-spotlight/solar-thermal-naked-energy/.

[7] Johntson, J. W. (2021). What Are the Social and Economic Benefits of Solar Thermal for Local Communities and Businesses. Retrieved from LinkedIn: https://www.linkedin.com/advice/0/what-social-economic-benefits-solar-thermal-local#:~:text=1-,3%20Social%20benefits,health%2C%20education%2C%20and%20productivity

[8] Renewables Now. (2023). Concentrating Solar Power Facts. Retrieved from REN21-Renewables in Energy Supply: https://www.ren21.net/gsr-2023/modules/energy_supply/02_market_developments/08_csp/#:~:text=The%20global%20installed%20capacity,a%20total%20of%206.3%20GW.&text=(See%20Figure%2028.),global%20CSP%20capacity%20in%202021.

[9] Charmala Suresh, R. P. (2020). Review on solar thermal energy storage technologies and their geometrical configurations. *International Journal of Energy Research*, 44(6), 4163–4195.

[10] Koohi-Fayegh, S. (2020). A review of energy storage types, applications and recent developments. *Journal of Energy Storage*, 27, 101047.

[11] T. M. Bhartendu, S. K. Shailendra, R. S. Pushpendra (2023). A comprehensive review on solar to thermal energy conversion and storage. *Journal of Energy Storage*, 72, 108280.

[12] Suresh, N. S., N. C. Thirumalai, B. S. Rao, M. A. Ramaswamy. (2014). Methodology for sizing the solar field for parabolic trough technology with thermal storage and hybridization. *Solar Energy*, 110, 247–259.

[13] Wong, B., L. Mesquita. (2019). Drake landing solar community: Financial summary and lessons learnt. IEA SHC International Conference on Solar Heating and Cooling for Buildings and Industry 2019 (p. 12). Ottawa: ISES Solar World Congress 2019.

[14] Wang, Z. (2022). Status and challenges for molecular solar thermal energy storage system based devices. *Chemical Society Reviews*, 51, 7313–7326.

[15] Loans Program Office. (2022). SOLANA. Retrieved from U.S Department of Energy - Loans Program Office: https://www.energy.gov/lpo/solana.

[16] Cabeza, L. F. (2020). 2-*Advances in Thermal Energy Storage Systems*. https://doi.org/10.1016/B978-0-12-819885-8.00002-4.

[17] Science Photo Library. (2019). Solar Thermal Power Diagram. Retrieved from Science Photo Library: https://www.sciencephoto.com/media/660311/view/solar-thermal-power-diagram.

[18] Clique Solar. (2018). ARUN 100. Retrieved from Clique Solar-Products: https://www.cliquesolar.com/Desalination.aspx.

[19] Sircar, A., N. Bist, K. Yadav. (2022). A comprehensive review on exploration and exploitation of offshore geothermal energy. Marine Systems & Ocean Technology, 17, 135–146.

[20] Gonzales, V. (2020). Geothermal Energy 101. Retrieved from Resources for the Future: https://www.rff.org/publications/explainers/geothermal-energy-101/.

[21] Alta Rock Energy. (2020). Enhanced Geothermal Systems. Retrieved from Alta Rock - Technology: https://altarockenergy.com/technology/enhanced-geothermal-systems/.

[22] Beckers, K. F., A. Kolker. (2021). Evaluating the feasibility of geothermal deep direct-use in the United States. *Energy Conversion and Management*, 243, 114335.

[23] Soltani, M., F. M. Kashkooli. (2021). Environmental, economic, and social impacts of geothermal energy systems. *Renewable and Sustainable Energy Reviews*, 140, 110750.

[24] The International Renewable Energy Agency. (2021). *Geo (World Economic Forum, 2023) Thermal: The Solution Underneath.* IRENA. https://www.irena.org/About

[25] Huttrer, G. W. (2020). Geothermal power generation in the world 2015-2020 update report. *World Geothermal Congress* 2020 (p. 17). Reykjavik: Proceedings World Geothermal Congress 2020.

[26] World Economic Forum. (2023). Explainer: This is How Geothermal Energy Works. Retrieved from World Economic Forum - Energy Transition: https://www.weforum.org/agenda/2023/01/geothermal-energy-infographics-data-renewable/.

[27] National Geographic. (2022). Geothermal Energy. Retrieved from National Geographic Education: https://education.nationalgeographic.org/resource/geothermal-energy/.

[28] Tasnuva Sharmin, N. R. (2023). A state-of-the-art review on geothermal energy extraction, utilization, and improvement strategies: Conventional, hybridized, and enhanced geothermal systems. *International Journal of Thermofluids*, 18, 100323.

[29] Prajapati, M., M. Shah, B. Soni. (2021). A review of geothermal integrated desalination: A sustainable solution to overcome potential freshwater shortages. *Journal of Cleaner Production*, 326, 129412.

[30] US Department of Energy. (2023). Hydrogen Storage. Retrieved from Office of the Energy Efficiency and Renewable Energy: https://www.energy.gov/eere/fuelcells/hydrogen-storage.

[31] US Department of Energy. (2019). Hydrogen Basics. Retrieved from Alternative Fuels Data Center: https://afdc.energy.gov/fuels/hydrogen_basics.html.

[32] Fuel Cell and Hydrogen Energy Association. (2020). Hydrogen as Energy Storage. Retrieved from Hydrogen as Energy Storage: https://www.fchea.org/hydrogen-as-storage.

[33] International Energy Agency. (2019). The Future of Hydrogen. Retrieved from IEA-Reports: https://www.iea.org/reports/the-future-of-hydrogen.

[34] IRENA. (2020). *Green Hydrogen Cost Reduction.* https://www.irena.org/-/media/Files/IRENA/Agency/Publication/2020/Dec/IRENA_Green_hydrogen_cost_2020.pdf .

[35] Amirthan, T., M. S. A. Perera. (2022). The role of storage systems in hydrogen economy: A review. *Journal of Natural Gas Science and Engineering*, 108, 104843.

[36] Agarwal, O. P., R. Pandit. (2021). The Path to a Hydrogen Economy. Retrieved from Indian Express: https://indianexpress.com/article/opinion/the-path-to-a-hydrogen-economy-7369262/.

[37] Raj, K., P. Lakhina, C. Stranger. (2022). *Harnessing Green Hydrogen*. New Delhi: NITI Aayog.

[38] Air Liquide. (2017). USA: Air Liquide Operates the World's Largest Hydrogen Storage Facility. Retrieved from Air Liquide - Sustainable Development: https://www.airliquide.com/group/press-releases-news/2017-01-03/usa-air-liquide-operates-worlds-largest-hydrogen-storage-facility.

[39] Kleperis, J., V. V. Fylenko, M. Vanags, A. Volkovs, et al. (2016). Energy storage solutions for small and medium-sized self-sufficient alternative energy objects. XV International Scientific Conference "Renewable Energy & Innovative Technologies" (pp. 290–296). Smolyan: Bulgarian Academy of Sciences, Union of Chemists in Bulgaria.

[40] US Department of Energy. (2020). Alternative Fuels Data Center. Retrieved from Fuel Cell Electric Vehicles: https://afdc.energy.gov/vehicles/fuel_cell.html.

[41] Abe, J. O., A. P. I. Popoola, E. Ajenifuja, O. M. Popoola. (2019). Hydrogen energy, economy and storage: Review and recommendation. *International Journal of Hydrogen Energy*, 44, 15072–15086.

[42] Kumar, A., V. P. Singh, C. S. Meena, N. Dutt. (2023). Thermal Energy Systems: Design, Computational Techniques and Applications. Boca Raton, FL: Taylor & Francis (CRC Press). https://doi.org/10.1201/9781003395768.

[43] Kundu, A., A. Kumar, N. Dutt, V. P. Singh, C. S. Meena. (2023). Modelling and simulation of thermal energy system for design optimization. In Kumar, A.; Singh, V.P.; Meena, C.S.; Dutt, N.; Editors; *Thermal Energy Systems: Design, Computational Techniques and Applications* (pp. 103–140). Boca Raton, FL: CRC Press; Chapter 7. https://doi.org/10.1201/9781003395768-7.

[44] Dewangan, A. K., S. Q. Moinuddin, M. Cheepu, S. K. Sajjan, A. Kumar. (2023). Thermal energy storage: Opportunities, challenges and future scope. In Kumar, A.; Singh, V.P.; Meena, C.S. ; Dutt, N.; Editors; *Thermal Energy Systems: Design, Computational Techniques and Applications* (pp. 17–28). Boca Raton, FL: CRC Press; Chapter 2. https://doi.org/10.1201/9781003395768-2.

[45] Kundu, A., A. Kumar, N. Dutt, C. S. Meena, V. P. Singh. (2023). Introduction to thermal energy resources and their smart applications. In Kumar, A.; Singh, V.P.; Meena, C.S.; Dutt, N.; Editors; *Thermal Energy Systems: Design, Computational Techniques and Applications* (pp. 1–15). Boca Raton, FL: CRC Press; Chapter 1. https://doi.org/10.1201/9781003395768-1.

[46] Dutt, N., A. J. Hedau, A. Kumar, M. K. Awasthi, V. P. Singh, G. Dwivedi. (2023). Thermo-hydraulic performance of solar air heater having discrete D-shaped ribs as artificial roughness. *Environmental Science and Pollution Research* 1–23. https://doi.org/10.1007/s11356-023-28247-9.

[47] Verma, V., C. S. Meena, S. Thangavel, A. Kumar, T. Choudhary, G. Dwivedi. (2023). Ground and solar assisted heat pump systems for space heating and cooling applications in the northern region of India - A study on energy and CO2 saving potential. *Sustainable Energy Technologies and Assessments*, 59, 103405. ISSN 2213-1388, https://doi.org/10.1016/j.seta.2023.103405.

Chapter 2

Technological advancements in sustainable and renewable solar energy systems

Mohammad Ahmadizadeh
Persian Gulf University, Bushehr

Milad Heidari, Sivasakthivel Thangavel,
Eman Al Naamani, and Morteza Khashehchi
Global College of Engineering and Technology, Muscat

Vikas Verma
Tezpur University Assam (Central
University, Government of India)

Ashwani Kumar
Technical Education Department Uttar Pradesh, Kanpur
(Under Government of Uttar Pradesh)

2.1 INTRODUCTION

The destructive effects of greenhouse gases resulting from the burning of fossil fuels and the consequences of electricity production with non-renewable energies have led people to think of a way to use them less. These destructive effects cause climate change and global warming [1]. Renewable energies are energies that are replaced naturally, and this replacement takes very little time compared to environmentally destructive and non-renewable energies [2]. In this procedure, researchers pay attention to new energies in all climates and regions.

Among the clean energy production methods, we can mention wind turbines, electricity generation from water, biofuels, geothermal energy, and solar energy [2]. The sun's radiation has played a significant role in human life since the past until now, to the extent that it is considered a high issue in the beliefs and thoughts of humans. According to past scientists, the sun converts millions of tons of hydrogen into helium per second. The earth receives part of this energy on its surface, which is different in different regions. The main reason for the difference in the amount of energy received

DOI: 10.1201/9781003472629-2

on the surface of the earth is the atmospheric conditions and climate of the earth, which controls life on the surface of the earth and help humans live better [3]. The ellipticity of the earth's orbit around the sun in a year causes significant differences in radiation in different parts of the globe. Maybe this difference is not a big amount in terms of percentage, but the energy of the sun is so great that paying attention to this small amount may be a promising promise for more use of this infinite source of energy for the comfort and convenience of humans. In 1996, the amount of energy received at sea level was 1 kw/m² [4].

Photovoltaic (pv) and csp (csp) methods are used to produce electricity with the help of the renewable energy of the sun. Arid climate areas in western China with 2000 kWh/m² provide a good opportunity to invest more in planting solar panels. In some areas, by strengthening CSP in the production of solar energy, it is possible to concentrate the sun's radiation at a point [5]. This helps to strengthen the csp network at night or cloudy times by taking advantage of concentrated thermal energy in environments containing molten salts [6]. In general, it is expected that the combined use of CSP and photovoltaics will bring more sustainable energy to us than before, which will reduce the price and costs of large-scale production [7].

Although solar panels have many advantages, batteries need to be used with them. These batteries are responsible for providing electricity at times such as night or in special weather conditions such as cloudy and rainy weather. A lot of battery usage causes this system to need more maintenance. Batteries are such that their charge should be kept above 50%. In other words, if this protection system is not managed properly, every time the battery charge reaches 0%, the battery life will decrease over time, and its pulp-shaped cells will be damaged. In countries such as Spain, India, Morocco, the United States, and China, photovoltaic, thermal, and combination technologies are used [2].

Installation of photovoltaic panels on the roofs is a huge improvement in the field of urban electricity generation. Over 90% of residential areas on the planet can install solar panels on the roof. The return on investment in this matter is estimated to be five years, which governments can implement in the form of a plan [8]. The implementation of this plan helps to release a lot of energy for the development of a country's industry or the sale of energy to neighbors [6].

After the nuclear accidents, Japan started to produce new energies, the most important of which were wind turbines and solar panels. Social political issues and developments in the electric energy industry and its entry into transportation have caused Japan to move toward being unique in the field of energy. Also, Japan has hot and humid summers, which helps pay special attention to solar panels [9,10]. After examining the power plants and thermal farms of power generation in India, we saw more longevity and efficiency of CSP, which encourages the development of CSP in India [11] According to World Bank data, Kuwait was the seventh country in terms of per capita electricity consumption in 2014. The growth and increase of the population in itself have caused governments to pay attention to providing affordable energy [7]. This is the requirement of Kuwait if, geographically, it has more than 2300 kWh/m^2 of solar energy production capacity [9]. After selecting and checking the cities in different geographical areas, such as Tunisia, Gaza, Alexandria, and Tripoli, were selected and the results of the research and analysis were as follows. The city of Tripoli has the best performance, accounting for 73% of the nominal power after Alexandria at about 66% and then Gaza at about 63%. The city of Tunis has the lowest solar fraction at about 59% according to the monthly and annual solar fraction data [3].

Solar power energy is free and permanent energy in coastal areas. Particularly those countries that are geographically an island or a large number of countries that are in the Gulf. Environmental factors such as wind, clouds, the position of the screens, etc., each in turn have a contribution to optimizing the production of electricity from solar energy. Most Gulf cities and countries have warm and sunny climates because they are level with the sea. This helps to ensure that solar panels generate electricity efficiently in these countries. The implementation of solar panels in coastal areas can be carried out on a large scale in the form of floating on the sea or beach roofs and roof coverings of commercial shopping centers near the sea. Covering some areas of the beach with the help of solar panels not only prevents excessive water vapor but also reduces the humidity of the beach and provides the required energy for coastal water purifiers in the Gulf regions.

2.2 SOLAR ENERGY HARVESTING SYSTEM

The basic components of the electricity generation system with the help of solar energy are panels, a charge controller, a battery, and an inverter. Figure 2.1 shows panels are formed by connecting smaller modules, which are closed in series or parallel. Charge controllers are regulators that cut off the maximum charging time of the battery and prevent excessive current and voltage from entering the battery [12]. The controllers that control the minimization of the charge are called pwm and the controllers that control the maximum charging rate are called mppt. Table 2.1 shows the key technological advancements in solar energy systems. It is influenced by advances in energy storage, grid integration, and materials science, which are also critical for the development of sustainable solar energy systems.

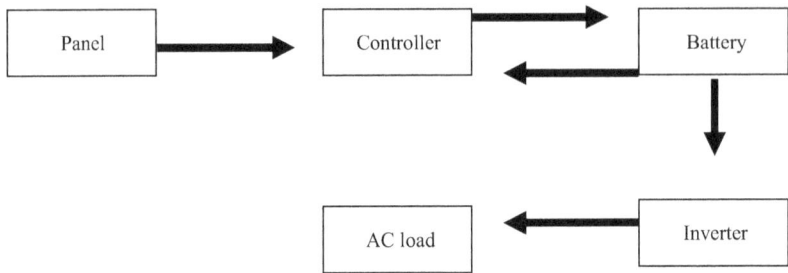

Figure 2.1 Solar panel system.

Table 2.1 Key technological advancements in solar energy systems [13–17]

Latest technology	Details	Advantages	Challenges
Photovoltaic (PV) panels	PV panels convert sunlight into electricity. Advancements include higher efficiency, flexible panels, and building-integrated designs	1. Clean and silent energy production 2. Low operating costs 3. Scalability	1. High initial costs 2. Intermittent power generation 3. Energy storage needed
Concentrated solar power (CSP)	CSP systems direct sunlight to heat a fluid, which powers a turbine to generate energy. Advances include better heat storage materials and higher efficiency systems	1. Continuous power generation 2. Energy storage capabilities 3. Suitable for large-scale power production	1. High upfront costs 2. Land and water usage 3. Environmental impact
Thin-film solar cells	Thin-film PV technologies are lighter and less resource-intensive than traditional PV panels. Advancements include improved materials and manufacturing processes	1. Lightweight and flexible 2. Lower production costs 3. Suitable for various surfaces	1. Lower efficiency compared to traditional PV 2. Degradation over time
Perovskite solar cells	Perovskite materials have shown great promise for high-efficiency solar cells. Advancements involve stability improvements and lead-free alternatives	1. High efficiency potential 2. Lower manufacturing costs 3. Flexibility in design	1. Durability and stability challenges 2. Toxicity concerns with lead-based perovskites

(Continued)

Table 2.1 (Continued) Key technological advancements in solar energy systems [13–17]

Latest technology	Details	Advantages	Challenges
Bifacial solar panels	Bifacial panels catch sunlight from both sides, resulting in higher energy production. Advancements focus on optimizing rear-side reflection and improving durability	1. Increased energy output 2. Flexibility in installation 3. Lower levelized cost of electricity	1. Relatively higher upfront costs 2. Installation complexity
Solar tracking systems	Tracking systems monitor the movement of the sun to maximize energy capture. Advances include better algorithms and materials for durability	1. Increased energy generation 2. Higher efficiency 3. Improved energy capture during all daylight hours	1. Higher maintenance requirements 2. Increased system complexity

2.3 EFFICIENCY OF SOLAR PV SYSTEM

The efficiency of a solar photovoltaic (PV) system depends on various factors. The efficiency of a solar photovoltaic (PV) system hinges on a multitude of factors that play a crucial role in determining its overall performance. One of the primary influencers is the geographical location where the PV system is installed. Solar radiation levels, which vary with location, significantly impact the system's efficiency. Regions with abundant sunlight, such as those near the equator, tend to generate more power from their PV systems compared to areas with less sunlight. This geographical variability underscores the importance of careful site selection when planning solar installations, aiming to maximize energy production.

Furthermore, the orientation and tilt angle of the PV panels are pivotal factors in optimizing efficiency. Panels need to be positioned to capture the maximum amount of sunlight during the day. In the Northern Hemisphere, panels are typically oriented to face south to receive the most sunlight, while in the Southern Hemisphere, they face north. The tilt angle is adjusted to match the latitude of the installation site, ensuring that the panels are optimally aligned with the sun's path. These considerations, along with factors like temperature, shading, and proper maintenance practices, collectively affect the overall efficiency of a solar PV system. By addressing and optimizing these factors, it is possible to enhance the system's energy production, making it a more sustainable and cost-effective source of renewable energy. Table 2.2 shows the key factors governing the efficiency of a solar PV system. During the design of a photovoltaic system, the following factors are reviewed and implemented.

Table 2.2 Description of key factors governing the solar PV system efficiency [18–23]

Key factors	Details
Solar panel efficiency	The efficiency of the PV panels in converting sunlight into electricity. Higher-efficiency panels produce more power
Solar insolation	The amount of sunlight the location receives, which varies with geography, weather, and time of day/year. Higher insolation results in greater energy production
Tilt and orientation	The angle and orientation of the PV panels affect the angle at which they receive sunlight. Proper alignment optimizes energy capture
Shading	Shadows from trees, buildings, or other obstructions can reduce panel efficiency. Minimizing shading is crucial for optimal performance
Temperature	High temperatures can reduce the efficiency of solar panels. Cooler panels generally operate more efficiently
Dust and dirt	Accumulated dust and dirt on panels can reduce their efficiency. Cleaning and maintenance must be done on a regular basis
Inverter efficiency	The efficiency of the inverter, which converts DC electricity from the panels to AC for household use. Higher-efficiency inverters waste less energy
Wiring and connection losses	The quality of wiring and connections in the system can affect energy losses. Proper installation and maintenance are important
Angle of incidence	The angle at which sunlight strikes the panel affects energy capture. Tracking systems optimize this angle for maximum efficiency
Spectral mismatch	Mismatches between the solar spectrum and the panel's spectral response can reduce efficiency. Some panels are designed to minimize this effect
System degradation	Over time, solar panels can degrade in efficiency. Choosing high-quality panels and monitoring performance helps mitigate degradation
System design and sizing	Proper system design and sizing based on energy needs and available space are essential for efficient energy production
Reflection and albedo	The reflection of sunlight from surfaces and ground cover (albedo) can affect panel efficiency. Light-colored surfaces can increase reflection
Soiling and bird droppings	Accumulated soiling, such as bird droppings, can reduce panel efficiency and require regular cleaning
Component quality and aging	The quality of components used in the system, such as wires, connectors, and mounting structures, can impact long-term efficiency

These factors (Table 2.2) interact and vary depending on the specific location, system design, and maintenance practices. To achieve the highest efficiency in a solar PV system, it's essential to consider and optimize these factors during the planning, installation, and operation of the system. The current maximum efficiency of photovoltaic panels is 25%, which has increased in recent years. One of the most important factors in the efficiency of panels is their building modules. The maximum efficiency of solar panels reaches 43%, and countries that produce these panels, such as China, Germany, and Japan, provide a significant share. Factors such as direction, angle, temperature, radiation, etc. also affect the efficiency of modules. In addition to the quality of building modules, several environmental and situational factors come into play when optimizing solar panel efficiency. The orientation of panels, their tilt angle, and the geographical location all influence the amount of sunlight panels receive, impacting their overall efficiency. Moreover, temperature and radiation levels significantly affect the performance of solar panels. As researchers continue to delve into these factors, seeking ways to enhance solar panel efficiency, the solar energy sector is poised for even more remarkable advancements, making renewable energy a compelling and sustainable choice for the future.

2.3.1 Effect of irradiance

The amount of sunlight has a very important effect on the power produced by the panels. Sunlight is the source of energy that photovoltaic cells within the panels convert into electricity. When solar panels are exposed to ample sunlight, they can operate at their highest efficiency, producing the maximum amount of electrical power. The energy in sunlight is harnessed by the panels through the photoelectric effect, where photons of light knock electrons loose from atoms in the semiconductor material, generating an electric current. The more sunlight the panels receive, the greater the number of electrons set in motion, resulting in a higher power output. After checking the solar radiation in a city like Lahore at 10 a.m., it was seen that the highest amount of solar radiation is $558 \, W/m^2$ [18]. Reducing the amount of solar radiation causes a decrease in temperature, and almost every degree above 0.5% reduces the efficiency of the panel.

When solar radiation diminishes, it results in a drop in the temperature of the solar panels. Even a minor increase in temperature, exceeding just 0.5%, can lead to a noticeable reduction in the efficiency of the panels. Consequently, it is imperative to consider not only the geographical location but also the local climate and weather patterns when designing and installing solar panel systems. By carefully managing these factors, it is possible to maintain the highest possible efficiency of solar panels, ensuring a consistent and reliable source of clean energy even under varying weather conditions. This knowledge underscores the importance of accurate location-based planning and monitoring in the field of solar energy production.

2.3.2 Effect of shade and temperature

As already mentioned, the decrease in temperature reduces the efficiency of the panels. Shade is one of the factors that can significantly reduce the temperature of the panels. Shade can have a profound impact on the temperature of solar panels, which in turn affects their performance and efficiency. When solar panels are shaded, whether partially or completely, it disrupts the normal operation of the photovoltaic cells within the panels. As a result, the shaded areas of the panels tend to become cooler than the exposed areas. This temperature differential can lead to a condition known as "hot spots" on the panel's surface, where some areas are significantly warmer than others. These hot spots not only reduce the efficiency of the panel but can also damage the solar cells over time, ultimately shortening the lifespan of the entire system.

In addition to the negative effects on temperature, shade can also result in electrical losses. When solar cells are partially shaded, they may not generate electricity at all, or they may produce significantly less power than their unshaded counterparts. This is because the shaded cells create a bottleneck for the flow of electrical current within the panel. To mitigate the impact of shade, advanced technologies like bypass diodes are often integrated into solar panels. When the shadow phenomenon occurs, it causes some diodes to fail, which lowers the overall power of the remaining modules. This can happen in the form of two types of shadows: hard shade and soft shade. When soft shading occurs, the modules prevent high voltage drop, but when hard shading occurs, excessive voltage drop cannot be prevented [18].

2.3.3 Battery life protection

Preventing the battery from draining and increasing the battery life will ultimately help the system be more efficient. Increasing the load and increasing the life of the battery help ensure that our energy production does not stop completely when the sun is down [18]. By examining the regions of different countries, site selection, and evaluation of solar resources, it is possible to know which country has more capacity in the field of solar power plants. For example, India is located in a geographical location with an average level of sunlight. It is sunny for 8–10 hours a day. This radiation averages 4.5–6.5 kWh/m^2 during the day, which covers an area of more than 85% of the country's surface in India [11,24]. To establish a CSP (solar) power plant, the most basic issue is determining the area for the construction of the power plant. CSP power plants should be built in conditions where DNI≥ 5.5 kWh/m^2 per day (more than 1800 kWh/m^2) [5].

To evaluate and discover the best location, factors such as water, land, etc. are considered. Components such as water supply, transportation, internet (network), type of soil, cost of land, amount of investment, environmental protection, etc. are the most important things that need to be considered for the establishment of a power plant. Hence, in India Udaipur, Rajasthan is the best study area for the construction of the CSP plant. Because DNI during the year in this geographical location has a higher average than in other regions of India. The solar field in the region prepares a report on the amount of solar radiation based on absorption. It is possible to check the conditions of the solar thermal power plant based on the simulation of the hypothetical CSP power plant in any place. This simulation includes regional information such as hourly DNI, wind speed, ambient temperature, atmospheric pressure, sun angle, and solar azimuth angle for the whole year [11]. On average, annually 2248.17 kWh/m²/year DNI is charged in the Udaipur region.

2.4 ADVANTAGES AND DISADVANTAGES OF THE SOLAR ENERGY SYSTEM

It's important to note that the advantages and disadvantages (Table 2.3) can vary depending on factors such as location, system size, and technology used. In addition, ongoing advancements in solar technology and energy storage solutions continue to address some of the disadvantages, making solar energy a more viable and sustainable option.

Table 2.3 Key advantages and disadvantages of using renewable solar energy systems [25–28]

Advantages of renewable solar energy systems	Disadvantages of renewable solar energy systems
Clean and Sustainable: Solar energy is a renewable resource, and its generation produces no greenhouse gas emissions	Solar power is an intermittent energy source since it is affected by weather and daylight
Reduced Energy Costs: Once installed, solar systems can lead to significant cost savings on electricity bills	**High Initial Costs:** The upfront installation costs of solar panels and equipment can be substantial
Low Operating and Maintenance Costs: Solar systems require minimal maintenance and have low ongoing operational costs	**Land and Space Requirements:** Large-scale solar installations need significant land or space, which can be a limitation
Grid Independence: Solar panels can provide energy independence, reducing reliance on the grid and enhancing energy security	**Energy Storage Needs:** Energy storage solutions (e.g., batteries) are often required for reliable power during the night or cloudy days
Job Creation: Manufacturing, installation, and maintenance employment are all created by the solar sector	**Environmental Impact:** Manufacturing solar panels can have environmental consequences, and their disposal must be managed responsibly

(Continued)

Table 2.3 (Continued) Key advantages and disadvantages of using renewable solar energy systems [25–28]

Advantages of renewable solar energy systems	Disadvantages of renewable solar energy systems
Scalability: Solar energy systems may be readily scaled up or down to fulfill a variety of energy requirements	**Aesthetic Concerns:** Some people find solar panel installations unattractive or incompatible with certain architectural styles
Reduces Carbon Footprint: Solar energy helps reduce greenhouse gas emissions and combat climate change	**Geographic Limitations:** Solar power may not be as effective in areas with low sunlight or high cloud cover
Incentives and Rebates: Many countries provide rebates and tax credits to encourage solar adoption	**Energy-Intensive Manufacturing:** The production of solar panels can be energy-intensive

As we look to the future of sustainable renewable solar energy systems, there are several exciting research directions and potential areas for innovation. One key focus is enhancing energy storage solutions to address the intermittent nature of solar power. Researchers are exploring advanced battery technologies, such as next-generation lithium-ion batteries, solid-state batteries, and flow batteries, to improve energy storage capacity and efficiency. In addition, integrating renewable energy systems with energy storage and smart grid technologies is a promising avenue for achieving a more stable and reliable energy supply that can meet the demands of both urban and rural settings.

Another exciting research area involves developing innovative solar panel materials and designs. Scientists are working on novel materials that can boost the efficiency of solar panels, making them more cost-effective and environmentally friendly. Thin-film solar cells, perovskite solar cells, and other emerging technologies offer the potential for higher energy conversion rates and lower production costs. Moreover, exploring ways to integrate solar technology into building materials and infrastructure, such as solar-integrated windows, walls, and roads, presents an intriguing pathway for harnessing solar energy in urban environments. These innovative approaches aim to maximize energy generation while minimizing the environmental footprint of solar systems, ultimately contributing to a more sustainable and renewable energy future. Table 2.4 outlines some future directions and potential research areas in sustainable renewable solar energy systems.

These research areas represent ongoing efforts to advance and improve sustainable renewable solar energy systems, making them more efficient, cost-effective, and environmentally friendly.

These developments (Table 2.5) in solar energy storage technologies are crucial for ensuring a stable and reliable energy supply from renewable sources and for maximizing the utilization of solar power, regardless of weather conditions or time of day.

Table 2.4 Future directions and potential research areas in solar energy systems [29–33]

Future directions and research area	Details
Solar energy storage	Developing advanced energy storage technologies (e.g., next-gen batteries) for nighttime or cloudy day power supply. Research on improving energy density, cycle life, and cost-effectiveness
Solar panel efficiency	Enhancing the efficiency of photovoltaic panels through innovative materials, designs, and manufacturing processes to maximize energy capture
Perovskite solar cells	Continued research on perovskite solar cells to improve stability, toxicity concerns, and scale production for commercial use
Tandem solar cells	Tandem solar cells that combine multiple materials to collect a greater range of the light spectrum, thereby boosting efficiency, are being researched
Building-integrated PV	Integrating solar panels into building materials (e.g., solar roof tiles) to make solar power more accessible and aesthetically pleasing
Off-grid solutions	Developing decentralized solar systems for remote and off-grid areas, including microgrids and portable solar solutions
Solar farm optimization	Optimizing large-scale solar farms for increased energy output, reduced land use, and efficient grid integration
Grid integration	Research on smart grid technology and grid management to better integrate solar power, mitigate grid instability, and improve energy distribution
Environmental impact	Assessing and mitigating the environmental impact of solar panel production, disposal, and land use, including recycling solutions
Economic viability	Studying the economic feasibility of solar projects, including cost-benefit analyses, financing models, and market development strategies
Policy and regulation	Research on supportive policies, incentives, and regulatory frameworks to encourage solar adoption and sustainable practices
Hybrid solar systems	Integrating solar with other renewable sources (e.g., wind, hydro) for hybrid systems that offer reliable and consistent power generation
AI and data analytics	Implementing artificial intelligence and data analytics for improved solar system performance monitoring, predictive maintenance, and energy optimization
Energy internet	Exploring the concept of an "energy internet" where distributed solar and renewable energy sources are connected seamlessly for efficient energy sharing
Research area	Description

Table 2.5 Latest developments in solar energy storage technologies [34–36]

Latest solar energy storage technologies	Details
Advanced lithium-ion batteries	Ongoing research has led to improvements in lithium-ion battery technology, resulting in higher energy density, longer cycle life, and reduced costs
Solid-state batteries	Solid-state batteries offer increased safety, faster charging, and improved energy density compared to traditional liquid electrolyte batteries. Development is ongoing to bring them to market
Flow batteries	Flow batteries, including redox flow and solid oxide fuel cells, have seen advancements in materials, which improve efficiency and lower costs for large-scale energy storage
Sodium-ion batteries	With recent advancements in cathode materials and performance, sodium-ion batteries are being explored as a lower-cost alternative to lithium-ion batteries
Supercapacitors	Supercapacitors, or ultra-capacitors, are being explored for their ability to store and release energy rapidly, making them suitable for short-term energy storage applications
Thermal energy storage	Advances in thermal energy storage systems include innovative materials and designs, such as phase change materials and high-temperature molten salt systems
Hydrogen storage	Hydrogen can be produced using excess solar power and stored for later use in fuel cells or for power generation, and research is ongoing to improve hydrogen storage and release efficiency
Graphene-based storage	Graphene-based materials have shown potential for energy storage, with research focusing on improving capacitance and charge-discharge capabilities
Machine learning and AI	AI and machine learning are used for predictive maintenance and optimizing energy storage system performance. These technologies can maximize the value of stored solar energy
Grid-integrated storage	Grid-integrated storage systems, with improved communication and control, allow for more efficient and stable grid operation by providing power during peak demand and grid fluctuations
Vanadium redox flow batteries	Vanadium redox flow batteries (VRFBs) have advanced in terms of energy efficiency, reduced electrolyte costs, and system longevity, making them more competitive for large-scale storage

According to the United Nations, China now has the largest solar energy capacity (Table 2.5). This success is better understood by knowing that China is the largest producer of solar panels in the world, followed by Japan and then America. This success can be attributed to China's strategic focus on renewable energy development and its commitment to reducing carbon emissions. With its massive manufacturing capabilities, China has not only become the largest consumer of solar panels but also the world's top producer of these critical components. This vertically integrated approach has allowed China to drive down the cost of solar panel production while scaling up its installation of solar energy systems. Japan, following closely behind China in the global rankings, has also made significant strides in the solar energy sector. The nation's emphasis on clean energy was further intensified after the Fukushima nuclear disaster, which led to a major shift in its energy policies. Japan's geographical location and technological expertise make it well-suited for solar energy generation. In addition, its commitment to research and development has led to more efficient and cost-effective solar technologies. America, with its vast landmass and considerable solar potential, is not far behind in this global race for solar energy capacity. The United States has made substantial investments in renewable energy, and with its diverse geography, it has harnessed solar power across various regions, from sunny deserts to urban environments. The competition among these major players reflects a global trend toward cleaner and more sustainable energy sources, demonstrating the importance of solar energy in mitigating climate change and meeting the world's growing energy demands. In Table 2.6, three Asian countries are among the top five countries in the world in this field. Asia is leading with 274,866 MW with 56.58% of the total capacity. That is more than half of the total solar energy installed in the world [12].

Table 2.6 Solar energy installed capacity region wise

Region	Solar energy installed capacity (2018)	Country	S/N
Asia	175.03	China	1
Asia	55.50	Japan	2
North America	51.45	USA	3
Europe	45.93	Germany	4
Asia	27.98	India	5
Europe	20.13	Italy	6
Europe	13.11	United Kingdom	7
Oceania	9.77	Australia	8
Europe	9.48	France	9
Asia	7.86	Korea Republic	10

2.5 CONCLUSION

So far, a lot of work has been done to optimize and increase the efficiency of solar panels. Many things are done in the study of optimization and planning for the implementation of panels including the type of panel, how the modules are placed, the effect of the direction of the panels, the effect of the angle of the panels, the battery system, the regulator modules, and the effect of the radiation point on the pane. Some of the effective factors in the productivity of panels are related to the geographical properties of the environment such as the shadow of the clouds, the length of the day, the temperature of the environment, etc. Radiation power, ambient temperature, sun power, absence of shadow and clouds, etc., are the most basic parameters in installing solar panels. The Gulf countries, which are at sea level, are a good place to install solar panels due to their geographical conditions and desert state. The surge in population and the rapid expansion of the tourism and construction sectors have resulted in a heightened need for energy. In this context, regions with desert beaches, like Qatar, Oman, Kuwait, and the United Arab Emirates, offer substantial potential for increased solar energy production. These areas, characterized by their sunny and arid conditions, are particularly well-suited to meet the growing electricity demands driven by factors such as population growth, tourism, and construction activities. Consequently, they present an opportunity for enhanced productivity in harnessing solar energy compared to other regions. In some cases, the sea level can be installed with solar or floating panels such as the lakes. Coastal shopping centers and tall buildings in these places are very strong points for the installation and efficiency of solar panels.

REFERENCES

1. Soomar, A.M., et al., Solar photovoltaic energy optimization and challenges. *Frontiers in Energy Research*, 2022. 10: p. 879985.
2. Behura, A.K., et al., Towards better performances for a novel rooftop solar PV system. *Solar Energy*, 2021. 216: pp. 518–529.
3. Almusaied, Z., B. Asiabanpour, and S. Aslan, Optimization of solar energy harvesting: An empirical approach. *Journal of Solar Energy*, 2018: 8 pp. https://doi.org/10.1155/2018/9609735
4. Adelakun, N.O. and B.A. Olanipekun, A review of solar energy. *Journal of Multidisciplinary Engineering Science and Technology (JMEST)*, 2019.6(12)
5. Wang, W.-Q., R. Jiang, Y.-L. He, and D. Li, Optical-thermal-mechanical analysis of high-temperature receiver integrated with gradually sparse biomimetic heliostat field layouts for the next-generation solar power tower. *Solar Energy*, 2022. 232: pp. 35–51.
6. Zhang, X., X. Dong, and X. Li, Study of China's optimal concentrated solar power development path to 2050. *Frontiers in Energy Research*, 2021. 9: p. 724021.

7. Alotaibi, S., F. Alotaibi, and O.M. Ibrahim, Solar-assisted steam power plant retrofitted with regenerative system using parabolic trough solar collectors. *Energy Reports*, 2020. 6: pp. 124–133.

8. Radosevic, N., et al., Solar energy modeling and mapping for the sustainable campus at Monash University. *Frontiers in Sustainable Cities*, 2022. 3: pp. 745197.

9. Li, A., Y. Xu, and H. Shiroyama, Solar lobby and energy transition in Japan. *Energy Policy*, 2019. 134: pp. 110950.

10. Bishoyi, D. and K. Sudhakar, Modeling and performance simulation of 100 MW LFR based solar thermal power plant in Udaipur India. *Resource-Efficient Technologies*, 2017. 3(4): pp. 365–377.

11. Assadeg, J., K. Sopian, and A. Fudholi, Performance of grid-connected solar photovoltaic power plants in the Middle East and North Africa. *International Journal of Electrical and Computer Engineering*, 2019. 9(5): pp. 3375–3383.

12. Ferahtia, A., Surface water quality assessment in semi-arid region (El Hodna Watershed, Algeria) based on water quality index (WQI). *Studia Ubb Chemia*, 2021. LXVI(1): pp. 127–142. DOI:10.24193/subbchem.2021.1.10

13. Meena, C.S., A.N. Prajapati, A. Kumar, and M. Kumar, Utilization of solar energy for water heating application to improve building energy efficiency: An experimental study. *Buildings*, 2022. 12: p. 2166. https://doi.org/10.3390/buildings12122166.

14. Meena, C.S., A. Kumar, S. Jain, A. Ur Rehman, and S. Mishra, Innovation in green building sector for sustainable future. *Energies*, 2022. 15: p. 6631. https://doi.org/10.3390/en15186631.

15. Dutt, N., Hedau, A. J., Kumar, A., Awasthi, M. K., Singh, V. P., and Dwivedi, G., Thermo-hydraulic performance of solar air heater having discrete D-shaped ribs as artificial roughness. *Environmental Science and Pollution Research*, 2023. doi: 10.1007/s11356-023-28247-9.

16. Kushwaha, P.K., N.K. Sharma, A. Kumar, and C.S. Meena. Recent advancements in augmentation of solar water heaters using nanocomposites with PCM: Past, present, and future. *Buildings*, 2023. 13: p. 79. https://doi.org/10.3390/buildings13010079.

17. Singh, V.P., Jain, S., Karn, A. et al., Experimental assessment of variation in open area ratio on thermohydraulic performance of parallel flow solar air heater. *Arabian Journal for Science and Engineering*, 2023. 48: pp. 11695–11711. https://doi.org/10.1007/s13369-022-07525-7

18. Mills, D. and G. Morrison, Modelling study for compact Fresnel reflector power plant. *Le Journal de Physique IV*, 1999. 9(PR3): pp. Pr3-159–Pr3-165.

19. Singh, V.P., S. Jain, A. Kumar, and S. Mishra, and N.K. Sharma, Heat transfer and friction factor correlations development for double pass solar air heater artificially roughened with perforated multi-V ribs. *Case Studies in Thermal Engineering*, 2022. 39: p. 102461. https://doi.org/10.1016/j.csite.2022.102461.

20. Singh, V.P., S. Jain, A. Karn, A. Kumar, G. Dwivedi, and C.S. Meena, Recent developments and advancements in solar air heaters: A detailed review. *Sustainability*, 2022. 14: p. 12149. https://doi.org/10.3390/su141912149.

21. Singh, V.P., S. Jain, A. Karn, A. Kumar, G. Dwivedi, C.S. Meena, and R. Cozzolino, Mathematical modeling of efficiency evaluation of double-pass parallel flow solar air heater. *Sustainability*, 2022. 14: p. 10535. https://doi.org/10.3390/su141710535.

22. Singh, V.P., S. Jain, and A. Kumar, Establishment of correlations for the thermo-hydraulic parameters due to perforation in a multi-V rib roughened single pass solar air heater. *Experimental Heat Transfer*, 2022. 89(2): pp. 1–21. https://doi.org/10.1080/08916152.2022.2064940.

23. Saxena, A., A.N. Prajapati, G. Pant, C.S. Meena, A. Kumar, and V.P. Singh, Water consumption optimization of hybrid heat pump water heating system. In: Shukla, A. K., Sharma, B. P., Arabkoohsar, A., Kumar, P. (eds) *Recent Advances in Mechanical Engineering. FLAME 2022. Lecture Notes in Mechanical Engineering*. Springer, Singapore, 2023. https://doi.org/10.1007/978-981-99-1894-2_61.

24. Kaygusuz, K., Prospect of concentrating solar power in Turkey: The sustainable future. *Renewable and Sustainable Energy Reviews*, 2011. 15(1): pp. 808–814.

25. Verma, V., C.S. Meena, S. Thangavel, A. Kumar, T. Choudhary, and G. Dwivedi, Ground and solar assisted heat pump systems for space heating and cooling applications in the northern region of India – A study on energy and CO2 saving potential. *Sustainable Energy Technologies and Assessments*, 2023. 59: p. 103405. https://doi.org/10.1016/j.seta.2023.103405.

26. Awasthi, M.K., N. Dutt, A. Kumar, S. Kumar, Electrohydrodynamic capillary instability of Rivlin-Ericksen viscoelastic fluid film with mass and heat transfer. *Heat Transfer*, 2023. 53(1): pp. 1–19. https://doi.org/10.1002/htj.22944.

27. Meena, C.S., A. Kumar, S. Roy, A. Cannavale, and A. Ghosh, Review on boiling heat transfer enhancement techniques. *Energies*, 2022. 15: p. 5759. https://doi.org/10.3390/en15155759.

28. Kundu, A., A. Kumar, N. Dutt, C.S. Meena, and V.P. Singh, Introduction to thermal energy resources and their smart applications. In: Kumar, A., Singh, V. P., Meena, C. S., Dutt, N. (eds) *Thermal Energy Systems: Design, Computational Techniques and Applications*. CRC Press, Boca Raton, FL, 2023, pp. 1–15; Chapter 1. https://doi.org/10.1201/9781003395768-1.

29. Dewangan, A.K., S.Q. Moinuddin, M. Cheepu, S.K. Sajjan, and A. Kumar, Thermal energy storage: Opportunities, challenges and future scope. In: Kumar, A., Singh, V. P., Meena, C. S., Dutt, N. (eds) *Thermal Energy Systems: Design, Computational Techniques and Applications*. CRC Press, Boca Raton, FL, 2023, pp. 17–28; Chapter 2. https://doi.org/10.1201/9781003395768-2.

30. Kundu, A., A. Kumar, N. Dutt, V.P. Singh, and C.S. Meena, Modelling and simulation of thermal energy system for design optimization. In: Kumar, A., Singh, V. P., Meena, C. S., Dutt, N. (eds) *Thermal Energy Systems: Design, Computational Techniques and Applications*. CRC Press, Boca Raton, FL, 2023, pp. 103–140; Chapter 7. https://doi.org/10.1201/9781003395768-7.

31. Pant, G., C.S. Meena, A. Saxena, A. Kumar, V.P. Singh, and N. Dutt, Study the temperature variation in alternate coils of insulated condenser cum storage tank: Experimental study. In: Sikarwar, B. S., Sharma, S. K., Jain, A., Singh, K. M. (eds) *Advances in Fluid and Thermal Engineering. FLAME 2022. Lecture Notes in Mechanical Engineering*. Springer, Singapore, 2023, Chapter 51. https://doi.org/10.1007/978-981-99-2382-3_52.

32. Kumar, A., V.P. Singh, C.S. Meena, and N. Dutt, *Thermal Energy Systems: Design, Computational Techniques and Applications*. Taylor & Francis (CRC Press, Boca Raton, FL, 2023. https://doi.org/10.1201/9781003395768.

33. Singh, S., A. Kumar, S.K. Behura, and K. Verma, Challenges and opportunities in nanomanufacturing. In: Singh, S., Behura, S. K., Kumar, A., Verma, K. (eds) *Nanomanufacturing and Nanomaterials Design: Principles and Applications.* CRC Press, Boca Raton, FL, 2022, pp. 17–30; Chapter 02. https://doi.org/10.1201/9781003220602-2.

34. Srivastava, S., D. Verma, S. Thusoo, A. Kumar, V.P. Singh, and R. Kumar, Nanomanufacturing for energy conversion and storage devices. In: Singh, S., Behura, S. K., Kumar, A., Verma, K. (eds) *Nanomanufacturing and Nanomaterials Design: Principles and Applications.* CRC Press, Boca Raton, FL, 2022, pp. 165–173; Chapter 10. https://doi.org/10.1201/9781003220602-10.

35. Singh, V.P., A. Dwivedi, A. Karn, A. Kumar, S. Singh, S. Srivastava, K. Srivastava, Nanomanufacturing and design of high-performance piezoelectric nanogenerator for energy harvesting. In: Singh, S., Behura, S. K., Kumar, A., Verma, K. (eds) *Nanomanufacturing and Nanomaterials Design: Principles and Applications.* CRC Press, Boca Raton, FL, 2022, pp. 241–242; Chapter 15. https://doi.org/10.1201/9781003220602-15.

36. Dada, M. and P. Popoola, Recent advances in solar photovoltaic materials and systems for energy storage applications: A review. *Beni-Suef University Journal of Basic and Applied Sciences*, 2023. 12: p. 66. https://doi.org/ 10.1186/s43088-023-00405-5.

Chapter 3

Thermal energy storage technologies and their applications

Naveen Sharma
Netaji Subhas University of Technology

Sunil Baloda
The Technological Institute of Textile and Sciences

Nazrul Islam Khan
Netaji Subhas University of Technology

3.1 INTRODUCTION

In general, energy sources fall into two groups: primary- and second-ary-energy sources [1]. Primary energy sources exist in nature and are typically found in their raw, unprocessed state. They have not been converted or transformed. Primary energy sources include coal, crude oil, biomass, solar energy, wind, tidal energy, natural uranium (used in nuclear energy), geothermal energy, flowing water (hydropower), natural gas, etc. To use energy contained in the primary energy sources for producing heat or mechanical work, various extraction, separation, cleaning, and grading processes must be undertaken. Secondary energy sources are derived from conversion of the primary energy sources, using diverse energy conversion techniques, thereby making them more suitable or usable forms of energy for a particular application. For example, electricity is a common secondary energy form that is generated from different primary energy sources. In our homes, industries, and transportation systems, we typically use secondary energy sources.

Understanding the distinction between primary- and secondary energy is essential in policymaking and energy planning, as it helps in managing the energy supply chain and optimizing the utilization of different energy sources to meet the diverse needs of society. A diagram indicating the link between primary and secondary forms of energy is shown in Figure 3.1 [1,2]. It clearly indicates that primary energy sources are the starting point, and secondary energy forms are the intermediate or final energy products derived from them using various conversion processes.

DOI: 10.1201/9781003472629-3

Primary energy   **Secondary energy**

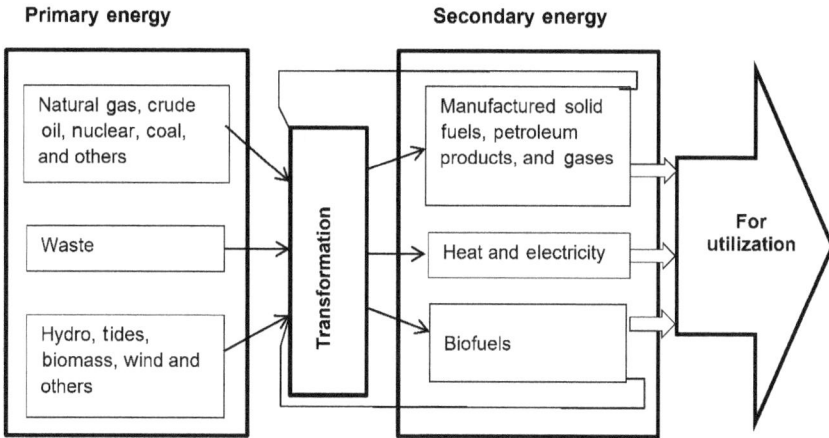

Figure 3.1 Relationship between primary and secondary energy [2].

The direct usability of secondary energy forms makes them particularly useful in the realm of energy. These secondary forms, sometimes called Energy Carriers, include electricity, diesel, gasoline, hydrogen, and heat. In Table 3.1, the various primary and secondary energy forms along with associated technologies are illustrated.

In power generation systems based on solar- and wind energy, the challenge of efficiently managing excess energy production and supplying it during peak demand hours has given rise to the need for advanced energy storage solutions known as hybrid ES systems. These systems are gaining increasing importance in the context of the growing imperative to combat global warming [3,4]. To meet the rising energy demands associated with variable energy sources, ongoing advancements in renewable energy production and their storage technologies are imperative [5].

Further, thermal energy storage (TES), a specific form of ES that focuses on storing thermal energy, allows for efficient storage and utilization of the end-use energy share that is available in the form of heat. It is often used for load leveling and helps in addressing the prominent issue of energy mismatch between when energy is needed and when it's generated, notably in systems with intermittent RESs.

In summary, ES, particularly TES, is a vital component of modern energy systems. It addresses multiple challenges in the energy sector, from grid stability to renewable energy integration and efficiency improvement, while also contributing to the reduction of fossil fuel usage and environmental impact. The role of TES will only become more prominent as the world shifts to a cleaner and more sustainable energy future.

Table 3.1 Conversion practices for primary energy to secondary energy [2]

S. n.	Primary energy	Conversion method	Secondary energy
1.	Uranium	Nuclear power plant	Heat, work, electricity
2.	Geothermal	Geothermal power plant	
3.	Coal	Thermal power plant	
4.	Solar	Solar thermal power plant	
5.		Photovoltaic power plant	Heat, electricity
6.	Wind	Wind farm	Work, electricity
7.	Tidal	Tidal power plant	
8.	Flowing or falling water	Hydro power plant	
9.	Crude oil	Oil refinery	Petrol, diesel
10.	Biomass	Biorefinery, thermal power plant	Heat, work, electricity, biofuel, biodiesel

3.2 MOTIVATIONS FOR ENERGY STORAGE

Energy storage stands as a vital pillar within the energy supply chain, addressing several critical imperatives in modern energy systems. Despite the ingenuity of engineering, no system achieves perfect thermodynamic efficiency. In most instances, energy losses manifest when heat dissipates into the surroundings. Remarkably, this waste heat represents a valuable source that, if harnessed and stored, can be repurposed to power other processes, fostering greater overall energy efficiency.

Beyond the realm of waste heat recovery, ES plays a pivotal role in the transition toward a low-carbon economy, where RESs take center stage. Notably, renewable resources like solar and wind power are characterized by being intermittent, rendering them unreliable for a consistent energy supply. However, through the implementation of ES solutions, this intermittent hurdle can be overcome, transforming RESs into dependable and steady energy sources [6].

The concept is simple yet transformative: surplus energy generated during periods of renewable resource abundance can be efficiently stored for later use when those resources are less available. This storage and subsequent discharge of energy allow for a more reliable and resilient energy supply, thereby smoothing out the inherent variability of renewable sources. As the world continues its journey toward sustainability and reduced carbon emissions, energy storage emerges as a linchpin technology, bridging the gap between intermittent renewables and uninterrupted energy provision.

3.2.1 Concept of ES

The fundamental idea underlying the storage of secondary energy is to collect energy generated at one point in time and reserve it for future use. Charging involves gathering energy while releasing it for practical

Production   **Storage**   **Regeneration**

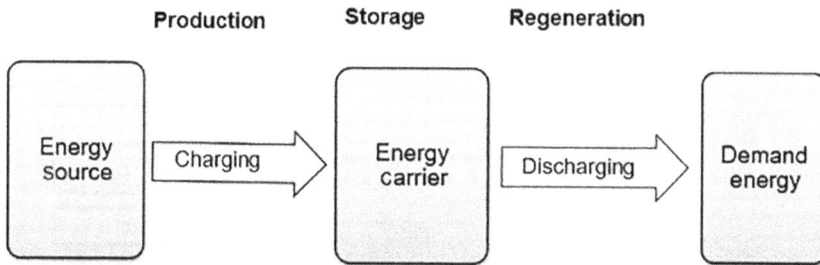

Figure 3.2 Concept of energy storage [7].

application, which is referred to as "discharging." Figure 3.2 provides a visual representation of the ES concept.

3.2.2 Advantages of ES

The importance of energy storage in today's energy landscape cannot be overstated. As mentioned, ES plays a key role in several aspects of the modern energy supply chain, and its significance is only growing. By optimizing the use of energy resources and reducing waste, ES helps in conserving fossil fuel resources, which is crucial from the perspective of addressing environmental change and transitioning to cleaner energy sources [8]. ES is indeed a crucial component in the field of energy management, and it offers several benefits, as outlined below:

- The ES system decreases the dependence on conventional fuels and reduces greenhouse gas emissions drastically, thereby indirectly improving the surrounding environment [8].
- The use of ES systems contributes to grid stability by storing excessive energy during off-peak hours and by distributing it when needed, thus overcoming variations in energy demand and supply [9].
- ES systems allow the capturing and storing of excessive energy, from RESs like solar and wind, and consequently, provide flexibility to use it when required, thereby increasing the penetration of renewables in the grid.
- By using energy storage, the overall efficiency of energy systems can be significantly improved.

3.3 CLASSIFICATION OF ES TECHNOLOGIES

There are numerous technologies employed for the purpose of storing energy. The choice of technique depends on factors like the application, duration of storage needed, and system size. The ES technologies can be broadly categorized based on the intended use of the stored energy and

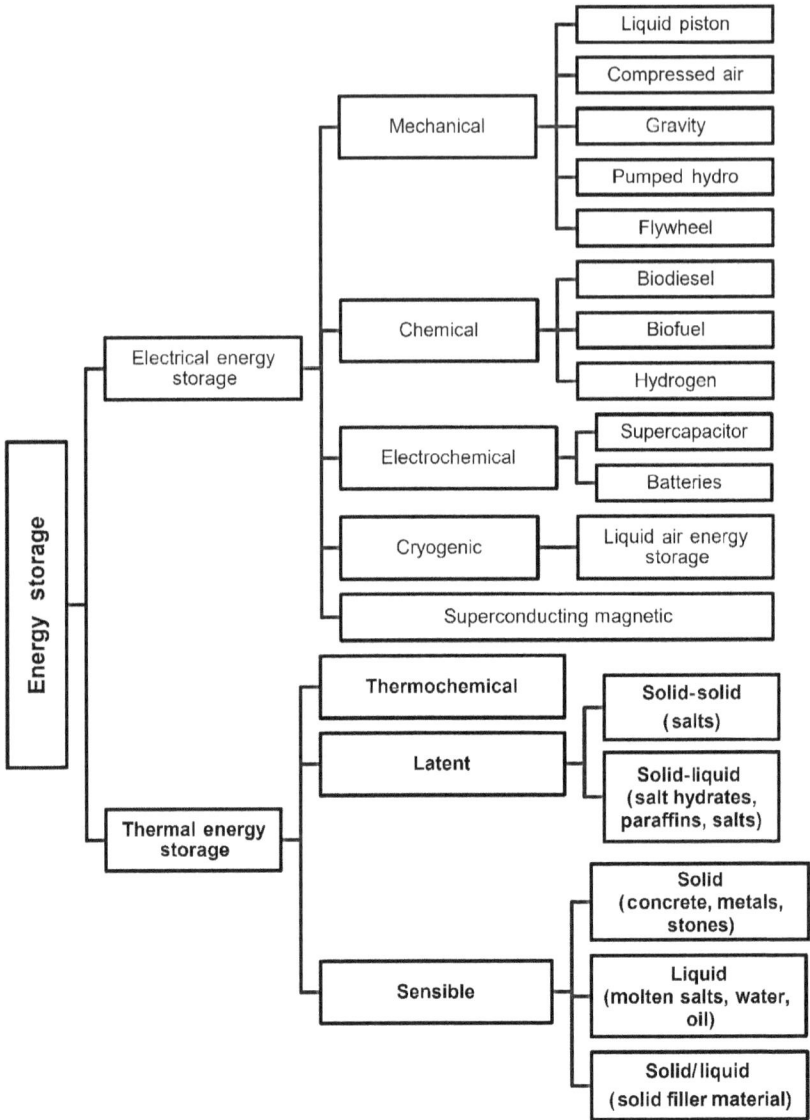

Figure 3.3 Classification of energy storage technologies [9].

their classification is illustrated in Figure 3.3. In the diagram, it's evident that technologies that store and subsequently release it in thermal form, like ice/chilled water storage, fall under the category of thermal energy storage technologies. Conversely, technologies that store energy in thermal form but release it as electrical energy, for instance liquid air energy storage, are categorized as electrical energy storage technologies.

3.4 THERMAL ENERGY STORAGE

This section focuses on TES technologies, materials used and their applications. TES indeed plays a key role in integrating high shares of solar and wind power into the energy grid. This is because both solar and wind power sources are intermittent and subject to fluctuations, as they depend on weather conditions and the time of day. Thermal energy storage helps address some of the challenges associated with these renewable energy sources by providing flexibility in several ways: energy shifting, grid stabilization, improved dispatchability, peak load management, reliability, and energy resilience [6].

In TES systems, there are three types: *sensible-, latent-, and thermochemical- heat storage.* Each has its benefits and is suitable for specific applications. By deploying a combination of these technologies strategically, it becomes possible to maximize the benefits of TES and integrate high shares of RESs effectively, reducing the reliance on fossil fuels and decreasing GHG emissions in the process.

3.4.1 Advantages of TES

In real-time scenarios, TES stands as a superior remedy for addressing issues related to energy management. TES systems play a crucial role in balancing seasonal demand variations and facilitating the evolution toward a predominantly renewable energy-based ecosystem [6]. The implementation of TES systems in various applications can yield numerous significant benefits, including [10]:

- Enhancement of the current system's efficiency.
- Reduction in the size and capacity of the available system.
- Decoupling the heating and cooling requirements from the power source.
- Efficient utilization of energy.
- Improved reliability and flexibility in operations.
- A reduction in overall costs results in economic benefits.
- An energy backup system that smooths out critical or partial loads in case of an emergency.
- Effective utilization of intermittent energy generated from RESs, thereby reducing the need for expensive grid upgrades.

3.5 TES METHODS

Thermal energy can be stored using a variety of techniques, depending on factors such as the material used for storage, the process for storing and retrieving the energy, and others. In the coming sub-sections, we will explore the three primary methods commonly used for TES: sensible, latent, and thermochemical storage.

3.5.1 Sensible heat storage

Sensible heat storage (SHS) is the most widely adopted and commercially developed form of TES, which involves energy storage by heating/cooling a storage medium without causing a change in its phase. Within the operational temperature range, energy stored depends directly on the temperature difference reached during charging/discharging and the thermal capacity of the storage medium [11].

Typically, the storage capacities of SHS systems range from 10 to 50 kWh per metric ton, with storage efficiencies varying from 50% to 98%. However, these values can vary, depending on the storage medium's thermal capacity and insulation technology used. Their operational temperature limit can extend from −160°C to over 1000°C [12,13].

In the SHS process, thermal energy is stored within a medium by altering its temperature, thereby leading to an increase in its internal energy. This stored energy is dependent on the mass (m), thermal capacity (C_p), and temperature change (ΔT) of the material and can be quantified using Eq. (3.1).

$$Q = \int_{T_i}^{T_f} mC_p\Delta T \tag{3.1}$$

where Q is the amount of energy stored, while T_i and T_f are the initial and final temperatures, respectively.

3.5.2 Latent heat storage (LHS)

In LHS, an ES material known as phase change material (PCM) is utilized. This material undergoes a physical phase transition during both the energy storage and discharge processes. These transformations can involve changing from a solid to a liquid state, a liquid to a gaseous state, or the reverse, depending on the specific application. Throughout both the energy storage and retrieval processes, the temperature remains relatively constant or very close to constant.

In hot storage, the solid PCM absorbs heat, causing it to transition from a solid to a liquid state. During discharge, the liquid PCM releases heat and transforms back into a solid. Conversely, cold storage causes the liquid PCM to turn solid by absorbing cold energy and vice versa while releasing the stored cold energy. LHS energy storage can be quantified using Eq. (3.2).

$$Q = \int_{T_i}^{T_m} mC_p\Delta T + mH + \int_{T_m}^{T_f} mC_p\Delta T \tag{3.2}$$

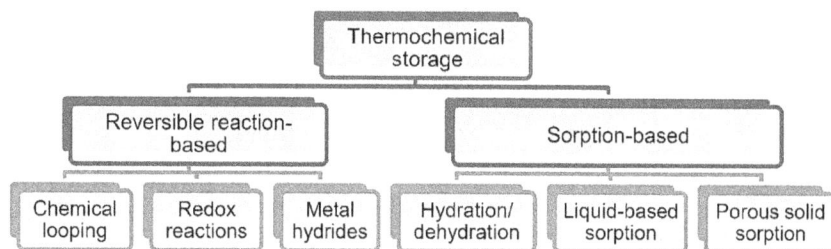

Figure 3.4 TCS methods [14].

where Q denotes amount of energy stored, H denotes latent heat, and T_i, T_f and T_m denote initial, final, and melting temperature of the storage medium, respectively.

3.5.3 Thermochemical storage (TCS)

Compared to SHS and LHS, the TCS provides a higher energy density. TCS can be categorized into two main types: *reversible reaction-based and sorption-based energy storage*, as outlined in Figure 3.4. Thermochemical systems, excluding those relying on sorption, operate through a reversible chemical reaction between two distinct substances. This process results in the generation of a substantial amount of energy due to an exothermic synthesis reaction.

A sorption process stores heat by altering the chemical potential of the sorbent and sorbate by disrupting the binding force. Sorption storage, however, is limited to working at temperatures up to around 350°C. In contrast, thermochemical systems devoid of sorption can function at higher temperatures, offering increased energy densities.

3.6 TES MATERIALS

There are three ways in which heat can be stored in TES materials: sensible, latent, and thermochemical heat. Sensible heat is the form of energy stored in materials that results in a rise or fall in material temperature. LHS materials do not change the current temperature but change the current phase of the material (e.g., PCM). The energy stored is converted to enthalpy change in the materials during phase change. Thermochemical energy stored in materials is a reversible process. Energy is stored in the formation of a compound through an endothermic reaction and is released when the reversible reaction occurs in the compound through exothermic reaction. The energy in terms of heat stored or released is equivalent to the enthalpy of the reaction. Sensible and latent heat are stored as energy transfer without any mass transfer, and thermochemical energy is the total enthalpy of reaction of the materials stored in a container [15].

3.6.1 Physical properties of TES materials

The following physical properties should be present in the energy storage materials for effective operation of TES [16]:

1. Heat capacity, latent heat, and heat of reaction should be high in the materials.
2. Materials should have high density to have better volumetric storage capacity.
3. Service life should be very high.
4. The materials should be non-toxic.
5. The materials should be non-explosive.
6. It should be easy to handle, i.e., good hygroscopic.
7. It should be non-corrosive with the container.
8. It should not be costly.
9. Able to withstand a greater number of charging/discharging cycles without degrading.
10. Very low-density changes with temperature difference to counter thermal and mechanical stresses.

3.6.2 SHS materials

As discussed above, sensible heat only results in temperature rise or fall without any phase change. Different materials also in different phases have different SHS capacities as calculated using Eq. (3.1) [17]. In general, all materials have different sensible heat capacities, and due to this, the heat storage capacity also varies.

In the liquid phase, heat can be stored and transferred without the need for a separate heat transfer fluid (HTF). Most widely used liquids for SHS are water and thermal oil (synthetic liquid, polyethylene glycol mineral oil, etc.). The temperature and physical properties of some characteristic liquids are listed in Table 3.2. Water is mostly used as a SHS medium because it is abundant, non-toxic, non-explosive, etc. However, it has a low operating temperature, is corrosive, and has a high vapor pressure. Most widely used liquid TES medium is water as it is abundant, has low thermal diffusivity, storage capability, flow control capability, is non-toxic, non-combustible, and is mixable with additives. However, it has limitations such as low operating temperature, corrosiveness, high vapor pressure, etc.

For high operating temperature, mineral oil can be employed up to around 300°C. For above 400°C, synthetic oils are suitable. In addition, above 100°C, molten salts are also very good for SHS medium [15]. However,

Table 3.2 Physical properties of liquid SHS materials

Material	T (°C)	ρ (kg/m³)	C_p (kJ/ (kg.K))	λ (W/ (m.K))	$10^6 \times a$ (m²/s)	$10^{-3} \times b$ (J/ (m²Ks^{1/2}))
Molten salt	230	1950	1.57	0.5	0.16	1.24
Silicone oil	25	970	1.47	0.17	0.12	0.49
Transformer oil	60	842	2.09	0.12	0.07	0.46
Water	20	998	4.18	0.598	0.14	1.58
Sodium	100	927	1.39	85.8	66.85	10.5
Paraffin	20	900	2.13	0.26	0.14	0.71

they are very costly; very low vapor pressure results in freezing; they are very corrosive; they have difficulty in handling or storage; etc. The solid SHS materials have the advantage of a very wide and high-temperature range. For example, magnesium bricks can store heat in Cowper regenerators up to 1000°C. Moreover, materials in solid form remain inert and have low vapor pressure.

A few selected solid storage media and their thermal properties are reported in Table 3.3 [16]. Moreover, natural materials like rocks and soil are available in abundance and at low cost and can be used as ground storage materials. Most widely used solid SHS natural materials are granite, basalt, and quartzite when thermomechanical stress is the major requirement. Ceramics are also used as solid SHS materials.

Table 3.3 Physical properties of solid SHS materials at 20°C

Material	ρ (kg/m³)	C_p (kJ/ (kg.K))	λ (W/ (m.K))	$10^6 \times a$ (m²/s)	$10^{-3} \times b$ (J/ (m²Ks^{1/2}))
Aluminum (99.99%)	2700	0.945	238.4	93.3	24.66
Sodium chloride	2165	0.86	6.5	3.5	3.5
Copper	8300	0.419	372	107	35.97
Granite	2750	0.89	2.9	1.18	2.67
Iron	7850	0.465	59.3	16.3	14.7
Lead	11,340	0.131	35.25	23.6	7.24
Graphite	2200	0.61	155	120	14.41
Brick	1800	0.84	0.50	0.33	0.87
Concrete	2200	0.72	1.45	0.94	1.52
Sandstone	2200	0.71	1.8	1.15	1.68
Soil (clay)	1450	0.88	1.28	1.0	1.28
Soil (gravelly)	2040	1.84	0.59	0.16	1.49
Limestone	2500	0.74	2.2	1.19	2.02
Slag	2700	0.84	0.57	0.25	1.13

3.6.3 LHS materials

The LHS materials are also called phase change material (PCM) as they store energy in the material when they undergo physical phase change from solid to liquid and release the energy when they convert from liquid to solid phase. Whenever the surrounding temperature reaches the melting point, the temperature does not change as the heat energy is absorbed for phase transition. Therefore, the selection of PCMs is mainly governed by the operating temperature. The classification of PCM is shown in Figure 3.5. For low temperatures, below 120°C, organic PCMs are used and for high temperatures from 120°C to 1000°C, inorganic anhydrous salt PCMs (e.g., Na_2SO_4) are used. Several organic and inorganic PCM properties are presented in Table 3.4 with their melting points and heat of fusion [15,18,19].

The PCMs capacity of heat storage is dependent on the following properties:

1. Suitable operating temperature.
2. Enthalpy of phase change is extremely high.
3. Improved thermal stability at low vapor pressures and high temperatures.
4. During phase changes, there is a small change in volume.
5. No supersaturation during melting or subcooling during freezing.
6. The ability to store large amounts of heat when coupled with sensible.

Paraffin wax is a popular organic PCM whose melting points increase with molar mass, thus giving a wide range of operating temperature in PCM selection. They are widely available, inexpensive, non-corrosive, non-toxic, chemically stable, low density, fast phase transformation, and negligible supercooling. However, the major limitations are low volumetric energy

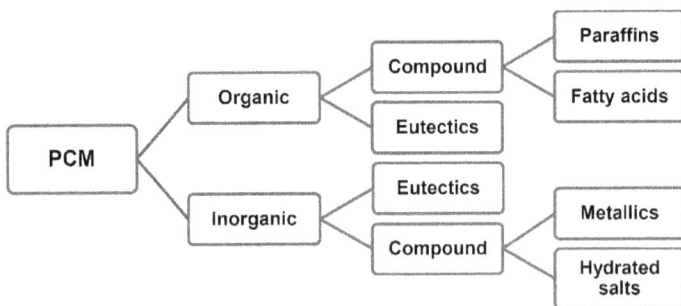

Figure 3.5 Classification of PCMs [18].

Table 3.4 Organic and inorganic PCM properties

Compounds		Melting points (°C)	Heat of fusion (kJ/kg)
Paraffins			
Paraffin	14-carbons	05.50	228.0
	15-carbons	10.00	205.0
	16-carbons	18.20	237.1
	17-carbons	21.70	213.0
	18-carbons	21.70	243.5
	19-carbons	32.00	222.0
	20-carbons	36.70	246.0
	21-carbons	40.20	215.0
Metallics			
Gallium		30.00	80.30
Caesium		28.65	16.40
Rubidium		38.85	25.74
Mercury		38.87	11.40
Cerrolow eutectic		59.00	91.00
Inorganic substance			
$LiClO_3 . 3H_2O$		08.30	253.0
$KF. 4H_2O$		18.70	231.0
$Mn(NO_3)_2 . 6H_2O$		26.00	125.9
$Ca(Cl)_2 . 6H_2O$		29.20	190.8
$Li NO_3 . 3H_2O$		30.20	296.0
$Na_2SO_4 . 10H_2O$		32.60	254.0
$Zn(NO_3)_2 . 6H_2O$		36.20	246.5
$Na_2CO_3 . 10H_2O$		34.20	146.9

density, high volumetric variation during phase transition, flammability, low thermal conductivity and diffusivity, and non-compatibility with other materials.

3.6.4 TCS materials

The materials that store energy as enthalpy change during reversible chemical reactions are called TCS materials. The energy density of TCS materials is higher than that of other TES materials. A system without storage capacities is called a heat transformer, which stores energy at low temperatures and releases it at high temperatures, or a heat pump, which stores energy at higher temperatures and releases it at lower temperatures. The list of materials used as TCS materials is presented in Table 3.5 [18,20,21].

Table 3.5 List of TCS materials

TCS material	Solid reactant	Working fluid	Energy density (GJ/m^3)	Charging reaction temperature (°C)
$MgSO_4.7H_2O$	$MgSO_4$	H_2O	2.8	122
$CaCO_3$	CaO	CO_2	3.3	837
$Ca(OH)_2$		H_2O	1.9	479
$CaSO_4.2H_2O$	$CaSO_4$		1.4	89
$Fe(OH)_2$	FeO		2.2	150
$FeCO_3$		CO_2	2.6	180

Table 3.6 Several applications of TES

Applications	TES material	Energy system
Power generation	Solar salt as SHS ($NaNO_3$ (60%) – KNO_3 (40%) + Sodium nitrate as PCM	Steam generation using dish type solar thermal power [22]
	Concrete	TES and solar driven organic Rankine cycle [23]
	Natural rocks	Concentrated solar power plants+TES [24]
	Ice, water tank and battery banks	Photovoltaic- thermal ES [25]
Solar ponds	Floating rings, gels, porous and PCMs, salts	Long term heat storage system [26]
Solar air heater	PCM	Solar air heater with PCM for crop drying, building and space heating [27]
Low temperature applications	Sensible heat	Packed bed solar TES [28]
	Metal foam+heat pipe+PCM	LHS system [29]
Heating and cooling	SHS+LHS+TCS	TES a heat reservoir for demand side management [30]
	Pit and tank TES	Energy system with TES [31]
Refrigeration systems	Organic and inorganic PCMs	Solar thermal water-lithium bromide vapor absorption system with PCM [32]
Water heating and desalination	Clay pots	Solar still with SHS [33]
	Salt hydrate PCMs	Solar LHS system [34]

3.7 APPLICATIONS OF TES

In the field of energy storage, TES refers to temporary energy storage because of the heating or cooling of a storing medium. This temporary energy storage allows it to be used later for diverse applications such as solar water heating and desalination, solar ponds, space heating and cooling, buildings, cold chain logistics, sustainable power supply, solar air heater, sustainable electric vehicles, grid applications, etc. [22–34]. Table 3.6 summarizes some of the latest research articles related to the use of TES in different applications.

3.8 CONCLUSION

TES serves a prominent role in load leveling scenarios, where disparities between energy demand and generation arise. Various TES techniques are currently in practice, each chosen based on factors like application type, duration, and scale. Conclusively, TES is a versatile and essential technology that plays a significant role in optimizing energy systems, reducing environmental impact, decarbonizing heat, saving consumers money, and enhancing the reliability and efficiency of energy utilization. There are many uses for it, including integrating RESs, improving grid stability, and managing energy consumption. Through the adoption of TES, our energy future can be more sustainable and resilient. In addition, it summarizes recent advancements in the use of TES in energy systems for different applications.

REFERENCES

1. Øvergaard, S. (2008). Definition of primary and secondary energy, In: *Standard International Energy Classification (SIEC) in the International Recommendation on Energy Statistics (IRES)*, Oslo Group on Energy Statistics, Statistics Norway, Oslo, pp. 1–7.
2. Aneke, M., & Wang, M. (2016). Energy storage technologies and real life applications – A state of the art review. *Applied Energy*, 179, 350–377. https://doi.org/10.1016/j.apenergy.2016.06.097.
3. Sakai, M., & Amano, M. (2013). *Energy Storage Devices and Systems*, Hitachi Chemicals Technical Report No.55, Hitachi Chemicals-Working on Wonders.
4. Kodama, H. (2014). *Electrical Energy Storage Devices and Systems*, Hitachi Chemicals Technical Report No.57, Hitachi Chemicals-Working on Wonders.
5. Lakshmi, G. S., Olena, R., Divya, G., & Oleksandr, R. (2020). Battery energy storage technologies for sustainable electric vehicles and grid applications. *Journal of Physics: Conference Series*, 1495(1), 012014. https://doi.org/10.10 88/1742-6596/1495/1/012014.
6. IRENA. (2020). *Innovation Outlook: Thermal Energy Storage*, International Renewable Energy Agency, Abu Dhabi. ISBN 978-92-9260-279-6.
7. Jonathan, R. (2013). *Energy Storage Technologies*, Ingenia, London, pp. 27–32.
8. Mahlia, T. M. I., Saktisahdan, T. J., Jannifar, A., Hasan, M. H., & Matseelar, H. S. C. (2014). A review of available methods and development on energy storage; technology update. *Renewable and Sustainable Energy Reviews*, 33, 532–545. https://doi.org/10.1016/j.rser.2014.01.068.
9. Tan, X., Li, Q., & Wang, H. (2013). Advances and trends of energy storage technology in microgrid. *International Journal of Electrical Power & Energy Systems*, 44(1), 179–191. https://doi.org/10.1016/j.ijepes.2012.07.015.
10. SBC Energy Institute Analysis Based on US DOE Energy Storage Program Planning Document. (2011). https://www.energy.gov/oe/articles/energy-storage-program-planning-document-2011.

11. Cabeza, L. F. (2014). *Advances in Thermal Energy Storage Systems: Methods and Applications*, Elsevier, Cambridge.
12. European Association for Storage of Energy and European Energy Research Alliance. (2013). *European Energy Storage Technology Development Roadmap towards 2030 - Technical Annex*, EERA, Belgium.
13. European Association for Storage of Energy and European Energy Research Alliance. (2017). *Energy Storage Technology Development Roadmap 2017*, EERA, Belgium.
14. Scapino, L., Zondag, H. A., Van Bael, J., Diriken, J., & Rindt, C. C. M. (2017). Sorption heat storage for long-term low-temperature applications: A review on the advancements at material and prototype scale. *Applied Energy*, 190, 920–948. https://doi.org/10.1016/j.apenergy.2016.12.148.
15. Bauer, T., Steinmann, W. D., Laing, D., & Tamme, R. (2012). Thermal energy storage materials and systems. *Annual Review of Heat Transfer*, 15, 131–177. https://doi.org/10.1615/AnnualRevHeatTransfer.2012004651.
16. Jarimi, H., Aydin, D., Yanan, Z., Ozankaya, G., Chen, X., & Riffat, S. (2019). Review on the recent progress of thermochemical materials and processes for solar thermal energy storage and industrial waste heat recovery. *International Journal of Low-Carbon Technologies*, 14(1), 44–69. https://doi.org/10.1093/ijlct/cty052.
17. Agyenim, F., Hewitt, N., Eames, P., & Smyth, M. (2010). A review of materials, heat transfer and phase change problem formulation for latent heat thermal energy storage systems (LHTESS). *Renewable and Sustainable Energy Reviews*, 14(2), 615–628. https://doi.org/10.1016/j.rser.2009.10.015.
18. Mitali, J., Dhinakaran, S., & Mohamad, A. A. (2022). Energy storage systems: A review. *Energy Storage and Saving*, 1(3), 166–216. https://doi.org/10.1016/j.enss.2022.07.002.
19. Sharma, A., Tyagi, V. V., Chen, C. R., & Buddhi, D. (2009). Review on thermal energy storage with phase change materials and applications. *Renewable and Sustainable Energy Reviews*, 13(2), 318–345. https://doi.org/10.1016/j.rser.2007.10.005.
20. Donkers, P. A. J., Pel, L., & Adan, O. C. G. (2016). Experimental studies for the cyclability of salt hydrates for thermochemical heat storage. *Journal of Energy Storage*, 5, 25–32. https://doi.org/10.1016/j.est.2015.11.005.
21. Abedin, A. H. (2011). A critical review of thermochemical energy storage systems. *The Open Renewable Energy Journal*, 4(1), 42–46. https://doi.org/10.2174/1876387101004010042.
22. Bian, R., Deng, Y., Feng, C., Yu, B., Sun, D., & Zhang, W. (2023). Performance and optimization study of graded thermal energy storage system for direct steam generation dish type solar thermal power. *Case Studies in Thermal Engineering*, 49, 103369. https://doi.org/10.1016/j.csite.2023.103369.
23. Shakeri, A., Eshghi, H., Salek, F., & Babaie, M. (2023). Energy assessment for integration of concrete thermal energy storage with low-grade solar power generation system. *Renewable Energy*, 218, 119249. https://doi.org/10.1016/j.renene.2023.119249.
24. El Alami, K., Asbik, M., & Agalit, H. (2020). Identification of natural rocks as storage materials in thermal energy storage (TES) system of concentrated solar power (CSP) plants – A review. *Solar Energy Materials and Solar Cells*, 217, 110599. https://doi.org/10.1016/j.solmat.2020.110599.

25. Ndwali, P. K., Kusakana, K., Numbi, P. B., Liu, S., Sun, W., & Cai, J. (2022). A review of multistage solar driven photovoltaic-thermal components with cascade energy storage system for tri-generation. *Energy Reports*, 8, 14–20. https://doi.org/10.1016/j.egyr.2022.09.096.

26. Rghif, Y., Colarossi, D., & Principi, P. (2023). Salt gradient solar pond as a thermal energy storage system: A review from current gaps to future prospects. *Journal of Energy Storage*, 61, 106776. https://doi.org/10.1016/j.est.2023.106776.

27. Pathak, S. K., Tyagi, V. V., Chopra, K., Kalidasan, B., Pandey, A. K., Goel, V., Saxena, A., & Ma, Z. (2023). Energy, exergy, economic and environmental analyses of solar air heating systems with and without thermal energy storage for sustainable development: A systematic review. *Journal of Energy Storage*, 59, 106521. https://doi.org/10.1016/j.est.2022.106521.

28. Gautam, A., & Saini, R. P. (2020). A review on sensible heat based packed bed solar thermal energy storage system for low temperature applications. *Solar Energy*, 207, 937–956. https://doi.org/10.1016/j.solener.2020.07.027.

29. Jaisatia Varthani, A., Shasthri, S., Baljit, S., & Kausalyah, V. (2022). A systematic review of metal foam and heat pipe enhancement in Latent Heat Thermal Energy Storage system. *Journal of Energy Storage*, 56, 105888. https://doi.org/10.1016/j.est.2022.105888.

30. Guelpa, E., & Verda, V. (2019). Thermal energy storage in district heating and cooling systems: A review. *Applied Energy*, 252, 113474. https://doi.org/10.1016/j.apenergy.2019.113474.

31. Sifnaios, I., Sneum, D. M., Jensen, A. R., Fan, J., & Bramstoft, R. (2023). The impact of large-scale thermal energy storage in the energy system. *Applied Energy*, 349, 121663. https://doi.org/10.1016/j.apenergy.2023.121663.

32. Raut, D., & Kalamkar, V. R. (2022). A review on latent heat energy storage for solar thermal water-lithium bromide vapor absorption refrigeration system. *Journal of Energy Storage*, 55, 105828. https://doi.org/10.1016/j.est.2022.105828.

33. Yarramsetty, N., Sharma, N., & Narayana, M. L. (2023). Experimental investigation of a pyramid type solar still with porous material: Productivity assessment. *World Journal of Engineering*, 20, 178–185. https://doi.org/10.1108/WJE-02-2021-0096.

34. Jin, X., & You, S. (2023). Experimental investigation of the solar latent heat thermal energy storage system integrated with salt hydrate phase-change materials. *Energy Reports*, 9, 502–511. https://doi.org/10.1016/j.egyr.2023.09.076.

Chapter 4

Solar energy scenario in India

Opportunities and challenges

Ranjana Singh and Puneet kumar
IIMT University Meerut

4.1 INTRODUCTION

In recent years, the world has witnessed an increasing global focus on renewable energy sources as a means to address pressing environmental concerns, mitigate climate change, and foster sustainable development [1]. Solar power stands out as a prominent and promising renewable energy source, thanks to its widespread availability and immense potential for producing clean electricity. While developed nations have made considerable strides in adopting solar energy, its utilization in developing countries poses distinct opportunities and obstacles. This book chapter aims to explore the opportunities and challenges associated with harnessing solar energy in developing countries. By shedding light on this topic, we seek to contribute to the understanding of how solar power can empower these nations in their pursuit of sustainable energy systems, economic growth, and improved quality of life [1–4].

4.2 CURRENT STATUS OF SOLAR ENERGY IN INDIA

Rapid Capacity Growth: In recent years, India has experienced a remarkable expansion in its solar energy capacity installation. The nation has established ambitious goals, aiming to attain 175 gigawatts (GW) of renewable energy capacity by 2022, with a significant portion, 100 GW, designated for solar power. Impressively, India has surpassed its initial objective of achieving 20 GW of installed solar capacity, accomplishing this milestone four years ahead of schedule.

Utility-Scale Solar Projects: India has seen a surge in the development of large-scale solar power projects. Solar parks and ultramega solar power projects have been established across various states to attract investments and facilitate the deployment of grid-connected solar farms. These projects contribute significantly to India's solar capacity and help in achieving economies of scale.

DOI: 10.1201/9781003472629-4

Rooftop Solar Systems: The adoption of rooftop solar systems in India has gained momentum, driven by government incentives and net metering policies. Residential, commercial, and industrial customers are progressively adopting rooftop solar panel installations to produce environmentally friendly energy and lower their electricity expenses. The government has established a goal of achieving 40 GW of rooftop solar installations by 2022.

Policy Support: The Government has implemented several policy actions to promote solar energy. The Jawaharlal Nehru National Solar Mission (JNNSM), launched in 2010, aims to establish India as a global leader in solar power generation.[5] Diverse financial incentives, including subsidies, tax advantages, and viability gap funding, have been implemented to incentivize investments in the industry.

International Collaborations: India has actively participated in cooperative initiatives and alliances with international organizations and nations to advance the cause of solar energy [2]. A notable example is the launch of the International Solar Alliance (ISA) in 2015, where India played a pivotal role as a founding member. The ISA aims to tackle shared challenges and expedite global solar energy adoption.

Job Creation: The growth of the solar industry in India has contributed to job creation and skill development. The installation, operation, and maintenance of solar projects have generated employment opportunities, particularly in rural areas, thereby supporting India's socio-economic development [6]. The installed solar energy capacity in India has surged from 2.63 GW in March 2014 to reach 49 GW by December 2021 (Figure 4.1).

During the initial three months of FY23, non-hydro renewable energy capacity witnessed an augmentation of 4.2 GW, a notable increase compared

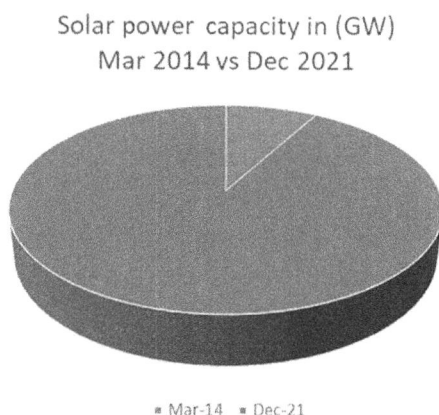

Solar power capacity in (GW)
Mar 2014 vs Dec 2021

▪ Mar-14 ▪ Dec-21

Figure 4.1 Installed solar energy capacities in India.(Source: Union Budget 2022–2023.)

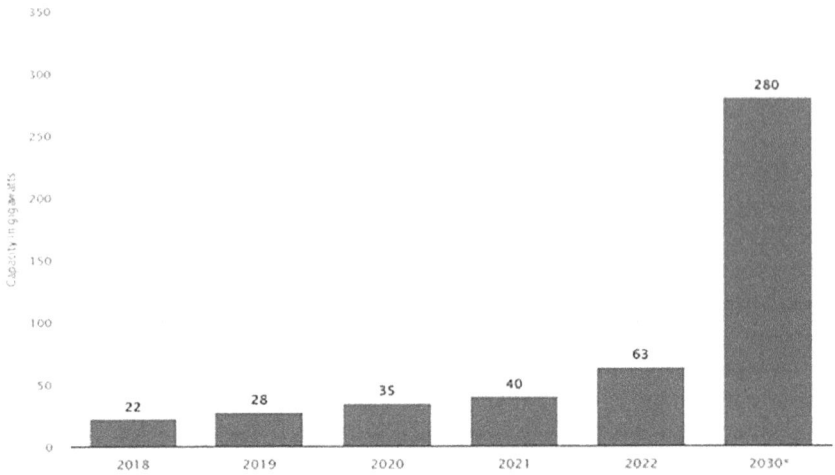

Figure 4.2 Installations and goals for solar power capacity in India from 2018 to 2030. (Source: Statista 2023.)

to the 2.6 GW added during the same period in FY22 [3]. Impressively, the installed capacity of solar power has surged by over 18-fold, progressing from 2.63 GW in March 2014 to a substantial 49.3 GW by the conclusion of 2021.

From 2018 to 2021, India experienced nearly a twofold increase in its installed solar power capacity, reaching approximately 40 GW by the end of the latter year (Figure 4.2). By 2022, the country would havefurther expanded its solar capacity to 63 GW [7]. Achieving the projected capacity for 2030 necessitates a growth of over fourfold in the installed capacity compared to the existing levels [1].

4.3 OPPORTUNITIES

Energy Access: Solar energy provides a unique opportunity for developing countries to expand access to electricity in remote and underserved areas [8]. Since solar systems can be installed off-grid, they offer a decentralized solution for communities that are far from the traditional power grid.

Cost Reduction: The cost of solar panels and related technologies has been declining steadily over the years. This cost reduction, combined with the abundance of sunlight in many developing countries, makes solar energy an increasingly affordable option for power generation.

Job Generation: The establishment of solar energy infrastructure has the potential to generate employment opportunities across different phases of the value chain, encompassing manufacturing, installation, operation, and maintenance. This can play a role in fostering local economic growth and mitigating unemployment.

Energy Independence: Solar energy diminishes reliance on foreign-imported fossil fuels, which is particularly beneficial for countries with limited domestic energy resources. By tapping into their solar potential, developing nations can enhance energy security and reduce vulnerability to fluctuating fuel prices.

Environmental Advantages: Solar power is a clean and renewable energy source, generating electricity without harmful emissions or greenhouse gases. Embracing solar energy can help developing countries mitigate the effects of climate change, enhance air quality, and minimize their carbon emissions.

Government Policy Support: The Indian government has implemented a series of significant policies and initiatives aimed at promoting the adoption and growth of solar energy within the nation [9]. These regulations aim to create a conducive environment for solar power generation, spur investments, and assist in achieving the country's renewable energy goals. Notable among these government policies are:

National Solar Mission (NSM): Established in 2010, NSM's objective is to achieve a solar power capacity of 100 GW by 2022 through a phased approach and various projects. This mission encompasses solar applications that are connected to the grid as well as those operating off-grid.

Ultra Mega Solar Power Projects and Solar Parks: The government has selected ideal places for solar power generation and built solar parks, which offer the essential infrastructure while lowering project development risks [10]. These parks are home to large-scale solar arrays known as Ultra Mega Solar Power Projects.

Renewable Purchase Obligation (RPO): This policy requires electricity distribution firms and big users to obtain a specific amount of their power from renewable sources, including solar energy.

Solar Rooftop Subsidies and Incentives: A variety of subsidies and incentives are available to promote solar rooftop installations for residential, commercial, and industrial clients. These include capital subsidies, net metering, and tax breaks.

Solar Energy Corporation of India (SECI): SECI, a government-owned organization, enables the development of solar projects through competitive bidding and auctions, assuring transparency and efficiency [11].

4.4 CHALLENGES

Upfront Costs: Despite the declining costs, the initial investment required for installing solar systems can still be a significant barrier, especially for low-income communities and governments with limited financial resources. Access to affordable financing options and innovative funding mechanisms is crucial to overcome this challenge.

Technical Expertise: Developing countries may lack the necessary technical expertise and skilled workforce to design, install, and maintain solar energy systems [12]. Building local capacity through training programs and knowledge transfer initiatives is essential to address this challenge.

Infrastructure Limitations: The integration of solar energy into existing power grids can be challenging, particularly in regions with inadequate transmission and distribution infrastructure. Upgrading and expanding the grid infrastructure to accommodate solar power is essential for its effective deployment.

Policy and Regulatory Frameworks: The absence of supportive policies and clear regulatory frameworks can hinder the growth of solar energy in developing countries. Governments need to establish favorable policies, such as feed-in tariffs, tax incentives, and streamlined permitting processes, to encourage investment and promote solar energy deployment.

Intermittency and Storage: Solar power generation is dependent on sunlight, which is intermittent and subject to variations throughout the day and seasons [13]. Developing effective energy storage solutions, such as batteries, is crucial to ensure a stable and reliable electricity supply from solar sources, especially during non-sunlight hours.

4.5 CONCLUSION

India's solar energy landscape offers great prospects for both sustainable growth and energy security. The country's tremendous solar potential, falling solar technology costs, and supporting legislative frameworks have resulted in a significant increase in solar installations. However, difficulties suchas intermittency, grid integration, land availability, and financial restrictions must be solved in order to fully realize the promise of solar energy [14]. Continued investment in research, technology innovation, and infrastructure development will be critical in overcoming these obstacles. Through the broad use of solar power, India can shift to a cleaner and more resilient energy future with a coordinated effort from government, business, and society.

REFERENCES

1. Murthy, V. (2014). *India's Solar Energy Future: Policy and Institutions.* Washington, DC: CSIS.
2. Sharma,A.(2011). A comprehensive study of solar power in India and world. *Renewable and Sustainable Energy Reviews*, 15(4), 1767–1776.
3. Deambi, S. (2015). *From Sunlight to Electricity: A Practical Handbook on Solar Photovoltaic Applications* (3rd Edn.). New Delhi: TERI.

4. Bhaskar, B. (2013). *Energy Security and Economic Development in India: A Holistic Approach.* New Delhi: TERI.
5. Kapoor, T. (2015). Fulfilling India's RE dreams. *Energy Next*, Hyderabad, Vol. 5.
6. REN21. 2023. *Renewables 2023 Global Status Report Collection.*
7. National Solar Mission. *Annual Report 2022-23.* Ministry of New and Renewable Energy. Government of India.
8. Report of Energy_Statistics_2022. Ministry of Statistics & Programme Implementation.
9. Ministryof New and Renewable Energy Government of India.(2022). *Annual Reports* 2020-2021. Ministry of New and Renewable Energy.
10. Ministry of New and Renewable Energy Government of India. *Vision and Mission.*
11. SolarEnergy Corporation of India Ltd. (SECI) Government of India. https://www.seci.co.in/
12. Ministry of New and Renewable Energy. www.mnre.gov.in
13. The Make in India initiative. www.makeinindia.com (sector/renewable energy).
14. www.reconnectenergy.com

Chapter 5

Applications of IoT in renewable energy sector

Smart energy management, opportunities, challenges and applications

Manish D Mahale
G H Raisoni College of Engineering and
Business Management Jalgaon

Bipasa B Patra
Padmashri Dr. V. B. Kolte College of Engineering Malkapur

Sushma M Mahale
NMKC College of Engineering Jalgaon

5.1 INTRODUCTION

Internet of Things (IoT) and Global Energy Interaction Technologies are two interconnected concepts that have the potential to revolutionise the energy sector. The network of physical appliances which are embedded with sensors, software, and connectivity that enable the system to collect or exchange the data, this system is known as Internet of Things (IoT). This system communicates with each other using internet which allows remote monitoring, controlling the various processes to make the system automated.

In the energy sector the IoT can have various applications:

Smart Grids: IoT enables the creation of smart grids, which are intelligent electrical grids capable of monitoring, analysing, and optimising energy distribution. IoT devices installed on power lines, transformers, and other grid components can accommodate the data of energy consumption, analyse inefficiencies, and enable real-time adjustments for better energy management.

Energy Monitoring and Management: To monitor the energy consumption of residential, commercial, and industrial sectors, the IoT system can be used. Advanced metres and sensing technology provide real-time data on energy consumption. It allows customers and energy suppliers to analyse patterns, reduce the rate of energy consumption, and make informed decisions.

DOI: 10.1201/9781003472629-5

Demand Response: The IoT can support demand response programmes, in which energy users modify their consumption in response to price signals or grid circumstances. With IoT-enabled devices, consumers can receive notifications and automate the control of appliances, optimising energy consumption during peak and off-peak periods.

Renewable Energy Integration: IoT can play a significant role in integrating renewable energy sources into the grid, such as solar panels and wind turbines. IoT devices can monitor the energy generated, predict output, and optimise energy flows to ensure efficient utilisation and minimise grid instability [1–3].

5.2 GLOBAL ENERGY INTERACTION TECHNOLOGIES

The term "global energy interaction technologies" refers to sophisticated systems and technologies that allow the fusion and interaction of multiple energy sources, including non-renewable and renewable sources, to build a more effective and sustainable energy ecosystem. Some examples of global energy interaction technologies include:

Energy Storage Systems: These technologies involve the development of efficient and scalable energy storage solutions. With the help of batteries, pumped hydro storage, compressed air energy storage (CAES), and other cutting-edge technologies, excess energy produced by renewable sources can be stored and used at times of high demand or low generation [4,5].

Smart Energy Management Systems: These systems utilise advanced software, data analytics, and automation to optimise energy generation, distribution, and consumption. By collecting and analysing data from various sources, including IoT devices, these systems can optimise energy flows, reduce losses, and improve overall efficiency.

Interconnected Grids: Interconnected grids enable the exchange of energy between different regions or countries. With the use of this technology, the supply and demand of energy may be balanced, allowing excess energy from one area to be used in another. The integration of various energy sources is made possible through interconnected grids, which also improve grid stability.

5.3 INTERACTION BETWEEN IoT AND GLOBAL ENERGY INTERACTION TECHNOLOGIES

IoT can greatly enhance the effectiveness and efficiency of global energy interaction technologies. Data may be gathered and analysed in real-time by integrating IoT devices into energy infrastructure, facilitating better decision-making and the optimisation of energy systems. For example, IoT sensors installed in energy storage systems can monitor their performance and provide valuable insights for maintenance and optimisation. A more dependable and sustainable energy ecosystem can be achieved by utilising IoT to provide seamless communication and coordination across different energy system components, including smart grids, renewable energy sources, and energy storage systems.

Global Energy Interaction Technologies and IoT have a lot of potential to alter the energy industry, increase energy efficiency, promote the integration of renewable energy sources, and make it possible for a more resilient and sustainable global energy infrastructure [6–8].

5.4 IMPORTANCE OF IoT IN ENERGY TECHNOLOGY

By enabling improved monitoring, control, optimisation, and administration of energy systems, the IoT plays a vital role in the field of energy technology. IoT devices, sensors, and connectivity enable the collection and analysis of real-time data, enabling more efficient and sustainable energy practices. Here are some key areas where IoT is applied in energy technology [9,10].

Smart Grid Management: IoT devices and sensors can be deployed throughout the grid infrastructure to collect data on energy generation, transmission, and consumption. This data enables grid operators to monitor grid conditions, identify faults or inefficiencies, optimise energy distribution, and respond to demand fluctuations in real time. It supports load balancing, fault detection, and predictive maintenance, ultimately improving grid reliability and efficiency.

Energy Monitoring and Management: IoT enables real-time monitoring of energy consumption in homes, buildings, and industrial facilities. Smart metres and energy monitoring systems collect data on energy usage patterns, identify energy-intensive areas, and provide actionable insights to optimise energy consumption. This empowers consumers and facility managers to make informed decisions to reduce energy waste, adjust usage during peak demand periods, and implement energy-saving strategies.

Demand Response and Energy Flexibility: Demand response programmes, where energy users alter their usage in response to grid problems or price signals, are made possible by IoT. Connected devices and smart thermostats can receive signals from utilities to automatically adjust energy consumption during peak demand periods or in response to dynamic pricing. This helps to balance supply and demand, reduce stress on the grid, and potentially lower energy costs for consumers.

Energy Efficiency and Conservation: IoT devices, such as smart appliances and lighting systems, enable energy-efficient practices by utilising sensors, automation, and data analytics. For example, smart lighting systems can adjust brightness based on occupancy or daylight levels, optimising energy use. Connected devices can also provide energy-saving recommendations, real-time feedback, and alerts to promote energy-conscious behaviour.

Renewable Energy Integration: IoT is essential for the grid's integration of renewable energy sources like solar and wind. IoT devices and sensors can monitor renewable energy generation, forecast weather conditions, and optimise the usage of renewable energy based on real-time data. This ensures efficient utilisation of renewable energy, minimises curtailment, and supports grid stability.

Energy Storage Optimisation: IoT enables smarter management and optimisation of energy storage systems, such as batteries or thermal storage. Connected energy storage systems can be integrated with IoT platforms to optimise charging and discharging cycles, respond to grid signals or demand response events, and ensure optimal use of stored energy.

Predictive Maintenance and Asset Management: IoT sensors can keep an eye on the condition and functionality of energy equipment and infrastructure, including power plants, transformers, and distribution systems. Real-time data from these sensors can be analysed to detect anomalies, predict failures, and schedule maintenance activities proactively. This improves asset management, reduces downtime, and enhances overall system reliability.

Overall, IoT technology enables the energy industry to gather real-time data, optimise energy usage, integrate renewable sources, improve grid management, and enhance overall energy efficiency. It empowers consumers, businesses, and utilities to make data-driven decisions, reduce energy waste, and contribute to a more sustainable and resilient energy future [11–14].

5.5 HYBRID SYSTEM

A hybrid system refers to a combination or integration of two or more different technologies, methodologies, or approaches to create a unified and enhanced system that leverages the strengths of each component. In the context of energy systems, a hybrid system often refers to the integration of multiple energy sources or technologies to improve efficiency, reliability, and sustainability [15–18].

Here are a few examples of hybrid systems in the energy sector:

5.5.1 Hybrid renewable energy systems

These systems combine multiple renewable energy sources to overcome the intermittent nature of individual sources and ensure a more reliable power supply. For instance, a hybrid renewable energy system may integrate solar panels, wind turbines, and energy storage technologies like batteries or pumped hydro storage. This allows for a more balanced and consistent power generation profile, as different sources can compensate for each other's variations in output.

5.5.2 Solar and wind hybrid renewable energy system

To harvest renewable energy from two separate sources, a hybrid solar and wind energy system, sometimes referred to as a hybrid renewable energy system, integrates both solar and wind power producing technology. This system takes advantage of the complementary nature of solar and wind resources, as they often have different patterns of availability throughout the day and year [19–21]. Here are the key components and benefits of a hybrid solar and wind energy system:

Solar Photovoltaic (PV) Panels: Sunlight is immediately converted into power by solar PV panels. They are set up to collect solar energy during the day. When exposed to sunlight, the panels' many solar cells produce DC electricity.

Wind Turbines: The kinetic energy of the wind is captured by wind turbines and turned into electricity. When the wind blows, the turbines' blades rotate, pushing a generator to generate power. In locations with reliable and ample wind resources, wind turbines are effective.

Battery Storage: To ensure continuous power supply, a hybrid system incorporates battery storage. Excess electricity generated from solar and wind sources during periods of high availability can be stored in batteries for later use when solar and wind generation is low. Batteries help balance the intermittent nature of renewable energy sources and provide a reliable power supply.

Charge Controller and Inverter: The charging and discharging of batteries are controlled by a charge controller, assuring their optimum performance and longevity. The DC electricity produced by solar panels and wind turbines is converted into AC electricity by an inverter so that it can be used in houses or connected to the electrical grid.

Energy Management System: The hybrid system is monitored and operated at its best by an energy management system. It optimises the use of renewable energy, makes sure that energy production from solar and wind sources is balanced, and effectively controls battery charging and discharging (Figure 5.1).

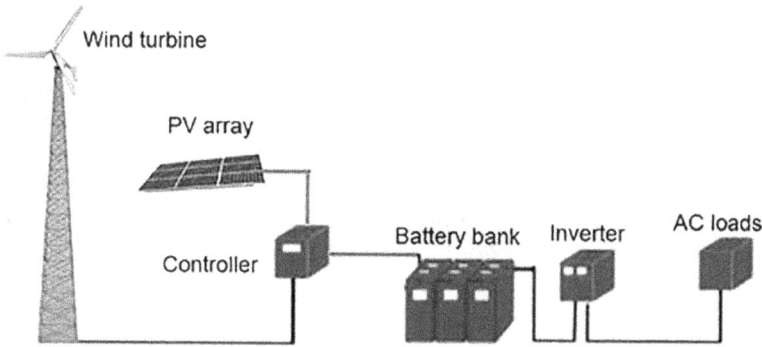

Figure 5.1 Hybrid of solar and wind energy generation system.

5.6 BENEFITS OF SOLAR AND WIND HYBRID SYSTEM

Increased Energy Production: The system's overall energy output can be enhanced by combining solar and wind energy sources. Solar power generation is typically highest during the day, while wind power generation can be more consistent throughout the day and year, including nighttime and winter months.

Enhanced System Reliability: The reliability of the power supply can be improved by a hybrid system by including multiple renewable energy sources. It reduces the dependence on a single energy source and provides more consistent power output, even during periods of low solar or wind resource availability.

Better Utilisation of Resources: Solar and wind resources often complement each other. Solar power generation tends to be higher during sunny days, while wind power generation can be more significant during cloudy or low-sunlight conditions. By combining both, a hybrid system optimises the utilisation of available renewable energy resources.

Flexibility in Location: Hybrid systems can be deployed in a variety of locations, as they can harness solar energy in sunnier regions and wind energy in areas with favourable wind conditions. This flexibility allows for the adoption of renewable energy solutions in diverse geographical locations.

Reduced Carbon Footprint: Utilising solar and wind energy reduces reliance on fossil fuels, resulting in a lower carbon footprint and decreased greenhouse gas emissions. Hybrid systems contribute to mitigating climate change and promoting a cleaner and more sustainable energy future.

Hybrid solar and wind energy systems offer a flexible and efficient approach to harnessing renewable energy. They provide a reliable and sustainable power supply, reduce dependence on conventional energy sources, and contribute to a greener and more resilient energy infrastructure

Hybrid Power Systems: Several distinct power production methods, such as traditional fossil fuel-based generators and renewable energy sources, are combined in hybrid power systems. Hybrid power systems can take advantage of the dispatchability and dependability of conventional power sources while minimising their environmental impact by incorporating cleaner, renewable energy sources [22–26].

5.7 WIND AND HYDROGEN-BASED HYBRID ENERGY SYSTEM

A wind hydrogen system, often referred to as a wind-to-hydrogen system, combines the generation of wind power with the production and storage of hydrogen. By using extra electricity produced by wind turbines, this system electrolyses water to make hydrogen. Here are the key components and benefits of a wind hydrogen system:

Wind Turbines: The kinetic energy of the wind is captured by wind turbines and turned into electricity. The main source of energy in a wind hydrogen system is these turbines. When the wind blows, it provides electricity that can be utilised to run electrolysers for the creation of hydrogen.

Electrolysers: Electrolyses are machinery that electrolyse water (H_2O) to produce hydrogen (H_2) and oxygen (O_2) using electricity. The electrolyses, which separate hydrogen from water molecules, receive the extra electricity produced by wind turbines. The generated hydrogen is collected for further usage and storage.

Hydrogen Storage: It is necessary to store the hydrogen created for later use. Different kinds of hydrogen can be kept, such as compressed gas, liquid hydrogen, or as a component of a chemical complex. System needs, available infrastructure, and safety considerations are only a few examples of the variables that influence the storage method selection.

Fuel Cells or Hydrogen Combustion: Fuel cells or hydrogen combustion systems can be used to turn the stored hydrogen back into power as necessary. Fuel cells electrochemically combine hydrogen and oxygen to generate electricity and produce water as a byproduct. Hydrogen combustion systems burn hydrogen in an engine or turbine to drive a generator and produce electricity.

Hydrogen Utilisation: The electricity generated from the stored hydrogen can be used to power various applications, such as grid stabilisation, transportation (fuel cell vehicles), heating systems, industrial processes, or as a backup power source (Figure 5.2).

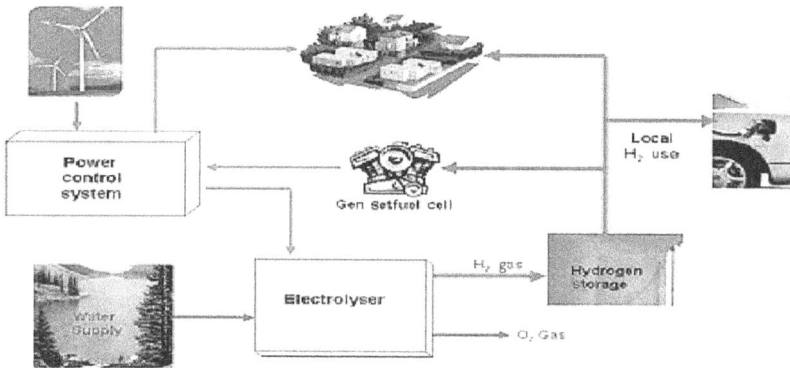

Figure 5.2 Wind hydrogen hybrid system.

5.8 BENEFITS OF A WIND HYDROGEN SYSTEM INCLUDE

Energy Storage: Wind hydrogen systems offer a way to store excess electricity generated by wind turbines, which would otherwise go unused or be curtailed. The stored hydrogen acts as a form of energy storage, providing a reliable and dispatchable energy source when wind power generation is low or intermittent.

Renewable Energy Integration: Wind hydrogen systems make it possible to incorporate renewable energy sources into a variety of industries. They offer a way to use wind energy for things other than making electricity, such as transportation or industrial activities that often depend on fossil fuels.

Carbon-Free Energy: Carbon-free hydrogen is defined as that produced from renewable energy sources like wind energy. Wind hydrogen systems aid in lowering greenhouse gas emissions and climate change mitigation by using hydrogen as an energy carrier.

Energy Sector Decentralisation: Wind hydrogen systems can be deployed in various locations, including remote or off-grid areas. This decentralisation of energy production and storage reduces reliance on centralised power plants and transmission infrastructure, increasing energy resilience and promoting energy independence.

Versatile Applications: Wind energy can be used to produce hydrogen, which has a variety of uses, including powering fuel cell cars, heating homes or businesses, assisting industrial processes, and serving as a backup power source in emergency situations.

In the shift to a greener and more sustainable energy supply, wind hydrogen systems are important. They provide a mechanism to successfully store and use renewable energy, support decarbonisation efforts, and open the door for the widespread use of hydrogen as an energy source [27–32].

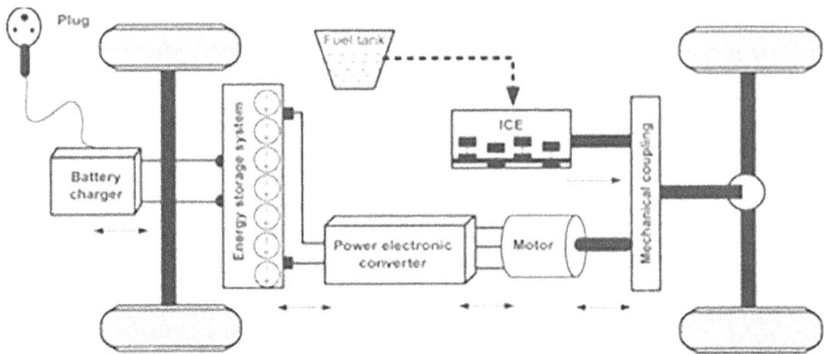

Figure 5.3 Hybrid electric vehicle.

Hybrid Electric Vehicles (HEVs): HEVs pair an electric motor with a battery system in addition to an internal combustion engine (ICE). These vehicles utilise the ICE for high-power demands or long-range driving, while the electric motor and battery provide energy efficiency and lower emissions for shorter trips and stop-and-go traffic. HEVs optimise fuel consumption and reduce greenhouse gas emissions by utilising both conventional and electric propulsion technologies. Benefits of HEVs include reduced fuel consumption, lower emissions, improved efficiency, and enhanced driving range compared to conventional vehicles. They offer a transitionary step towards electrification and can help reduce environmental impact while maintaining the convenience and range of traditional vehicles. Additionally, HEVs provide flexibility by utilising both gasoline and electric power sources, making them suitable for a wide range of driving conditions and infrastructure availability (Figure 5.3).

5.9 HYBRID ENERGY STORAGE SYSTEMS

To improve performance and flexibility, hybrid energy storage systems mix various energy storage methods. For example, a hybrid energy storage system may integrate lithium-ion batteries with ultra-capacitors. Lithium-ion batteries provide high energy density and capacity for longer-term storage, while ultra-capacitors offer fast charging and discharging capabilities for short-duration power demands. Hybrid systems often aim to optimise the use of resources, increase system resilience, reduce environmental impact, and improve overall efficiency. By combining complementary technologies, hybrid systems can provide more reliable, flexible, and sustainable energy solutions across various applications, including power generation, transportation, and energy storage. To diversify the generating mix and improve the efficiency and sustainability of energy production, a number of innovative technologies are being developed and deployed, including tidal power, wave energy, and improved energy storage systems (such as lithium-ion batteries and flow batteries) [33–37].

The choice of generation technology depends on factors such as resource availability, environmental impact, cost, scalability, grid requirements, and policy frameworks. Many regions and countries aim to transition to a more diverse and sustainable generation mix by incorporating an increasing share of renewable energy sources while gradually phasing out or reducing reliance on fossil fuel-based generation technologies.

5.10 GRID TECHNOLOGY

To generate, transmit, distribute, and manage electricity across an electrical grid, systems, equipment, and technologies are referred to as "grid technology." To guarantee the efficient, secure, and dependable distribution of electricity to consumers, grid technology is essential. The following are some crucial features of grid technology:

Power Generation: Grid technology includes the various methods and technologies used to generate electricity, as discussed in the previous response. This includes power plants, renewable energy sources, and emerging technologies that produce electrical energy.

Transmission Infrastructure: The transmission system consists of high-voltage power lines, transformers, substations, and associated equipment used to transmit electricity over long distances from power generation sites to distribution networks or major demand centres. Grid technology ensures the efficient and reliable transfer of electricity through the transmission infrastructure.

Distribution Infrastructure: Residential, commercial, and industrial end users receive power from transmission networks through distribution systems. To control the flow of electricity and maintain voltage levels, this infrastructure also consists of medium-voltage and low-voltage power lines, distribution transformers, switchgear, and other components.

Grid Monitoring and Control: Monitoring and control systems that continuously track the condition and efficiency of the grid infrastructure are included in grid technology. Real-time monitoring of grid characteristics, including voltage levels, power flows, and equipment status, is made possible by Supervisory Control and Data Acquisition (SCADA) systems, sensors, and automation technologies. Grid operators use these systems to optimise grid operation, prevent outages, and respond swiftly to faults or emergencies.

Grid Resilience and Security: Grid technology also addresses resilience and security aspects to protect the grid from disruptions, cyber-attacks, or natural disasters. This involves implementing measures such as redundancy, backup systems, grid hardening, and advanced security protocols to ensure the integrity and reliability of the grid infrastructure.

Smart Grid Technology: The idea of a "smart grid" entails the infrastructure of the grid being upgraded to incorporate cutting-edge technology, communication systems, and data analytics. Real-time data sharing between grid elements, grid operators, and end users is made possible by smart grid technology, which improves the monitoring, regulation, and optimisation of energy flows. Demand response programmes, grid integration of decentralised energy sources (such as rooftop solar panels), and the deployment of sophisticated energy management systems are all made possible by it.

Grid Modernisation: Grid technology includes ongoing efforts to modernise and upgrade the existing grid infrastructure to meet the evolving needs of the energy sector. This involves the incorporation of modern metering and invoicing systems, energy storage systems, electric vehicle charging infrastructure, and renewable energy sources. The development and adoption of grid technology are crucial for achieving a more reliable, efficient, and sustainable energy system. It enables the integration of diverse energy sources, improves grid stability, supports the transition to cleaner energy, and empowers consumers to actively participate in energy management. Grid technology continues to evolve as new innovations emerge, enabling a more decentralised, resilient, and intelligent grid infrastructure.

5.11 SMART GRID

Modern communication, control, and information technologies are incorporated into an improved electrical grid known as a "smart grid" to improve the efficiency, dependability, and sustainability of power generation, transmission, distribution, and consumption. To manage and operate the grid as efficiently as possible, it makes use of digital infrastructure, sensors, and real-time data analysis.

A smart grid has the following major characteristics and advantages:

Advanced Metering Infrastructure (AMI): Smart metres are used in smart grids to enable two-way communication between utility companies and customers. Smart metres gather and send real-time data on energy use, enabling correct invoicing and demand response initiatives as well as providing consumers with thorough data on their energy consumption.

Grid Monitoring and Control: To gather information on voltage levels, power quality, equipment performance, and grid conditions, smart grids place sensors and monitoring equipment throughout the grid infrastructure. Grid operators are able to quickly identify and address problems, reduce outages, and optimise energy flows thanks to the real-time analysis of this data.

Demand Response and Load Management: Demand response programmes are made possible by smart grids, allowing users to modify their energy usage in response to price signals or grid circumstances. Consumers can receive real-time information about energy pricing through linked devices and smart metres, and they can modify their usage accordingly to lower peak demand and increase system efficiency.

Integration of Renewable Energy Sources: Smart grids make it easier to integrate renewable energy sources like solar and wind power by keeping an eye on their sporadic nature and controlling their integration into the grid dynamically. This improves grid stability, lowers curtailment, and provides greater coordination between generation and demand.

Distributed Energy Resources (DERs): Rooftop solar panels, energy storage devices, and infrastructure for recharging electric vehicles can all be seamlessly integrated into smart networks. They give grid operators a framework for the effective management, control, and optimisation of DERs, allowing them to take use of their advantages and boost system performance as a whole.

Enhanced Grid Resilience and Reliability: Smart grids can more quickly locate faults, isolate impacted areas, and restore power more effectively since they have sophisticated monitoring and control capabilities. They provide self-healing methods so that grid elements may automatically detect and react to interruptions, reducing the impact on consumers.

Improved Energy Efficiency: Smart grids give customers access to real-time data on their energy usage, enabling them to make wise choices about their consumption patterns. This knowledge encourages energy conservation and efficiency, which helps consumers lower their energy costs and environmental impact.

Cyber Security and Grid Resilience: Strong cyber security measures are incorporated into smart grids to guard against online threats and guarantee the dependability and integrity of grid operations. In order to identify and reduce potential cyber security risks, this comprises secure communication methods, encryption, access controls, and constant monitoring.

In general, smart grids provide a number of advantages, such as increased grid resilience, improved grid efficiency, better integration of renewable energy sources, higher consumer participation, and lessening of environmental effects. Smart grids open the door for a more resilient, adaptable, and sustainable energy system by utilising cutting-edge technologies and data-driven insights.

5.12 ADVANTAGES OF SMART GRID

The adoption of a smart grid has numerous advantages for various parts of the power system. A smart grid offers the following main advantages:

Enhanced Grid Reliability and Resilience: Advanced monitoring, communication, and control capabilities built into smart grids allow for the quick isolation, identification, and restoration of outages. By enabling quick reaction to failures, enhancing real-time visibility into grid conditions, and decreasing downtime, they increase grid reliability. Additionally, self-healing methods are supported by smart grids, allowing grid elements to automatically identify and react to interruptions while minimising the impact on customers.

Improved Energy Efficiency: Smart grids enable more efficient energy generation, distribution, and consumption. Through real-time monitoring and data analysis, they identify and reduce energy losses, optimise power flows, and promote more efficient use of energy resources. Additionally, consumers now have access to real-time data on energy usage, which helps them make more educated choices, change their consumption habits, and increase the energy efficiency of their homes and businesses.

Integration of Renewable Energy Sources: Smart grids make it easier to integrate renewable energy sources into the electrical grid, such solar and wind power. They can keep an eye on and control how variable and dispersed renewable energy production is, ensuring efficient coordination between generation and demand. This promotes a more sustainable and low-carbon energy mix and increases the penetration of renewable energy sources while lowering curtailment.

Demand Response and Load Management: Demand response programmes are made possible by smart grids, allowing users to modify their energy usage in response to price signals or grid circumstances. Consumers may access real-time information about energy prices and modify their usage through linked devices and smart metres. As a result, peak demand is decreased, grid stability is increased, and energy resource efficiency is increased during periods of high demand.

Accommodation of Electric Vehicles (EVs) and Energy Storage: The integration of electric vehicles and energy storage technologies is supported by smart grids. They facilitate grid-friendly charging practices and provide effective monitoring and control of the EV charging infrastructure, optimising charging schedules. Energy storage devices can also be used by smart grids to balance supply and demand, store extra renewable energy, and support the grid during times of high demand or emergencies.

Grid Planning and Asset Management: Smart grids provide valuable data and insights for grid planning, expansion, and asset management. They enable utilities to analyse load patterns, identify capacity needs, and optimise infrastructure investments. By leveraging real-time data and analytics, utilities can prioritise maintenance activities, predict equipment failures, and optimise the performance of grid assets.

Improved Grid Security and Cyber Resilience: Smart grids incorporate robust cybersecurity measures to protect against cyber threats. They employ secure communication protocols, encryption, access controls, and continuous monitoring to ensure the integrity and reliability of grid operations. This enhances grid security and resilience in the face of evolving cyber risks.

Empowering Consumers: Smart grids empower consumers with greater visibility and control over their energy consumption. Real-time energy data, smart metres, and user-friendly interfaces enable consumers to monitor and manage their energy usage, adjust their behaviours, and save on energy costs. Additionally, smart grids facilitate the integration of DERs, allowing consumers to generate and sell their excess energy back to the grid.

Overall, smart grids offer benefits such as improved grid reliability, enhanced energy efficiency, better integration of renewables, demand response capabilities, optimised asset management, and increased consumer engagement. These benefits contribute to a more sustainable, resilient, and consumer-centric electricity system.

5.13 ZERO ENERGY BUILDING

An extremely energy-efficient building that generates as much renewable energy as it uses over the course of a year is referred to as a zero energy building (also known as a net-zero energy building, or NZEB). This results in a net energy consumption of zero. These structures use energy-efficient systems, materials, and architectural features to reduce their energy consumption, and they also have onsite renewable energy production to meet any lingering energy requirements. Here are some essential characteristics and tenets of zero energy structures:

Energy Efficiency: Zero energy buildings prioritise energy efficiency in all aspects of design and construction. This includes using high levels of insulation, efficient windows and doors, optimised Heating, Ventilation, and Air Conditioning (HVAC) systems, efficient lighting, and energy-efficient appliances and equipment. Energy modelling and analysis are typically performed during the design phase to identify and implement the most effective energy-saving measures.

Passive Design Strategies: Zero energy buildings often employ passive design strategies to reduce the need for mechanical heating, cooling, and lighting. Passive solar design, natural ventilation, day lighting, and shading techniques are utilised to optimise energy performance and occupant comfort.

Renewable Energy Generation: Zero energy structures use localised renewable energy production systems to balance their energy usage. This might include windmills, geothermal systems, solar photovoltaic (PV) panels, and other renewable energy sources. The objective is to produce enough renewable energy annually to satisfy or surpass the building's energy requirements.

Energy Monitoring and Controls: Zero energy buildings employ advanced energy monitoring and control systems to track energy consumption, renewable energy generation, and overall building performance. This allows building operators to monitor energy usage, identify areas for improvement, and optimise system operation for maximum energy efficiency.

Integration of Energy Storage: To store extra energy produced by renewable sources, some zero energy buildings incorporate energy storage technologies, such as batteries or thermal storage. Further reducing dependency on the grid, this stored energy can be used at times of high demand or low renewable energy production.

Optimised Water Efficiency: In addition to energy efficiency, zero energy buildings often prioritise water efficiency. This includes implementing low-flow fixtures, water-saving appliances, rainwater harvesting systems, and wastewater recycling technologies to minimise water consumption and support overall sustainability goals.

Life Cycle Approach: Zero energy buildings take a life cycle approach to sustainability. This involves considering the environmental impact of building materials, construction processes, and ongoing operations. Sustainable and locally sourced materials, recycling and waste management strategies, and long-term building durability are key considerations.

Zero energy buildings have several advantages, including reduced operating costs, improved occupant comfort and health, reduced greenhouse gas emissions, and increased resilience to energy price fluctuations. They play a vital role in the transition to a sustainable and low-carbon built environment by demonstrating the feasibility and benefits of highly efficient and renewable energy-integrated buildings.

5.14 ENERGY CONSERVATION BUILDING CODES

Energy conservation building codes, also known as energy codes or building energy codes, are regulations and standards that set minimum requirements for the energy efficiency of buildings. These codes aim to reduce energy consumption, promote sustainable building practices, and mitigate the environmental impact of buildings. Here are some key aspects of energy conservation building codes:

Minimum Performance Standards: Buildings must adhere to minimal energy performance standards set forth by energy codes. The criteria for renewable energy sources, air sealing, lighting efficiency, HVAC systems, and insulation of the building envelope are a few examples of the topics covered by these standards. To guarantee that buildings achieve a given level of energy efficiency, the codes provide specific criteria and requirements for each constituent.

Energy Modelling and Calculation: Energy codes often require energy modelling and calculation to assess a building's energy performance. Energy modelling involves using computer software to simulate and predict a building's energy use based on various design parameters. These calculations help evaluate the building's compliance with the code and identify opportunities for improving energy efficiency.

Envelope Requirements: Energy codes specify requirements for building envelopes, including insulation, windows, doors, and air leakage. These requirements ensure that buildings have proper insulation to reduce heat transfer, minimise air leakage to improve thermal comfort and optimise natural lighting while minimising solar heat gain.

Efficient Lighting and Appliances: Energy codes address the efficiency of lighting systems and appliances installed in buildings. They often mandate the use of energy-efficient lighting fixtures, such as LEDs (Light-Emitting Diodes), and may include requirements for automatic lighting controls, occupancy sensors, and daylighting strategies. Similarly, energy codes may set standards for the energy efficiency of appliances and equipment used in buildings.

HVAC Systems and Controls: Energy codes establish requirements for HVAC systems and controls to ensure optimal energy performance. These requirements may include efficiency standards for heating and cooling equipment, ventilation rates, energy recovery systems, and controls for temperature and airflow regulation.

Renewable Energy and Onsite Generation: Depending on the energy code, onsite generating and renewable energy technologies may be combined. They may include provisions for grid-connected systems, net metering, and mandates for the installation of solar panels, wind turbines, or other renewable energy technology.

5.15 GREEN BUILDING

The practice of planning, constructing, and managing buildings in a way that is resource- and environmentally conscious is known as green building, sometimes known as sustainable or eco-friendly building. Green buildings work to reduce their negative effects on the environment, improve tenant health and comfort, and encourage resource, energy, and water efficiency [38–40]. Here are some crucial elements of green construction:

Energy Efficiency: Energy efficiency is given priority in green buildings, which incorporate energy-saving technologies and architectural principles. The utilisation of renewable energy sources, such as solar or wind power, as well as energy-efficient insulation, high-performance windows, HVAC systems, and lighting controls that are state-of-the-art are all examples of this.

Water Conservation: Utilising water-saving plumbing fittings, rainwater collection systems, grey water recycling, and water-conserving landscaping practices, green buildings place a strong emphasis on water conservation. These actions aid in lowering water usage, relieving pressure on water supplies, and fostering sustainable water management.

Materials and Resources: Green buildings employ strategies to minimise the use of natural resources and reduce waste generation. This includes using sustainable and locally sourced materials, promoting recycling and waste management practices during construction and operation, and incorporating recycled content and low-emission materials.

Indoor Environmental Quality: By maintaining appropriate indoor air quality, adequate ventilation, and natural lighting, green buildings prioritise human health and well-being. To improve the sense of connection with nature, they might include elements like non-toxic materials, low-emission finishes, effective air filtering systems, and the use of biophilic design ideas.

Site Selection and Landscaping: Green buildings consider the site selection and landscaping practices to minimise environmental impact. They may incorporate features like green roofs, permeable surfaces, native landscaping, and preservation of open spaces to reduce heat island effect, manage storm water runoff, and support biodiversity.

Life Cycle Assessment: The life cycle impact of buildings, encompassing the extraction, manufacture, construction, operation, and end-of-life stages, is frequently taken into account in green building practices. Making educated judgements about materials, systems, and design techniques is made easier with the aid of life cycle assessment.

Certification and Rating Systems: LEED (Leadership in Energy and Environmental Design), BREEAM (Building Research Establishment Environmental Assessment Method), and Green Star are just a few examples of the several green building certification programmes and grading systems that are available globally. These initiatives offer criteria, benchmarks, and standards for evaluating and appreciating a building's sustainability performance.

Education and Awareness: Building occupants and stakeholders benefit from green buildings' promotion of education and awareness. To improve the building's total environmental performance, sustainable practices, energy-saving methods, and encouraging sustainable behaviours must all be promoted.

5.16 CONCLUSION

In order to solve environmental issues, reduce climate change, and create healthier and more sustainable built environments, it is becoming more and more crucial to adopt green building concepts and practices. Green buildings have many advantages, such as lessened energy and water use, cheaper running costs, increased productivity and health of occupants, and a less carbon footprint.

REFERENCES

1. B. Patra, M. Mahale, A. Mhaskar, *Madhur Chauhan "Smart Grid"*. Alpha Publication.
2. R. Beli, *Smart Grid Fundamentals: Energy Generation, Transmission and Distribution (Nano and Energy)*, Edition-1, ISBN-1482256673. CRC Press, Boca Raton, FL, 2022.
3. D. Baken, *Smart Grids*. CRC Press, Boca Raton, FL, ISBN-9781351831413, 2017.
4. S. Wilker, M. Meise, E. Piatkowska, T. Sauter, O. Jung, Smart grid reference architecture, an approached on a secure and model-driven implementation, 2018, https://doi.org/10.1109/ISIE.2018.8433754.
5. C. S. Meena, A. N. Prajapati, A. Kumar, M. Kumar, Utilization of solar energy for water heating application to improve building energy efficiency: An experimental study. *Buildings*, vol. 12, p. 2166, 2022. https://doi.org/10.3390/buildings12122166.
6. C. S. Meena, A. Kumar, S. Jain, A. Ur Rehman, S. Mishra, Innovation in green building sector for sustainable future. *Energies*, vol. 15, p. 6631, 2022. https://doi.org/10.3390/en15186631.
7. N. Dutt, A. J. Hedau, A. Kumar, M. K. Awasthi, V. P. Singh, G. Dwivedi, Thermo-hydraulic performance of solar air heater having discrete D-shaped ribs as artificial roughness. *Environ. Sci. Pollut. Res.*, 2023. https://doi.org/10.1007/s11356-023-28247-9.
8. P. K. Kushwaha, N. K. Sharma, A. Kumar, C. S. Meena, Recent advancements in augmentation of solar water heaters using nanocomposites with PCM: Past, present, and future. *Buildings*, vol. 13, p. 79, 2023. https://doi.org/10.3390/buildings13010079.
9. V. P. Singh, S. Jain, A. Karn, A. Kumar, Experimental assessment of variation in open area ratio on thermohydraulic performance of parallel flow solar air heater. *Arab. J. Sci. Eng.*, vol. 48, pp. 11695–11711, 2022. https://doi.org/10.1007/s13369-022-07525-7.
10. D. Mills, G. Morrison, Modelling study for compact Fresnel reflector power plant. *J. Phys. IV*, vol. 9(PR3), pp. Pr3-159–Pr3-165, 1999.
11. V. P. Singh, S. Jain, A. Kumar, S. Mishra, N. K. Sharma, Heat transfer and friction factor correlations development for double pass solar air heater artificially roughened with perforated multi-V ribs. *Case Stud. Therm. Eng.*, vol. 39, p. 102461, 2022, ISSN 2214-157X. https://doi.org/10.1016/j.csite.2022.102461.

12. V. P. Singh, S. Jain, A. Karn, A. Kumar, G. Dwivedi, C. S. Meena, Recent developments and advancements in solar air heaters: A detailed review. *Sustainability*, vol. 14, p. 12149, 2022. https://doi.org/10.3390/su141912149.

13. V. P. Singh, S. Jain, A. Karn, A. Kumar, G. Dwivedi, C. S. Meena, R. Cozzolino, Mathematical modeling of efficiency evaluation of double-pass parallel flow solar air heater. *Sustainability*, vol. 14, p. 10535, 2022. https://doi.org/10.3390/su141710535.

14. V. P. Singh, S. Jain, A. Kumar, Establishment of correlations for the thermo-hydraulic parameters due to perforation in a multi-V rib roughened single pass solar air heater. *Exp. Heat Transfer*, pp. 597–616, 2022. https://doi.org/10.1080/08916152.2022.2064940.

15. A. Saxena, A. N. Prajapati, G. Pant, C. S. Meena, A. Kumar, V. P. Singh, Water consumption optimization of hybrid heat pump water heating system. In: Shukla, A. K., Sharma, B. P., Arabkoohsar, A., Kumar, P. (eds) *Recent Advances in Mechanical Engineering. FLAME 2022. Lecture Notes in Mechanical Engineering.* Springer, Singapore, 2023, pp. 721–732. https://doi.org/10.1007/978-981-99-1894-2_61.

16. K. Kaygusuz, Prospect of concentrating solar power in Turkey: The sustainable future. *Renew. Sustain. Energy Rev.*, vol. 15(1), pp. 808–814, 2011.

17. V. Verma, C. S. Meena, S. Thangavel, A. Kumar, T. Choudhary, G. Dwivedi, Ground and solar assisted heat pump systems for space heating and cooling applications in the northern region of India – A study on energy and CO2 saving potential. *Sustainable Energy Technol. Assess.*, vol. 59, p. 103405, 2023, ISSN 2213-1388. https://doi.org/10.1016/j.seta.2023.103405.

18. M. K. Awasthi, N. Dutt, A. Kumar, S. Kumar, Electrohydrodynamic capillary instability of Rivlin-Ericksen viscoelastic fluid film with mass and heat transfer. *Heat Transfer*, vol. 53(1), pp. 115–133, 2024. https://doi.org/10.1002/htj.22944.

19. C. S. Meena, A. Kumar, S. Roy, A. Cannavale, A. Ghosh, Review on boiling heat transfer enhancement techniques. *Energies*, vol. 15, p. 5759, 2022. https://doi.org/10.3390/en15155759.

20. A. Kundu, A. Kumar, N. Dutt, C. S. Meena, V. P. Singh, Introduction to thermal energy resources and their smart applications. In: Kumar, A., Singh, V. P., Meena, C. S., Dutt, N. (eds) *Thermal Energy Systems: Design, Computational Techniques and Applications.* CRC Press, Boca Raton, FL, 2023, pp. 1–15; Chapter 1. https://doi.org/10.1201/9781003395768-1.

21. A. K. Dewangan, S. Q. Moinuddin, M. Cheepu, S. K. Sajjan, A. Kumar, Thermal energy storage: Opportunities, challenges and future scope. In: Kumar, A., Singh, V. P., Meena, C. S., Dutt, N. (eds) *Thermal Energy Systems: Design, Computational Techniques and Applications.* CRC Press, Boca Raton, FL, 2023, pp. 17–28; Chapter 2. https://doi.org/10.1201/9781003395768-2.

22. A. G. Olabi, M. Mahmoud, B. Soudan, T. Wilberforce, M. Ramadan, Geothermal based hybrid energy systems, toward eco-friendly energy approaches. *Renew. Energy*, vol. 147, pp. 2003–2012, 2020.

23. A. Kundu, A. Kumar, N. Dutt, V. P. Singh, C. S. Meena, Modelling and simulation of thermal energy system for design optimization. In: Kumar, A., Singh, V. P., Meena, C. S., Dutt, N. (eds) *Thermal Energy Systems: Design, Computational Techniques and Applications.* CRC Press, Boca Raton, FL, 2023, pp. 103–140; Chapter 7. https://doi.org/10.1201/9781003395768-7.

24. G. Pant, C. S. Meena, A. Saxena, A. Kumar, V. P. Singh, N. Dutt, Study the temperature variation in alternate coils of insulated condenser cum storage tank: Experimental study. In: Sikarwar, B. S., Sharma, S. K., Jain, A., Singh, K. M. (eds) *Advances in Fluid and Thermal Engineering. FLAME 2022. Lecture Notes in Mechanical Engineering.* Springer, Singapore, 2023, pp. 627–638. https://doi.org/10.1007/978-981-99-2382-3_52.

25. M. C. Bassetti, D. Consoli, G. Manente, A. Lazzaretto, Design and off-design models of a hybrid geothermal-solar power plant enhanced by a thermal storage. *Renew. Energy*, vol. 128, pp. 460–472, 2018.

26. K. Li, C. Liu, S. Jiang, Y. Chen, Review on hybrid geothermal and solar power systems. *J. Clean. Prod.*, vol. 250, p. 119481, 2020.

27. A. Kumar, V. P. Singh, C. S. Meena, N. Dutt, *Thermal Energy Systems: Design, Computational Techniques and Applications.* Taylor & Francis (CRC Press), Boca Raton, FL, 2023. https://doi.org/10.1201/9781003395768.

28. S. Singh, A. Kumar, S. K. Behura, K. Verma, Challenges and opportunities in nanomanufacturing. In: Singh, S., Behura, S. K., Kumar, A., Verma, K. (eds) *Nanomanufacturing and Nanomaterials Design: Principles and Applications.* CRC Press, Boca Raton, FL, 2022, pp. 17–30; Chapter 02. https://doi.org/10.1201/9781003220602-2.

29. S. Srivastava, D. Verma, S. Thusoo, A. Kumar, V. P. Singh, R. Kumar, Nanomanufacturing for energy conversion and storage devices. In: Singh, S., Behura, S. K., Kumar, A., Verma, K. (eds) *Nanomanufacturing and Nanomaterials Design: Principles and Applications.* CRC Press, Boca Raton, FL, 2022, pp. 165–173; Chapter 10. https://doi.org/10.1201/9781003220602-10.

30. V. P. Singh, A. Dwivedi, A. Karn, A. Kumar, S. Singh, S. Srivastava, K. Srivastava, Nanomanufacturing and design of high-performance piezoelectric nanogenerator for energy harvesting. In: Singh, S., Behura, S. K., Kumar, A., Verma, K. (eds) *Nanomanufacturing and Nanomaterials Design: Principles and Applications.* CRC Press, Boca Raton, FL, 2022, pp. 241–242; Chapter 15. https://doi.org/10.1201/9781003220602-15.

31. M. Dada, P. Popoola, Recent advances in solar photovoltaic materials and systems for energy storage applications: A review. *Beni-Suef Univ. J. Basic Appl. Sci.*, vol. 12, p. 66, 2023. https://doi.org/10.1186/s43088-023-00405-5.

32. M. El Haj Assad, M. H. Ahmadi, M. Sadeghzadeh, A. Yassin, A. Issakhov, Renewable hybrid energy systems using geothermal energy: Hybrid solar thermal-geothermal power plant. *Int. J. Low-Carbon Technol.*, vol. 16, no. 2, pp. 518–530, 2021.

33. W. Strydom, J. Pretorius, J. Hoffmann, Sensitivity analysis on the performance of a natural draft direct dry cooling system for a 50 MWe CSP application. *Proc. 8th World Congr. Momentum, Heat Mass Transf.,* Lisbon, pp. 1–8, 2023. https://doi.org/10.11159/enfht23.138.

34. B. Willemse, S. Nielsen, J. Pretorius, M. Owen, A. Rix, A comparative evaluation of heat dissipation factors for open-rack and floating solar photovoltaic installations. Available SSRN 4547936, 2023. http://dx.doi.org/10.2139/ssrn.4547936

35. S. A. Kalogirou, S. Lloyd, J. Ward, P. Eleftheriou, Design and performance characteristics of a parabolic-trough solar-collector system, *Appl. Energy*, vol. 47, no. 4, pp. 341–354, 1994.

36. S. J. Self, B. V Reddy, M. A. Rosen, Geothermal heat pump systems: Status review and comparison with other heating options. *Appl. Energy*, vol. 101, pp. 341–348, 2013.
37. O. Erdinc, M. Uzunoglu, Optimum design of hybrid renewable energy systems: Overview of different approaches. *Renew. Sustain. Energy Rev.*, vol. 16, no. 3, pp. 1412–1425, 2012.
38. J. J. Ding, J. S. Buckeridge, Design considerations for a sustainable hybrid energy system. *Trans. Inst. Prof. Eng. New Zeal. Electr. Eng. Sect.*, vol. 27, no. 1, pp. 1–5, 2000.
39. G. Leonzio, Solar systems integrated with absorption heat pumps and thermal energy storages: State of art. *Renew. Sustain. Energy Rev.*, vol. 70, pp. 492–505, 2017.
40. M. Kıyan, E. Bingöl, M. Melikoğlu, A. Albostan, Modelling and simulation of a hybrid solar heating system for greenhouse applications using Matlab/Simulink. *Energy Convers. Manag.*, vol. 72, pp. 147–155, 2013.

Chapter 6

Design and modelling of solar, geothermal and hybrid energy systems

Abhishek K. Sharma, Mohit Nayal,
Siddharth Jain, and Varun Pratap Singh
University of Petroleum and Energy Studies Dehradun

6.1 INTRODUCTION

Energy system design and modelling, including solar, geothermal, and hybrid designs, is critical to creating sustainable and efficient energy solutions. These systems use renewable energy to meet demand while reducing environmental effects and providing energy security. Solar, geothermal, and hybrid technology integration necessitates a comprehensive approach that considers resource availability, system layout, control strategies, and financial viability [1]. Solar energy systems, which use photovoltaic (PV) panels that transform sunlight into power, are a cornerstone of sustainable energy adoption. The design process includes determining the best sites, tilts, and orientations for PV arrays while considering local solar radiation patterns. To improve system output, advanced modelling tools simulate sun irradiance, shading effects, and panel performance. By storing excess energy for use during times of low sunlight, battery-powered integration in solar systems addresses intermittency difficulties [2]. Geothermal energy systems use the Earth's inherent heat reservoirs to heat, cool, and generate electricity. Subsurface resource evaluation, well drilling, and power plant setup are all part of the design. Simulation tools simulate ground heat transfer processes, assisting in the size and configuration of ground heat exchangers. Geothermal heat pumps are combined with solar or other energy sources in hybrid geothermal systems, increasing the efficiency of energy and flexibility [3].

To develop diverse and dependable solutions, hybrid energy systems integrate numerous sources of renewable electricity, energy storage, and typically conventional backup power. The combination of wind, solar, geothermal, and energy storage methods must be optimized while designing such systems [4]. To maintain optimal operation, complex control algorithms manage the flow of energy among sources, storage, and loads. Advanced modelling tools replicate dynamic interactions and assist system component and sizing decisions [5].

DOI: 10.1201/9781003472629-6

6.2 FINITE ELEMENT ANALYSIS AND MODELLING TECHNIQUES IN ENERGY SYSTEM DESIGN

Energy systems are at the heart of our modern world, providing power and sustainability for countless applications. The finite element analysis (FEA) and modelling techniques in designing and optimizing play various critical roles in analysing energy systems. By harnessing the power of FEA, engineers and researchers can assess structural integrity, evaluate thermal performance, and implement efficiency enhancements to create more effective and sustainable energy solutions [6].

6.2.1 FEA techniques for structural integrity assessment

Structural integrity is paramount in energy system design, ensuring safety, reliability, and longevity. This section delves into the application of FEA techniques for assessing and optimizing the structural components within energy systems. We will explore how FEA aids in simulating and analysing stresses, strains, and deformations in materials and structures, helping engineers identify weak points, predict failures, and develop solutions for enhanced structural integrity [7].

In the realm of energy system design, FEA techniques play a pivotal role in ensuring structural integrity. Engineers utilize FEA through methods such as static analysis, dynamic analysis, buckling analysis, and fatigue analysis to simulate and analyse stresses, strains, and deformations in components of energy systems [8]. This is critical for identifying potential weaknesses and optimizing designs, ultimately preventing costly structural failures. The advantages of FEA lie in its precision, cost-effectiveness, and versatility across various materials and configurations. However, it demands expertise due to its complexity, relies heavily on assumptions, and can be computationally intensive [9].

6.2.2 Thermal performance evaluation using FEA

Efficient heat management is a cornerstone of energy system design, impacting both energy production and conservation. In this section, we delve into how FEA is employed to evaluate and enhance thermal performance. Through advanced simulations and modelling, FEA allows engineers to understand heat transfer mechanisms, temperature distribution, and thermal stresses within energy systems. We'll explore case studies and methodologies that enable engineers to optimize thermal efficiency and ensure consistent performance [10].

Thermal performance evaluation is another vital aspect of energy systems, and FEA is instrumental in this domain. Techniques such as steady-state thermal analysis, transient thermal analysis, and conduction, convection, and radiation analysis enable engineers to assess temperature distribution and heat transfer mechanisms within components. This is essential for optimizing thermal behaviour, preventing overheating, ensuring consistent performance, and reducing energy losses [11]. FEA empowers engineers to fine-tune system components for peak thermal efficiency, identify hotspots, and enhance overall energy system efficiency. Nevertheless, it requires careful consideration of simplifications, accurate material properties, and boundary conditions [12].

6.2.3 Efficiency enhancement through different techniques

The quest for energy efficiency is a driving force behind energy system innovation. This section explores the diverse range of techniques that leverage FEA to enhance the efficiency of energy systems. From materials selection to innovative design approaches and advanced control strategies, FEA plays a pivotal role in achieving higher energy efficiency. We'll examine real-world examples of how FEA can be used to identify inefficiencies, reduce energy losses, and improve overall system performance [13].

Efficiency enhancement techniques in energy system design encompass material selection, design modifications, advanced control strategies, and the integration of renewable sources. These approaches aim to maximize energy output while minimizing waste, resulting in sustainable and cost-effective energy systems [14]. They offer benefits such as higher efficiency, sustainability, and cost savings over the system's lifetime. However, they can also introduce complexity, require specialized expertise, and incur integration challenges and costs, particularly in hybrid systems [15,16]. By comprehending these techniques and their applications, advantages, and limitations, engineers and researchers can make informed decisions to create energy systems that are both efficient and reliable, contributing to a greener and more sustainable future [17].

6.3 THERMAL PERFORMANCE EVALUATION

Thermal performance evaluation is an important step in determining the efficiency, efficacy, and general effectiveness of thermal systems, parts, and materials. This assessment includes a thorough examination of how these components manage heat, transfer thermal energy, and maintain desirable temperatures in a variety of applications ranging from buildings and industrial processes to systems for storing energy. A thorough investigation of heat transmission methods such as conduction, convection, and radiation is at the heart of thermal performance evaluation [18]. In the context of building envelopes, for example, analysing the thermal conductivity and ability of insulation materials to resist heat transfer aids in the design of energy-efficient buildings that limit heat loss during wintertime and heat gain during hot seasons. Understanding how heat is transmitted throughout equipment and systems is critical for increasing their performance and energy efficiency in industrial processes [19].

Thermal performance evaluation relies heavily on advanced modelling and simulation techniques. Simulations of computational fluid dynamics (CFD) provide extensive insights into heat distribution, fluid flow patterns, and temperature gradients within complex geometries, for example [20]. These simulations aid in the identification of possible hotspots, the optimization of designs for improved heat dissipation, and the refinement of thermal management systems. Energy modelling tools in buildings evaluate energy use and suggest the potential for energy savings by analysing heat flow through windows, walls, roofs, and HVAC systems. Real-world testing and experimentation supplement simulation efforts by allowing theoretical models to be validated and refined [21]. Thermal performance testing is exposing systems or parts to controlled temperatures and assessing their responses. Testing the thermal resistance of insulation materials, for example, gives empirical data that can be used to validate simulation results and make informed design decisions [22]. Table 6.1 shows various parameters used in the thermal performance evaluation, design, and modelling of solar, geothermal, and hybrid energy systems

6.3.1 Innovative approaches to thermal storage

Innovative thermal storage technologies have emerged as critical solutions for tackling energy intermittency concerns and maximizing waste heat utilization across multiple sectors. These methods include cutting-edge technologies and strategies that improve the effectiveness, capacity, and applicability of thermal storage devices. One novel approach is to include sophisticated phase change materials (PCMs) in thermal storage devices [23]. PCMs have the unusual capacity to collect and release considerable amounts of energy at practically constant temperature during phase transitions such as solid-to-liquid or liquid-to-gas transitions. Because of this feature, they can store and release thermal energy with minimum temperature variations. PCMs can be embedded in building materials like concrete or gypsum to provide thermal-enhanced building components that store heat during the day and release it at night, resulting in more energy-efficient structures [24]. Furthermore, PCMs can be integrated into systems such as solar water heaters, allowing for extended heat retention and utilization even when sunshine is not available [25].

Another novel technique entails the use of thermochemical storage devices. These systems make use of reversible chemical reactions that collect and

Table 6.1 Detail of various parameters used in the thermal performance evaluation, design, and modelling of solar, geothermal, and hybrid energy systems

Parameter	Description	Solar energy system	Geothermal energy system	Hybrid energy system
System type	The type of energy system (solar, geothermal, hybrid)	Solar PV, solar thermal, concentrated solar power (CSP)	Ground-source heat pump (GSHP), direct use, geothermal power plant	Combining solar and geothermal, etc.
Solar collector type	Type of solar collectors used (e.g., flat plate, parabolic trough, photovoltaic)	Flat plate, parabolic trough, PV panels, concentrated solar collectors, etc.	N/A	Depends on solar component
Geothermal heat source	Source of geothermal heat (e.g., ground, geothermal reservoir)	Ground, geothermal reservoir, aquifer, etc.	Ground, geothermal reservoir, etc.	Depends on the configuration
System size	Physical size or capacity of the energy system	kW, MW, GW, etc.	kW, MW, GW, etc.	kW, MW, GW, etc.
Efficiency	Energy conversion efficiency (e.g., solar panel efficiency, heat pump efficiency)	Solar panel efficiency (%), thermal efficiency (%), conversion efficiency (%)	Heat pump efficiency (%), conversion efficiency (%)	Combined efficiency (%)
Temperature range	Operating temperature range for the system	Low, medium, high	Low, medium, high	Varies

(Continued)

Table 6.1 (Continued) Detail of various parameters used in the thermal performance evaluation, design, and modelling of solar, geothermal, and hybrid energy systems

Parameter	Description	Solar energy system	Geothermal energy system	Hybrid energy system
Energy storage	Type of energy storage used (e.g., batteries, thermal storage)	Batteries, thermal storage, phase change materials, etc.	Thermal storage, aquifer storage, etc.	Combination of storage types
Location	Geographic location where the system is installed	Latitude, longitude	Geological characteristics	Latitude, longitude, geological characteristics
Insolation (solar)	Solar radiation available at the location	kWh/m²/day	N/A	kWh/m²/day
Ground temperature (geothermal)	Initial ground temperature at depth	°C or °F	°C or °F	°C or °F
Environmental factors	Environmental conditions affecting system performance	Ambient temperature, wind speed, humidity, dust, shading, etc.	Geological conditions, thermal gradients, etc.	Combination of environmental factors
Economic parameters	Cost-related parameters (e.g., cost per kWh, payback period)	LCOE (levelized cost of electricity), ROI, NPV, payback period	LCOE, ROI, NPV, payback period	LCOE, ROI, NPV, payback period
Modelling software	Software used for modelling and simulation	PVsyst, SAM, HOMER, TRNSYS, etc.	TRNSYS, COMSOL, TOUGH, etc.	Combination of modelling software

release heat as they transform. Thermochemical storage can provide high energy storage densities and long storage periods by selecting appropriate reaction pairings and developing efficient reactors [26]. Such technologies are especially promising for concentrated solar power (CSP) facilities, which may store solar energy as chemical potential energy and use it to generate electricity during overcast or night-time times. Innovative thermal storage techniques, such as high-temperature ceramic storage systems, can store waste heat generated during industrial processes. High-temperature ceramics are used as storage media in these systems, absorbing excess heat from manufacturing processes and then efficiently releasing it when needed [27]. This strategy not only improves industrial process energy efficiency but also reduces overall consumption of energy and emissions of greenhouse gases. Innovative storage solutions, such as composite materials with increased thermal conductivity, have resulted from advanced material research. These materials can efficiently disperse and transmit heat, surpassing the constraints of conventional storage systems' poor heat transfer. This development is especially useful for raising the charging and discharging rates of thermal storage systems, hence increasing their overall efficiency and performance [28].

6.3.2 Modelling of thermal energy storage systems

Thermal energy storage (TES) system modelling is a critical technique that allows for the study, optimization, and effective exploitation of these critical components in the field of sustainable energy management [29]. TES is critical for bridging the temporal gap between energy generation and demand, improving the flexibility and reliability of various energy systems such as renewable sources such as solar and wind power, industrial processes, and building heating and cooling. TES modelling, at its core, is the development of mathematical and computer representations that replicate the behaviour of complex systems under varied conditions [30].

These models encompass the complicated thermal dynamics, material characteristics, and transfer of heat processes that influence TES and release [31]. TES modelling can be done using a variety of approaches, including analytical, numerical, and empirical methods, each of which provides distinct insights into system behaviour. Analytical models are simplified yet insightful representations of TES systems that give quick solutions for initial system design and parameter estimates. These models frequently include assumptions that simplify calculations, making them appropriate for preliminary judgements [32].

They may, however, lack the precision required for comprehensive assessments of complicated systems. Numerical techniques, on the other hand, entail dividing the TES system into finite elements or cells and numerically solving governing equations using computer simulations. This method accommodates complex geometries, a wide range of materials, and changing working circumstances [33]. CFD techniques are frequently used to simulate heat transfer and fluid flow inside the storage medium, providing extensive insights into the distribution of temperatures, energy storage capability, and system performance [34].

Empirical models characterize the actions of TES systems using practical data and statistical methodologies. When dealing with real-world variability and uncertainty, these models are especially useful. They may be based on regression analysis or methods of machine learning, and they must capture the link between input parameters and system response [35]. Empirical models are useful for forecasting system behaviour based on previous data, which aids in real-time control and decision-making tactics. Integrating realistic TES models into energy system simulations allows for the optimization of TES operation techniques, which is especially important for maximizing the use of renewable energy sources that are intermittent. These models serve as the foundation for the creation of control algorithms that regulate both charging and discharging cycle while considering aspects such as energy demand, energy costs, and weather conditions. Optimal control provides efficient energy storage during periods of low demand and high availability of energy, followed by release when demand peaks or energy generation declines [36].

6.4 SOLAR AND RADIANT ENERGY SYSTEMS DESIGN

Solar and radiant energy system design is a critical activity in the field of renewable energy sources, with the goal of harnessing and utilizing the enormous potential of the sun's light. These systems are an important step toward reducing the environmental effects of traditional energy sources, lowering carbon emissions, and ensuring a more sustainable future for future generations. Such systems require a multidimensional strategy that includes technology innovation, structural integration, and efficient control of energy [37]. Solar energy system design is centred on absorbing and converting sunlight into useable forms of energy, primarily electricity or heat. The most prominent form of solar energy technology is photovoltaic (PV) systems, which consist of solar panels that absorb sunlight and generate direct current (DC) electricity via the photovoltaic effect. These panels, which are frequently made of semiconductor materials such as silicon, are strategically put on rooftops, solar farms, or even built into building facades. Solar thermal systems, like PV systems, capture solar radiation to generate heat, which is then utilized for water heating, space heating, or even powering thermal operations in enterprises. These systems are made up of solar collectors that absorb solar radiation and subsequently heat a fluid that can be used for a variety of purposes [38].

Radiant energy, which includes both solar and thermal radiation, presents a broad area for energy system designers. Radiant cooling systems, for example, use the radiative heat exchange principle to chill indoor spaces without relying only on energy-intensive air conditioning. These systems remove heat from the indoor environment by incorporating radiant cooling panels into ceilings or floors. This novel strategy improves energy efficiency while also providing a more comfortable and long-lasting cooling solution. The design of effective solar and radiant energy systems includes architectural concerns in addition to technological factors. Architects and designers are critical in fitting solar panels and collectors into structures without sacrificing aesthetic appeal [39]. Solar panels are combined with architectural elements in building-integrated photovoltaics (BIPV), making them into functional components of the building envelope. This combination not only creates clean energy but also improves the visual appeal and economic value of the structure. Furthermore, planning for radiant energy demands knowledge of thermal dynamics, material qualities, and occupant comfort. Architects and engineers must collaborate to optimize the location of radiant cooling and heating devices, ensuring efficient energy exchange while preserving a pleasant indoor environment [40]. Table 6.2 shows the various parameters used in the design of solar and radiant energy systems, which can be further discussed in detail.

6.4.1 Solar collector design and performance

The design and performance of solar collectors are key parts of capturing sunlight for warming and other applications. Solar panels are devices that turn sunlight into useful thermal energy. A solar collector's design entails a careful balancing of parameters that increase energy absorption, decrease losses, and improve overall efficiency [41].

The collector's glazing or cover is transparent, allowing sunlight to penetrate while reducing heat losses. It serves as an insulator, minimizing convective loss of heat and shielding the absorber from external variables. As glazing, several materials, like as glass or polymers, can be employed, each having its own set of benefits in terms of durability, transparency, and thermal qualities. Proper insulation around the collector's sides and rear reduces heat loss to the environment [42]. This guarantees that the heat received by the absorber is effectively retained and used. Low thermal conductivity insulation materials, such as foam or fibreglass, are often utilized to reduce heat transfer. To transport and transfer heat, solar collectors frequently use a fluid (typically water or a mixture of water and antifreeze). The efficiency of heat transmission from the absorber to the fluid is affected by the design of the fluid circulation system, including the flow rate and channel. Circulation systems that are well-designed ensure uniform heat transfer and prevent overheating or freezing. The collector's orientation and

Table 6.2 Details of various parameters used in the design of solar and radiant
energy systems

Parameter	Description	Solar energy system	Radiant energy system
System type	The type of energy system	Solar PV, solar thermal, concentrated solar power (CSP)	Radiant heating, radiant cooling, combined systems, etc.
Solar collector type	Type of solar collectors used	PV panels, solar thermal collectors, concentrated solar collectors, etc.	Radiant heating panels, radiant cooling panels, radiant tubing, etc.
Collector efficiency	Energy conversion efficiency	Solar panel efficiency (%), thermal collector efficiency (%)	Radiant panel efficiency (%), radiant tubing efficiency (%)
System size	Physical size or capacity of the system	kW, MW, GW, etc.	Square footage, length of tubing, etc.
Temperature control	Control mechanisms for temperature	Temperature sensors, controllers, tracking systems	Thermostats, zone controls, control valves, etc.
Heat source	Source of heat for the system	Solar radiation (sunlight), backup heat source (if applicable)	Hot water, electric resistance, heat pump, etc.
Heat transfer fluid	Fluid used for heat transfer (if applicable)	Water, thermal oil, air, etc.	Water, glycol-water mixture, etc.
Insolation	Solar radiation available at the location	$kWh/m^2/day$	N/A
Design temperature range	Desired temperature range for radiant system	Heating: $^\circ$C or $^\circ$F, cooling: $^\circ$C or $^\circ$F	Heating: $^\circ$C or $^\circ$F, cooling: $^\circ$C or $^\circ$F
Environmental factors	Environmental conditions affecting performance	Ambient temperature, shading, wind, humidity, etc.	Ambient temperature, indoor conditions, humidity, etc.
Economic parameters	Cost-related parameters (e.g., cost per kWh, payback period)	LCOE (levelized cost of electricity), ROI, NPV, payback period	Installation cost, operating cost, ROI, NPV, payback period
Control strategy	Strategies for system control and optimization	Tracking, temperature set points, on/off control, etc.	Thermostat control, zone control, modulation, etc.
Modelling software	Software used for modelling and simulation	PVsyst, SAM, HOMER, TRNSYS, etc.	Radiant heating/cooling design software, energy plus, etc.

tilt are critical factors in optimum energy extraction. Collectors are typically pointed toward the sun's direction (south in the northern hemisphere) and tilted at an angle to maximize sunlight exposure throughout the day and year [43].

Metrics including efficiency, heat gain, and overall system effectiveness are used to evaluate the performance of a solar collector. The efficiency of a collector is a measure of how well it converts incoming solar energy into useful heat. Factors such as absorber design, glazing qualities, and insulation all have an impact on it. The quantity of heat delivered by a collector to the fluid or air moving through it is referred to as its heat gain. All energy losses within the collector and its components are considered in the overall system efficacy [44].

6.4.2 Radiant heating systems

Radiant heating systems, a modern and energy-efficient solution to indoor comfort, function by actively heating things and surface in a space, including floors, walls, and even furniture. This type of heating works on the radiant heat transfer principle, which states that warmth radiates outward from a surface that is heated to cooler surfaces and objects, ensuring a comfortable and constant temperature distribution. Unlike standard forced-air systems, which heat the air and rely on convection currents to disperse warmth, radiant heating warms the surroundings directly, producing a comfortable and even warmth that is sometimes compared to sunbathing. A heat source, such as a heating system or a heat pump, heats a fluid, usually water, which then flows via a network of pipes installed in a building's floor, walls, or ceiling. This network of pipes, known as a "hydronic" system, progressively raises the temperature of the surfaces it encounters by releasing the stored heat into the environment. This warmth is subsequently radiated as infrared radiation, which is absorbed by the room's items and people, providing a mild and comfortable heat [45].

There are several advantages: first, because the entire area is heated consistently, it prevents drafts and stratification of temperatures that can occur with forced-air systems. Second, it reduces heat loss by focusing on warming surfaces instead of air. As a result, energy consumption is reduced, and utility expenses are cut. Furthermore, because low-temperature water heating is compatible with renewable energy sources such as solar thermal collectors, radiant systems are more conducive to using these technologies. Radiant heating systems are classified into several types [46]. The most prevalent are floor-based systems, which can be put beneath a variety of flooring materials such as tile, wood, and concrete. Wall-mounted systems, as the name implies, are built into walls to provide a discreet heating option. Ceiling-mounted devices are less popular, but they can be beneficial in certain settings. Radiant heating system installation necessitates precise planning and design. To guarantee optimal performance, factors such as the heat source, pipe architecture, insulation, and control systems must be considered. To avoid heat loss downhill, adequate insulation beneath the surface that is heated is critical. Furthermore, radiant systems have a slower response time than forced-air systems, which means they may take longer to attain the appropriate temperature [47].

6.5 GEOTHERMAL ENERGY SYSTEMS DESIGN

Geothermal energy system design entails an extensive approach to harnessing the Earth's heat for long-term energy production, heating, and cooling. Geothermal systems take advantage of the Earth's sub surface's stable temperature, offering a continuous, renewable source of power. The initial step in geothermal system development is a detailed assessment of the project site's geothermal resource potential. Subsurface gradients of temperature, rock characteristics, and fluid reservoirs are all evaluated. Geoscientists and engineers record the thermal features of the subsurface using techniques such as geological surveys, seismic studies, as well as well logging. Drilling geothermal wells allows access to the heat trapped in the outermost layer of the Earth's crust. The depth, temperature, and permeability of the geological formations are all considered in the design of these wells. Reservoir management is essential for avoiding overexploitation and ensuring long-term resource sustainability. Fluid pressure and flow dynamics inside the reservoir must be monitored and modelled for appropriate management [48].

Power Plant Configuration depends on the temperature as well as the quality of the geothermal resource, geothermal power plants can be configured in a variety of ways. Dry steam plants drive turbines with steam at high pressure directly from the reservoir. To power turbines, flash steam plants utilize high-temperature geothermal fluid that is flashed into steam. Heat is transferred from the geothermal liquid to a secondary fluid with a lower point of boiling, resulting in steam that powers turbines. The configuration of the power plant is determined by the individual resource characteristics [49]. The infrastructure of a geothermal power plant includes power producing machinery, exchangers for heat, and electrical components. Connecting the plant to the electrical grid allows extra power to be provided to the grid during periods of high production. The location of the power plant and its closeness to existing grid infrastructure influence the possibility of integration [50].

When compared to traditional power generation, geothermal energy systems are considered environmentally beneficial due to their low emissions and minimum land use needs. Potential environmental consequences, such as changes in subsurface pressure, subsidence of land, and the release of

Table 6.3 Description of various parameters used in the design of geothermal energy systems

Parameter	Description
System type	Type of geothermal energy system (e.g., direct use, geothermal power plant, ground-source heat pump)
Heat source	Source of geothermal heat (e.g., ground, geothermal reservoir, aquifer)
System size	Physical size or capacity of the geothermal system (kW, MW, GW, etc.)
Temperature range	Operating temperature range for the geothermal system (low, medium, high)
Heat exchanger type	Type of heat exchanger used for heat transfer (e.g., shell-and-tube, plate, coaxial)
Fluid type	Type of heat transfer fluid (e.g., water, brine, refrigerant)
Pumping system	Pumping mechanisms for fluid circulation (e.g., geothermal well pumps, circulation pumps)
Heat extraction rate	Rate at which heat is extracted from the geothermal source (kW, MW, GW)
Reservoir characteristics	Geological properties of the geothermal reservoir (e.g., permeability, depth, pressure)
Flow rate	Rate of fluid flow through the geothermal system (L/s, m³/s, GPM, etc.)
Pressure	Pressure conditions within the geothermal system (bar, psi, kPa, etc.)
Environmental factors	Environmental conditions affecting system performance (e.g., ambient temperature, seismic activity)
Economic parameters	Cost-related parameters (e.g., installation cost, operating cost, ROI, NPV)
Control strategy	Strategies for system control and optimization (e.g., temperature set points, flow control)
Modelling software	Software used for modelling and simulation (e.g., TOUGH, COMSOL, TRNSYS, etc.)

trace gases, must, however, be studied and addressed during the design phase. Effective planning and monitoring aid in reducing these consequences [51]. Table 6.3 shows the detailed description of various parameters used in the design of geothermal energy systems.

6.5.1 Geothermal heat pump systems

Geothermal heat pump systems, referred to as ground-source heat pump systems (GSHPs), are creative and energy-efficient technology utilized in homes, businesses, and factories for cooling, heating, and even water heating. These systems take advantage of the steady warmth of the subsurface of the earth to deliver dependable and long-term thermal comfort. The heat pumps unit, the earth's exchanger for the heat, and the distribution system are the three basic components of a geothermal heat pump system. Like a standard air-source heat pump, the heat pump unit distributes heat between the structure and the ground. However, because of the steady ground temperature, it is more efficient. The ground heat exchanger is made up of a series of pipes buried in the ground that circulate a heat-transfer fluid, transferring heat with the earth. Depending on available space, this loop can be put horizontally in pits or vertically in boreholes. Finally, throughout the structure, the distribution system distributes air that is heated or cooled or water. Geothermal heat pump systems have substantial environmental benefits. They can save up to 45% more energy than traditional HVAC systems, resulting in lower emissions of greenhouse gases and less dependency on fossil fuels. GSHPs are regarded by the United States EPA (Environmental Protection Agency) as one of the most effective and ecologically benign heating and cooling systems available. Although geothermal heat pumps have significant advantages, they are not without limitations. Because of the excavation or drilling needed for ground loop installation, initial installation costs may be greater than for conventional systems. These expenses, however, can be offset by reduced energy consumption during the system's lifetime [52].

6.6 HYBRID ENERGY SYSTEMS DESIGN

Hybrid energy systems combine different green energy sources, technologies for storing energy, and, in many cases, traditional sources of electricity to offer a more dependable, effective, and environmentally friendly power solution. These systems take advantage of the complimentary properties of many energy sources to improve efficiency, maximize energy utilization, and alleviate the intermittency difficulties that are frequently associated with sources of renewable energy. Solar, wind, hydro, biomass, and geothermal energy sources are frequently combined in hybrid systems. Solar panels may provide electricity during the day, whereas wind turbines can produce electricity at night or during cloudy conditions. Combining various sources increases overall energy availability and decreases dependency on a single source [53].

Batteries, pumped hydro storage, and TES are important energy storage technologies in hybrid systems. They store extra energy created during peak production periods and release it during high demand or low generation periods. This assures a consistent and uninterrupted power supply, even when renewable energy sources are not supplying energy [54]. Advanced control and management algorithms are used in hybrid energy systems to maximize the use of available energy sources and storage. To meet demand efficiently, these algorithms calculate when to charge and discharge storage systems, as well as when to transition between different energy sources [55].

Natural gas generators or diesel engines, for example, can be used as backup power in hybrid systems. They ensure dependability in the event of extended periods of low renewable energy generation or unexpected interruptions. Connecting hybrid systems to the grid allows surplus energy to be fed into the grid during periods of high generation. When renewable energy is inadequate, they can pull power from the grid. Both the system's owner and the grid profit from this bidirectional energy flow. Hybrid systems are especially useful in rural or off-grid areas when access to a steady grid is limited. They offer a self-sufficient energy option that eliminates reliance on fossil fuels while also lowering operating costs [56].

6.6.1 Design considerations for hybrid systems

To ensure optimal performance, dependability, and cost-effectiveness, hybrid energy systems must be designed with certain critical aspects in mind. To produce a coherent and efficient system, these considerations include the integration of diverse renewable energy sources, energy storage technology, backup power alternatives, and control systems. The availability of resources at the installation site should be considered while selecting renewable energy sources. Because of their variable daily and seasonal rhythms, solar and wind resources complement each other. Biomass and geothermal energy can provide constant power output, compensating for the intermittent nature of solar and wind energy [57]. Understanding the energy need profile is critical for appropriately sizing the system's hybrid components. Analysing peak demand periods and recognizing load fluctuations aids in evaluating the capacity of energy storage devices and backup power sources required for continuous delivery. The technology used to store energy, such as batteries, pumped hydro storage, or thermal storage, is determined by parameters such as discharge length, cycle life, efficiency, and available space. The technology chosen should be compatible with the system's needs and provide dependable energy storage and discharge capabilities. Control algorithms and management strategies that are effective play a critical role in optimizing the operation of hybrid systems [58].

When to charge or drain energy storage, when to switch between various sources of energy, and how to respond to shifting demand are all determined by these techniques. Incorporating backup power sources such as diesel generators or natural gas turbines ensures a consistent energy supply during periods of low renewable energy generation or unexpected interruptions. Backup power sources should be sized to match key demands and serve as a backup measure. The seamless integration of grid-connected systems with the grid is critical [59]. Bidirectional energy flow enables surplus energy to be delivered to the grid in exchange for credits or money, while grid electricity can be pulled when needed. Smart grid technologies enable the hybrid system and the grid to communicate and control each other more effectively. The design approach should be guided by life cycle assessments and cost-benefit studies. Evaluating the environment's impact, reduction of greenhouse gas emissions, and economic feasibility aids in making informed judgements about system parts and topologies. Compliance with local legislation, codes, and standards is critical to ensuring hybrid systems operate safely and legally. Grid codes may specify grid connectivity, frequency regulation, and voltage stability requirements. Implementing a maintenance plan that includes frequent inspections, equipment service, and component replacement is critical for the hybrid energy system's longevity and reliability [60].

6.7 CONCLUSION

In conclusion, this chapter has provided a comprehensive overview of the design and modelling of solar, geothermal, and hybrid energy systems. We have explored a range of topics, from the use of FEA tools to the optimization of TES systems. Through our discussions on solar and radiant systems, geothermal energy systems, and hybrid energy solutions, we have highlighted the versatility and potential of renewable energy technologies.

The key takeaways are as follows:

Diverse Design Techniques: Various design techniques can be applied to assess the structural integrity, thermal performance, and overall efficiency of renewable energy systems. By tailoring these techniques to specific projects, researchers and engineers can optimize system designs for maximum effectiveness.

Energy Storage Optimization: The exploration of energy storage systems, including batteries, thermal storage, and hydrogen technologies, underscores their crucial role in renewable energy integration. Efficient energy storage is vital for managing intermittent energy sources and ensuring a reliable power supply, making it a fundamental aspect of renewable energy systems.

Sustainability and Innovation: The topics covered here are at the forefront of sustainable energy innovation. By employing the knowledge and techniques presented in this chapter, we can advance the transition to clean and renewable energy sources, reducing our dependence on fossil fuels and mitigating the effects of climate change.

The design and modelling of renewable energy systems are central to the future of sustainable energy production and utilization. A person equipped with the tools, knowledge, and insights will lead and be needed to contribute to this critical field. As we strive for a greener and more sustainable future, the principles and practices outlined here will play an essential role in achieving our renewable energy goals.

REFERENCES

[1] M. El Haj Assad, M. H. Ahmadi, M. Sadeghzadeh, A. Yassin, and A. Issakhov, "Renewable hybrid energy systems using geothermal energy: Hybrid solar thermal-geothermal power plant," *Int. J. Low-Carbon Technol.*, vol. 16, no. 2, pp. 518–530, 2021.

[2] M. C. Bassetti, D. Consoli, G. Manente, and A. Lazzaretto, "Design and off-design models of a hybrid geothermal-solar power plant enhanced by a thermal storage," *Renew. Energy*, vol. 128, pp. 460–472, 2018.

[3] K. Li, C. Liu, S. Jiang, and Y. Chen, "Review on hybrid geothermal and solar power systems," *J. Clean. Prod.*, vol. 250, p. 119481, 2020.

[4] M. Saini, A. Sharma, V. P. Singh, G. Dwivedi, and S. Jain, "Solar thermal receivers – A review," *Adv. Mater. Manuf. Energy Eng. Vol. II*, vol. II, pp. 310–325, 2022, https://doi.org/10.1007/978-981-16-8341-1.

[5] L. Gong, Y. Zhang, and Z. Bai, "Geothermal-solar hybrid power with the double-pressure evaporation arrangement and the system off-design evaluation," *Energy Convers. Manag.*, vol. 244, p. 114501, 2021.

[6] S. Rajamand, "Analysis of effect of physical parameters on the performance of lead acid battery as efficient storage unit in power systems using new finite-element-method-based model," *J. Energy Storage*, vol. 47, p. 103620, 2022.

[7] B. Shen, Y. Chen, C. Li, S. Wang, and X. Chen, "Superconducting fault current limiter (SFCL): Experiment and the simulation from finite-element method (FEM) to power/energy system software," *Energy*, vol. 234, p. 121251, 2021.

[8] V. P. Singh and S. Jain, "Economic analysis of a large scale solar updraft tower power plant," *Sustain. Energy Technol. Assess.*, vol. 58, p. 103325, 2023, https://doi.org/10.1016/j.seta.2023.103325.

[9] N.-H. Kim, B. V Sankar, and A. V Kumar, *Introduction to Finite Element Analysis and Design*. Hoboken, NJ: John Wiley & Sons, 2018.

[10] V. P. Singh et al., "Recent developments and advancements in solar air heaters: A detailed review," *Sustainability*, vol. 14, no. 19, pp. 1–57, 2022, https://doi.org/10.3390/su141912149.

[11] V. P. Singh, A. Karn, G. Dwivedi, T. Alam, and A. Kumar, "Experimental assessment of variation in open area ratio on thermohydraulic performance of parallel flow solar air heater," *Arab. J. Sci. Eng.*, vol. 41, no. 12, pp. 1–17, 2022, https://doi.org/10.1007/s13369-022-07525-7.

[12] C. Huynh, L. Zheng, and P. McMullen, "Thermal performance evaluation of a high-speed flywheel energy storage system," In *IECON 2007-33rd Annual Conference of the IEEE Industrial Electronics Society*, Taipei, 2007, pp. 163–168.

[13] V. P. Singh, S. Jain, and A. Kumar, "Establishment of correlations for the Thermo-Hydraulic parameters due to perforation in a multi-V rib roughened single pass solar air heater," *Exp. Heat Transf.*, vol. 35, no. 5, pp. 1–20, 2022, https://doi.org/10.1080/08916152.2022.2064940.

[14] N. Dutt, A. Binjola, A. J. Hedau, A. Kumar, V. P. Singh, and C. S. Meena, "Comparison of CFD results of smooth air duct with experimental and available equations in literature," *Int. J. Energy Resour. Appl.*, vol. 1, no. 1, pp. 40–47, 2022, https://doi.org/10.56896/IJERA.2022.1.1.006.

[15] J. P. Pretorius, "Optimization and control of a large-scale solar chimney power plant" (Ph.D. Thesis). Univ. Stellenbosch, South Africa, 2007.

[16] 3rd International Conference on Advanced Engineering Materials and Technology, AEMT 2013, *Adv. Mater. Res.*, vol. 750–752, 2013 [Online]. Available: https://www.scopus.com/inward/record.uri?eid=2-s2.0-84884890 430&partnerID=40&md5=07286d2f8bc2ab307507c1732f3f3cf1.

[17] V. P. Singh, S. Jain, A. Karn, A. Kumar, and G. Dwivedi, "Mathematical modeling of efficiency evaluation of double pass parallel flow solar air heater," *Sustainability*, vol. 14, no. 17, pp. 1–22, 2022, https://doi.org/10.3390/ su141710535.

[18] J. Pretorius, "Incorporating vegetation under the collector," no. January 2008, 2023. https://www.semanticscholar.org/paper/Incorporating-Vegetation-(-lnder-the-Collector-Roqf-Johannes-P./89f76bb880d13065458117672b66474c378 5c0eb

[19] V. P. Singh, S. Jain, and J. M. L. Gupta, "Analysis of the effect of perforation in multi-v rib artificial roughened single pass solar air heater: Part A," *Exp. Heat Transf.*, vol. 36, no. 2, pp. 163–182, 2021, https://doi.org/10.1080/0891 6152.2021.1988761.

[20] J. P. Pretorius and D. G. Kröger, "Basic theory and numerical simulation of large scale solar updraft power plants," In *Proceedings of the Second International Conference on Solar Chimney Power Technology.* Boca Raton, FL: Taylor & Francis; CRC Press, 2010, pp. 45–54. https://doi. org/10.1201/9781003220602

[21] R. K. Shubham Srivastava, D. Verma, S. Thusoo, A. Kumar, V. P. Singh, "Nanomanufacturing for energy conversion and storage devices," In *Nanomanufacturing and Nanomaterials Design: Principles and Applications,* 2022, pp. 165–174.

[22] A. Kundu, A. Kumar, V. P. Singh, C. S. Meena, and N. Dutt, "Chapter 1: Introduction to thermal energy resources and their smart applications," In *Thermal Energy Systems:Design, Computational Techniques, and Applications.* Boca Raton, FL: CRC Press, 2023, pp. 1–15.

[23] F. D. Boshoff, S. J. van der Spuy, J. P. Pretorius, and C. J. Meyer, "Design of an axial flow fan for a unique cooling application," 2022, https://doi. org/10.1115/GT2022-80256.

[24] J. P. Pretorius, S. J. Van Der Spuy, and M. Strümpfer, "Tip vortex effects on air-cooled condenser axial flow fan performance," 2022, https://doi. org/10.1115/GT2022-82283.

[25] A. Anderson, B. Rezaie, and ,M. A. Rosen, "An innovative approach to enhance sustainability of a district cooling system by adjusting cold thermal storage and chiller operation," *Energy*, vol. 214, p. 118949, 2021.

[26] F. Boshoff, S. J. van der Spuy, J. Pretorius, and C. J. Meyer, "Investigation into the predicted performance of a cooling fan for an sCO2 CSP plant," Available SSRN 4528886, 2023. http://dx.doi.org/10.2139/ssrn.4528886

[27] A. Datta, A. Kumar, A. Kumar, A. Kumar, and V. P. Singh, "Advanced materials in biological implants and surgical tools," In *Advanced Materials for Biomedical Applications*, 1st ed., N. D. Ashwani Kumar, Y. Gori, A. Kumar, C. S. Meena, Eds. Boca Raton, FL: CRC Press, 2022, pp. 283–298.

[28] L. Roccamena, M. El Mankibi, and N. Stathopoulos, "Development and validation of the numerical model of an innovative PCM based thermal storage system," *J. Energy Storage*, vol. 24, p. 100740, 2019.

[29] C. McGregor, J. Pretorius, A. Attieh, and J. Hoffmann, "Techno-economic assessment of electricity generation from a medium-scale CSP-PV hybrid plant using long-duration storage," Available SSRN 4571684, 2023. https://www.researchgate.net/publication/374447688_Techno-economic_Assessment_of_Electricity_Generation_from_a_Medium-Scale_CSP-PV_Hybrid_Plant_Using_Long-Duration_Storage

[30] A. Kundu, A. Kumar, N. Dutt, V. P. Singh, and C. S. Meena, "Chapter 7: Modelling and simulation of thermal energy system for design optimization," In *Thermal Energy Systems: Design, Computational Techniques, and Applications*, A. Kumar, N. Dutt, V. Singh, C. Meena, Eds. Boca Raton, FL: CRC Press, 2023, pp. 103–137.

[31] W. Strydom, J. Pretorius, and J. Hoffmann, "Sensitivity analysis on the performance of a natural draft direct dry cooling system for a 50 MWe CSP application," *Proc. 8th World Congr. Momentum, Heat Mass Transf.*, Lisbon, pp. 1–8, 2023, https://doi.org/10.11159/enfht23.138.

[32] A. Kumar, V. P. Singh, C. S. Meena, and N. Dutt, *Thermal Energy Systems: Design, Computational Techniques, and Applications*, 1st ed. Boca Raton, FL: CRC Press, 2023.

[33] J. P. Pretorius and J. A. Erasmus, "Effect of tip vortex reduction on air-cooled condenser axial flow fan performance: An experimental investigation," *J. Turbomach.*, vol. 144, no. 3, 2021, https://doi.org/10.1115/1.4052234.

[34] V. P. Singh et al., "Heat transfer and friction factor correlations development for double pass solar air heater artificially roughened with perforated multi-V ribs," *Case Stud. Therm. Eng.*, vol. 39, no. September, p. 102461, 2022, https://doi.org/10.1016/j.csite.2022.102461.

[35] M. Owen, J. Pretorius, B. Buitendag, and C. Nel, "Heat dissipation factors for building-attached Pv modules," Available SSRN 4119337, 2022, https://doi.org/10.2139/ssrn.4119337.

[36] D. Fernandes, F. Pitié, G. Cáceres, and J. Baeyens, "Thermal energy storage: 'How previous findings determine current research priorities,'" *Energy*, vol. 39, no. 1, pp. 246–257, 2012.

[37] J. P. Pretorius and A. F. Du Preez, *ESKOM Cooling Technologies*, 2009.

[38] V. P. Singh, S. Jain, and J. M. L. Gupta, "Performance assessment of double-pass parallel flow solar air heater with perforated multi-V ribs roughness – Part B," *Exp. Heat Transf.*, vol. 35, no. 7, pp. 1059–1076, 2022, https://doi.org/10.1080/08916152.2021.2019147.

[39] V. P. Singh, S. Jain, and J. M. L. Gupta, "Analysis of the effect of variation in open area ratio in perforated multi-V rib roughened single pass solar air heater – Part A," *Energy Sources, Part A Recover. Util. Environ. Eff.*, vol. 44, no. January, pp. 1–21, 2022, https://doi.org/10.1080/15567036.2022.2029976.

[40] R. B. Lee III *et al.*, "The Clouds and the Earth's Radiant Energy System (CERES) sensors and preflight calibration plans," *J. Atmos. Ocean. Technol.*, vol. 13, no. 2, pp. 300–313, 1996.

[41] Y. G. Ashwani Kumar, A. K. S. Gangwar, A. Kumar, C. S. Meena, V. P. Singh, N. Dutt, and A. Prasad, "Biomedical study of femur bone fracture and healing," In *Advanced Materials for Biomedical Applications*. Boca Raton, FL: CRC Press, 2022, pp. 283–298.

[42] B. Willemse, S. Nielsen, J. Pretorius, M. Owen, and A. Rix, "A comparative evaluation of heat dissipation factors for open-rack and floating solar photovoltaic installations," Available SSRN 4547936, 2023. http://dx.doi.org/10.2139/ssrn.4547936

[43] S. A. Kalogirou, S. Lloyd, J. Ward, and P. Eleftheriou, "Design and performance characteristics of a parabolic-trough solar-collector system," *Appl. Energy*, vol. 47, no. 4, pp. 341–354, 1994.

[44] V. P. Singh et al., "Nanomanufacturing and design of high-performance piezoelectric nanogenerator for energy harvesting," In *Nanomanufacturing and Nanomaterials Design: Principles and Applications*. Boca Raton, FL: CRC Press, 2022, pp. 241–272.

[45] S. Jain, N. Kumar, V. P. Singh, S. Mishra, and N. K. Sharma, "Transesterification of algae oil and little amount of waste cooking oil blend at low temperature in the presence of NaOH," *Energies*, vol. 16, no. 1, pp. 1–12, 2023, https://doi.org/10.3390/en16031293.

[46] J. Pretorius, *Eskom Perspective on Specifications for Large ACC's*. Eskom, 2012.

[47] M. Bojić, D. Cvetković, V. Marjanović, M. Blagojević, and Z. Djordjević, "Performances of low temperature radiant heating systems," *Energy Build.*, vol. 61, pp. 233–238, 2013.

[48] J. P. Pretorius and D. G. Kröger, *Regulating Solar Chimney Power Plant Output According to Demand*, 2009 [Online]. Available: https://hdl.handle.net/10019.1/44278.

[49] I. W. Johnston, G. A. Narsilio, and S. Colls, "Emerging geothermal energy technologies," *KSCE J. Civ. Eng.*, vol. 15, pp. 643–653, 2011.

[50] J. P. Pretorius and D. G. Kröger, Critical evaluation of solar chimney power plant performance, *Solar Energy*, vol. 80, no. 5, pp. 535–544, 2006. https://doi.org/10.1016/j.solener.2005.04.001.

[51] I. Lee, J. W. Tester, and F. You, "Systems analysis, design, and optimization of geothermal energy systems for power production and polygeneration: State-of-the-art and future challenges," *Renew. Sustain. Energy Rev.*, vol. 109, pp. 551–577, 2019.

[52] S. J. Self, B. V Reddy, and M. A. Rosen, "Geothermal heat pump systems: Status review and comparison with other heating options," *Appl. Energy*, vol. 101, pp. 341–348, 2013.

[53] N. Dutt, A. Jageshwar, H. Ashwani, K. Mukesh, K. Awasthi, and V. Pratap, "Thermo - hydraulic performance of solar air heater having discrete D - shaped ribs as artificial roughness," *Environ. Sci. Pollut. Res.*, 2023, https://doi.org/10.1007/s11356-023-28247-9.

[54] V. P. Singh and G. Dwivedi, "Technical analysis of a large-scale solar updraft tower power plant," *Energies*, vol. 16, no. 1, p. 103325, 2023, https://doi.org/10.3390/en16010494.

[55] K. Chauhan and V. P. Singh, "Proceedings prospect of biomass to bioenergy in India: An overview," *Mater. Today Proc.*, no. xxxx, 2023, https://doi.org/10.1016/j.matpr.2023.01.419.

[56] O. Erdinc and M. Uzunoglu, "Optimum design of hybrid renewable energy systems: Overview of different approaches," *Renew. Sustain. Energy Rev.*, vol. 16, no. 3, pp. 1412–1425, 2012.

[57] J. Pretorius and D. Kröger, "Incorporating vegetation under the collector," *R D Journal, South African Inst. Mech. Eng.*, vol. 24, no. 1, pp. 3–11, 2008.

[58] V. P. Singh, C. S. Meena, A. Kumar, and N. Dutt, "Double pass solar air heater: A review," *Int. J. Energy Resour. Appl.*, vol. 1, no. 2, pp. 22–43, 2022, https://doi.org/10.56896/IJERA.2022.1.2.009.

[59] A. R. Singh, S. K. Singh, and V. P. Singh, "Process parameters optimization of carbon nano tube based catalytic transesterification of algal oil," *Mater. Today Proc.*, no. xxxx, 2023, https://doi.org/10.1016/j.matpr.2023.01.418.

[60] J. J. Ding and J. S. Buckeridge, "Design considerations for a sustainable hybrid energy system," *Trans. Inst. Prof. Eng. New Zeal. Electr. Eng. Sect.*, vol. 27, no. 1, pp. 1–5, 2000.

Chapter 7

Optimization techniques of solar thermal and hybrid energy systems

Mohit Nayal, Abhishek K. Sharma,
Siddharth Jain, and Varun Pratap Singh
University of Petroleum and Energy Studies

7.1 INTRODUCTION TO OPTIMIZATION IN SOLAR THERMAL AND HYBRID ENERGY SYSTEMS

An innovative method for utilizing the entire potential of various energy sources while meeting the rising needs for sustainability and dependability is hybrid energy system optimization. These systems seamlessly combine two or more different energy generation methods, such as fossil fuels, fossil wind, hydro, solar, and storage of energy, with the main objective of improving overall performance in terms of resilience, efficiency, and the environment. The improvement of hybrid systems has gained popularity as the world's energy landscape shifts to cleaner and more varied sources, providing a multifaceted answer to the problems of intermittency, grid integration, and emissions of carbon reduction [1]. Creating a symphony of sources of energy to produce a more dependable and resilient energy supply is the core of a hybrid energy system optimization. By combining them with reliable sources like natural gas or energy storage systems, the inherent variability of renewable energy sources like solar and wind power can be reduced. The grid is less susceptible to outages thanks to this synergy's ability to produce more steadily fluctuating power that can satisfy different demand profiles [2].

The development of solar thermal systems reflects a crucial frontier in the search for effective and sustainable energy sources on a global scale. By transforming sunlight into heat, solar thermal technology captures the sun's plentiful and sustainable energy. Heat can then be used for a variety of purposes, such as the production of electricity, heating space, and industrial processes. Solar thermal systems present a promising route to produce clean, renewable energy as the world struggles with global warming and attempts to move away from fossil fuels [3]. In this case, optimization entails a multifaceted strategy aimed at improving the effectiveness, economy, and sustainable development of these systems. Maximizing the

DOI: 10.1201/9781003472629-7

capture of energy and utilization is at the core of solar thermal optimization [4]. To ensure that a greater portion of the incident solar energy is efficiently converted into usable heat, this necessitates improvements in collector designs, transfer of heat fluids, and storage systems. Higher absorption rates and lower thermal losses are made possible by advances in collector material technology, such as developed coatings and selective surfaces. This increases system efficiency overall [5].

7.2 MATHEMATICAL MODELLING AND SIMULATION FOR SYSTEM OPTIMIZATION

Mathematical modelling and simulation are vital tools for system optimization. They involve creating mathematical representations of real-world systems and using computer simulations to analyze and improve their performance [6]. By formulating systems mathematically, we gain a deep understanding of their behaviour and can explore various scenarios to identify optimal solutions [7]. This approach is widely used in fields such as engineering, economics, and science to solve complex problems, make informed decisions, and optimize processes. It provides a cost-effective and efficient way to improve systems, but its accuracy depends on the quality of the model and the data used [8].

7.2.1 System equations and constraints

Mathematical modelling and simulation play a pivotal role in system optimization. This optimization technique involves the creation of mathematical models that represent real-world systems [9]. These models encapsulate the relationships between various system components, parameters, and constraints. By formulating the system's behaviour mathematically, it becomes possible to analyze and optimize its performance efficiently [10].

Mathematical modelling helps identify the underlying equations and constraints governing a system's behaviour. These equations describe how different variables interact with one another. Constraints, on the other hand, define limitations within which the system must operate [11]. They can be related to physical, economic, or environmental factors. System equations and constraints provide a solid foundation for optimization, as they enable researchers and engineers to mathematically represent and understand complex systems [12].

7.2.2 Dynamic simulation approaches

Dynamic simulation is an essential aspect of mathematical modelling for system optimization. It involves running computer-based simulations to predict the system's behaviour over time. Dynamic simulation considers how the system evolves and responds to various inputs and disturbances. This approach is particularly valuable for systems with time-dependent components, such as manufacturing processes, chemical reactors, and traffic systems [13].

Dynamic simulation enables researchers to explore different scenarios and test the impact of changes in real time. By simulating a system's behaviour under various conditions, it becomes possible to identify optimal operating points, strategies, and control parameters. This leads to more informed decision-making and improved system performance [14].

7.3 EVOLUTIONARY ALGORITHMS AND GENETIC OPTIMIZATION

Evolutionary algorithms and genetic optimization are powerful techniques used in the field of optimization and search. They draw inspiration from the process of biological evolution and genetics to find optimal solutions to complex problems [15].

7.3.1 Genetic algorithm fundamentals

Genetic algorithms are a powerful optimization technique inspired by the process of natural selection. These algorithms mimic the principles of evolution to find optimal solutions to complex problems. Genetic algorithms work by generating a population of potential solutions represented as "genomes" or "chromosomes." These solutions undergo a process of selection, crossover (recombination), and mutation, just like genes in biological organisms [16].

Genetic algorithms are used in various fields, including engineering, finance, and machine learning. They are particularly useful when dealing with complex, non-linear, and multi-modal optimization problems. Genetic optimization techniques excel in finding near-optimal solutions in a wide range of applications [17].

7.3.2 Multi-objective optimization

One significant advantage of genetic optimization is its applicability to multi-objective optimization problems. In many real-world scenarios, decision-makers need to consider multiple conflicting objectives simultaneously. Genetic algorithms can efficiently handle such situations by evolving a population of solutions that represent trade-offs between different objectives. This results in a Pareto front, which represents the best possible compromises among conflicting goals [18].

While genetic optimization has numerous advantages, it's not without limitations. One limitation is the computational complexity associated with large-scale problems. Additionally, genetic algorithms may require a substantial number of function evaluations to converge to a solution, making them computationally expensive in some cases [19].

7.4 MACHINE LEARNING-BASED OPTIMIZATION

Machine learning-based optimization (MLBO) is a powerful approach that leverages machine learning techniques to automate and improve the process of optimization. It combines the principles of optimization with the capabilities of machine learning algorithms to find optimal solutions in complex and dynamic environments.

7.4.1 Neural networks for system optimization

Machine learning techniques, such as neural networks, have gained prominence in system optimization. Neural networks are computational models inspired by the human brain's structure and function. They are capable of learning complex relationships from data and can be used to optimize systems in various domains, including image processing, natural language processing, and control systems [20].

Neural networks can approximate complex, non-linear functions, making them suitable for optimization tasks. They can learn from historical data and adapt their parameters to optimize a specific objective. For example, in manufacturing, neural networks can be used to optimize production processes by predicting equipment failures or quality issues [21].

7.4.2 Support vector machines in energy system design

Support vector machines (SVMs) are another machine learning tool employed in system optimization, especially in energy system design. SVMs are used for classification and regression tasks and have found applications in predicting energy consumption, optimizing renewable energy integration, and fault detection in electrical grids.

One of the advantages of using SVMs is their ability to handle high-dimensional data and provide robust results. They are effective in situations where data may not be linearly separable. However, SVMs can be computationally intensive for large datasets, and their performance heavily relies on proper parameter tuning.

In conclusion, mathematical modelling and simulation provide a foundational understanding of systems, while evolutionary algorithms and genetic optimization offer powerful tools for finding optimal solutions to complex problems. MLBO, with neural networks and SVMs, expands the scope of optimization to data-driven applications. Each technique has its unique advantages and limitations, making them suitable for specific scenarios and problem domains. The choice of optimization technique depends on the nature of the problem, available data, and computational resources [22].

7.5 DESIGN CONSIDERATIONS AND COMPONENT SIZING

For maximum energy efficiency and effectiveness, designing a solar thermal system requires careful consideration of several factors and the appropriate sizing of its components. First, the intended application and regional climate conditions influence the choice of collector's type, whether flat-plate or concentrating. While focusing collectors are appropriate for larger, industrial setups, flat-plate collectors are perfect for residential and tiny systems. To meet energy demands, it is crucial to size the collector area [23]. This calculation is influenced by variables like location, solar radiation, and desired temperature rise. For minimal heat loss, the collectors must have selective coatings, glazing, and sufficient insulation. The type of heat transfer fluid (HTF) used is crucial to the effectiveness of the system. It needs to be thermally stable, with high capacity for heat and a low freezing point. Pumps and piping must be the right size to ensure effective heat transfer throughout the system. Another important factor is energy storage. To meet energy needs during cloudy periods, it is essential to size thermal storage tanks or materials like phase-change materials effectively. For the system to operate as efficiently as possible and to avoid overheating or freezing, controls like sensors for temperature and automatic valves are crucial [24].

To achieve seamless integration and optimum performance, the design of an energy system that is hybrid, which typically integrates two or more conventional and/or renewable energy sources, calls for careful consideration of several important factors. First and foremost, the choice of energy sources needs to be in line with the availability of local resources and energy demands. For grid stability, common combinations include fossil fuel generators and renewable energy sources, solar and wind, and solar and battery storage [25]. To balance energy production and consumption, it is crucial to size the various components. This entails figuring out the right number of batteries, backup generators, wind turbines, and solar panels to use while considering the load profiles, weather, and desired resilience of the system. Conversion of energy and storage technologies must be efficient. To reduce energy losses and increase storage capacity, superior inverters, charge controllers, and battery packs must be used. Additionally, the depth of discharge, cycle duration, and capacity to withstand peak loads should all be considered when sizing battery banks [26].

7.5.1 Optimal collector and storage sizing

For a system powered by sunlight to be efficient and effective, collector and storage sizing must be ideal. Location, sunlight, and the desired rise in temperature must all be carefully considered when determining the collector size. To avoid oversizing, which can result in unnecessary costs, the collector area must be balanced with the system's power requirements. A bigger collector space can capture more sunlight and produce more heat [27]. To size collectors accurately, tools for simulation and past weather information can be used. Regarding storage, the system's ability to deliver a steady and dependable heat supply, even on overcast days or at night, depends on the size of thermal storage tanks or materials. Undersized storage could cause a shortage of heat while oversized storage could result in excessive heat losses. The intended length of energy storage should be taken into consideration when sizing, as well as daily variations in energy production and consumption patterns. It's important to consider the type of heat storage methods, whether it's a form of latent heat storage using phase-change substances or sensible heat storage in molten salt or water. Based on its unique properties and intended use, each has unique sizing considerations [28].

A vital step in making sure that energy is produced, used, and resilient in a hybrid energy system is determining the ideal collector and storage sizing. The collector component, which frequently consists of solar and wind turbines, should be sized to meet the energy demand while taking resource availability variations into account. To find the right balance, a thorough analysis of historical weather data and load profiles is necessary. Under sizing collectors could result in insufficient power generation while oversizing them could result in excess production of energy and potential waste [29]. Another crucial component of hybrid systems is storage capacity, which is frequently provided by batteries. The size of the energy storage system must consider both the short-term nature of renewable sources and the daily energy demand. It should be built to store extra energy when production is high and release it when demand is at its highest or when renewable energy production is at its lowest. To ensure durability and effectiveness, the storage capacity should be sized considering the depth of discharge, both charge and discharge rates, and the system's expected cycle life [30]. The best collector and storing size can be determined with the help of tools for optimization and modelling software. These tools calculate the most cost-effective sizing strategy by considering a number of variables, including weather patterns, energy prices, and system constraints. In the end, it's important to find a balance that maximizes the use of renewable resources, reduces reliance on backup sources of information, and ensures the system's dependability and financial viability over the course of its operational lifespan [31].

7.5.2 Integration of heat exchangers and storage units

A key tactic for improving energy capture, storage, and utilization in solar thermal systems is the integration of exchangers for heat and storage components. To maximize the transfer of heat from the solar collector to the storage medium, heat exchangers act as a link between the collector and the thermal storage unit. Heat exchangers effectively transfer thermal energy from the collector's working fluid, frequently a HTF, to the storing medium, which can be water, melting salt, or other phase-change materials, during solar radiation periods. For solar thermal systems to be useful outside of sunny days, a thermal storage unit must be included. With the help of this storage component, extra heat produced during the height of the day can be saved for use later, like at night or on cloudy days. To reduce heat losses and increase system efficiency, the storage unit's effective size and insulation are essential. To ensure that the thermal energy is efficiently delivered to the application, whether it be for space heating, the production of hot water, or industrial processes, thermal exchangers can play a dual role by obtaining heat from the place of storage when necessary [32].

For solar thermal systems to be useful outside of sunny days, a thermal storage chamber must be included. With the help of this storage component, extra heat produced during the height of the day can be saved for use later, like at night or on days with clouds. To reduce heat losses and increase system efficiency, the storage unit's effective size and insulation are essential. To ensure that the thermal energy is effectively delivered to the application, whether it be for space heating, the production of hot water, or industrial processes, thermal exchangers can play a dual role by obtaining heat from the place of storage when necessary. Energy supply and demand in a hybrid system must be balanced, and storage units—often in the form of batteries or thermal storage systems—are essential for this. Batteries can be used to store extra electricity produced by renewable sources like solar or wind power, while phase-change materials or insulated tanks can be used to store extra thermal energy. To ensure a steady supply of electricity, this stored energy can then be used during times of high demand or when sustainable resources are scarce [33].

7.6 CONTROL STRATEGIES FOR ENHANCED PERFORMANCE

To guarantee effective operation and effortless integration of various energy sources, control methods for hybrid energy systems are crucial. These systems can adjust to shifting conditions thanks to sophisticated control algorithms and continuous tracking. Increased system reliability is achieved through load prioritization, where vital loads are powered up first. Predictive control models also optimize the production and storage of energy by using forecasts for the weather and load predictions. Energy exchange with the grid is facilitated by demand-side administration and grid interaction control, which lowers costs and improves system robustness. By balancing energy demand with supply, these techniques make hybrid systems a dependable and sustainable source of energy. Advanced control methods, storage of energy optimization, and a variety of energy sources are just a few ways to improve the efficiency of a hybrid energy system. The system effectively manages the balance between renewable and conventional energy sources, optimizing the production and consumption, by utilizing smart controls and predictive analytics. Energy storage systems with increased capacity and efficiency guarantee a consistent power supply under changing circumstances. Enhancing reliability through the integration of complementary energy sources, like solar and wind, reduces reliance on a single source. Hybrid energy systems are a formidable answer to our energy needs because improved performance ultimately translates into higher energy reliability, long-term viability, and cost-effectiveness.

A solar energy system can perform better by implementing efficient control strategies. First, to ensure that the solar panels follow the trajectory of the sun throughout the day and maximize energy absorption, advanced tracking devices are essential. Whether single-axis or dual-axis, these tracking systems allow collectors to keep their ideal angle in relation to the sun, optimizing solar radiation absorption. Controlling the temperature is also essential. The system can control the functioning of auxiliary parts like pumps and valves as well as the flow rate of the HTF by using intelligent control algorithms. This makes sure the HTF doesn't overheat and damage the collection devices or the storage system in the process of reaching the desired temperature. Precise modifications based on the surroundings are possible thanks to the integration of sensors and real-time monitoring. Continuous monitoring of cloud cover, the outside temperature, and the amount of solar radiation allows the system to adjust and improve its performance. For instance, the system might direct extra heat to storage tanks during cloudy days so that it can be used later [34].

7.6.1 On/off control strategies

To operate solar thermal systems efficiently and consistently, on/off control strategies are essential. Based on conditions or requirements, these tactics involve enabling and removal of various system components. The operation of circulation pumps is a fundamental On/Off control method in solar thermal systems. These pumps oversee transferring the HTF through the heat exchanger's collectors. On/Off control is used to avoid irrational energy use and heat losses during times of low sunlight or when the desired temperature is reached. The pump starts to circulate the HTF when there is enough sunlight and heat is required, and it stops when the ideal temperature is reached or when there is insufficient sunlight. Controlling auxiliary heating elements, like electrical heaters or backup boilers, is another crucial component. When the solar thermal system is unable to achieve the necessary temperature or load demand on its own, these components are turned on. By ensuring that auxiliary heating is only used when necessary and not when it is not, On/Off control prevents energy waste and lowers operating costs [35].

The effective operation of hybrid power systems, which frequently integrate multiple energy sources and storage components, depends on On/Off control strategies. Based on system conditions, energy demand, and source availability, these strategies involve activating or deactivating particular components. The application of set-point thresholds is a typical on/off control strategy. For instance, a solar power system will turn on the solar panels and begin producing electricity when it determines that the amount of sunlight is greater than a set threshold. On the other hand, if the amount of sunlight drops below a certain point, the system might use another energy source or draw power from storage to meet demand. Real-time data and weather forecasting are essential components of these control strategies. The system can make proactive choices to switch between energy sources, charge or discharge storage, or activate backup generators as necessary by anticipating weather patterns and solar irradiance. This anticipatory strategy maximizes the use of green energy while minimizing reliance on fossil fuels [36].

7.6.2 Advanced model predictive control

Advanced model predictive control (MPC) is a sophisticated optimization technique widely employed in the field of control systems engineering and various industrial processes. It sets itself apart from conventional control methods by dynamically computing optimal control actions based on a system's current state and an established objective function. This adaptability allows MPC to thrive in scenarios where system conditions change or uncertainties are prevalent, making it a valuable tool across a diverse range of industries. MPC finds applications in controlling chemical processes, automotive systems, robotics, energy management, and autonomous systems like drones and self-driving vehicles. Its primary strength lies in its ability to effectively manage constraints, ensuring system safety and optimal performance. For instance, in industrial settings, MPC can simultaneously regulate multiple variables while adhering to safety limits, minimizing energy consumption, and maximizing productivity [37].

However, MPC does come with its set of limitations, including computational complexity, model accuracy requirements, tuning challenges, and sensitivity to sampling rates. These considerations should be carefully weighed against its advantages when choosing MPC over other control techniques for a particular application [36].

7.7 APPLICATIONS OF OPTIMIZATION TECHNIQUES

Techniques for solar thermal optimization are used in many different fields and industries. These methods are used to raise the effectiveness, performance, and affordability of solar thermal heating systems. In concentrated solar power (CSP) plants, where light is focused onto a receiver to generate high-temperature heat, solar thermal optimization is essential. The layout of receivers, heat-transferring fluids, and thermal storage systems are improved using optimization techniques, which leads to higher energy efficiency and lower costs [38]. For uses like steam generation, drying, and desalination, solar thermal systems are incorporated into industrial processes. By making these systems more efficient at supplying the necessary temperatures and energy, optimization helps industrial operations rely less on fossil fuels. Buildings use solar thermal systems for space cooling and heating. Designing effective solar thermal systems that are integrated into buildings and provide comfortable temperatures while consuming the least amount of energy is made possible by optimization techniques [39].

Techniques for energy system hybrid optimization are used in a variety of fields and industries, offering solutions to improve energy efficiency, lower costs, and support sustainability. When incorporating renewable energy sources like solar and wind into the electrical grid, optimization is essential. Hybrid systems can maintain grid stability while balancing energy generation and demand. To efficiently and sustainably meet localized energy needs, hybrid systems are used in microgrid design to combine different sources of energy (solar, wind, and diesel generators) and storage [40]. The best mix of internal combustion engine and electric propulsion for fuel economy and lower emissions is determined using optimization techniques for hybrid electric vehicles (HEVs) and plug-in hybrid electric vehicles (PHEVs). To reduce energy consumption and costs, fleets of hybrid or electric vehicles can be routed and scheduled with the aid of optimization models in logistics and transportation. To manage energy consumption, businesses use hybrid systems with methods of optimization, integrating conventional and renewable energy sources while reducing energy costs and environmental impact. Industrial processes are made more energy-efficient by combined heat and power (CHP) systems, which use optimization to generate electricity and useful heat simultaneously [41].

7.7.1 Solar thermal collector array optimization

Designing an effective and affordable solar thermal system requires careful consideration of how to optimize a solar thermal collector array. Maximizing solar energy harvesting while reducing system costs are the objectives [42].

To ensure the greatest energy input, pick a location with high sunlight (solar radiation). To maximize the capture of energy throughout the year, determine the collector array's ideal azimuth and tilt angles [43]. Frequently, this entails pointing the collectors in the direction of the sun. According to the purpose (e.g., flat plate, concentrating, evacuated tube) and efficiency requirements, choose the best type of collector. Select coatings and collector materials that maximize absorption and reduce thermal losses [44]. Reduce shading and increase the collector's exposure to sunlight by strategically placing collectors within the array. To prevent shading and airflow interference, determine the ideal distance between collectors. Consider factors like heat load and necessary temperature levels when sizing the collector array to meet the application's energy needs [45]. Utilize an effective thermal storage system (such as a water tank or phase-change material) to store any excess thermal energy so that it can be used later when there is less sunlight or more demand. Make sure the storage system is the right size to accommodate both the energy needs of the application and the output of the collector array [46].

7.7.2 Hybrid PV-thermal system optimization

By simultaneously producing both electricity and heat from a single configuration, hybrid photovoltaic-thermal (PV-T) systems represent a possible route for maximizing the use of solar energy. These systems combine thermal collectors and solar panels, combining their individual capabilities to increase total efficiency and energy output. To maximize the benefits of combined heat and electricity production while considering technical, financial, and environmental factors, hybrid PV-T systems must be optimized [47]. The primary goal of the optimization procedure is to increase the resource sharing between the photovoltaic and heating components. To ensure that the thermal collection doesn't impact the PV efficiency and vice versa, rigorous design and engineering are required [48]. The primary goal of the optimization process is to increase the resource sharing among the solar power and thermal components. To ensure that the heat collection does not compromise the photovoltaic productivity and vice versa, careful planning and construction are required. To determine the most effective layout, various system arrangements, such as co-located or combined panels, must be assessed [49].

An essential aspect of hybrid PV-T system optimization is thermal energy storage. By including effective storage systems, intermittency can be reduced, and energy availability increased. Surplus thermal energy can be stored and used later. To enable effortless energy, transfer and storage, phase-change substances, water-based storage, and other cutting-edge approaches are investigated. For hybrid PV-T systems to operate as efficiently as possible, control strategies are essential. The simultaneous generation of electricity and heat is optimized with the aid of intelligent control algorithms, which are informed by real-time data and predictive analytics. To ensure optimum system performance, these strategies also adjust to shifting weather patterns and energy requirements [50].

7.8 FUTURE TRENDS AND CHALLENGES IN OPTIMIZATION OF SOLAR THERMAL AND HYBRID ENERGY SYSTEMS

Optimization of solar thermal and combined energy sources is expected to experience revolutionary developments and confront formidable obstacles. Solar thermal systems have drawn attention for their potential in effective heat generation and storage, which is becoming more and more important as the world places more focus on sustainable energy. One noticeable development is the incorporation of innovative substances into collector designs with the goal of increasing absorption rates and enhancing heat transfer effectiveness [51]. This development has the potential to improve solar thermal systems overall performance and make them more cost-effective for large-scale applications. This path to optimization faces significant obstacles. Even in thermal systems, the unpredictable nature of solar energy continues to be a major barrier. To close the gap between energy production and consumption, researchers are working diligently to develop novel thermal energy storage techniques, such as phase-change materials and high-capacity storage devices. A significant task that calls for advanced control strategies and system designs is integrating these storage mechanisms into hybrid energy systems, which mix solar thermal with other renewable sources like biomass or geothermal.

The optimization process is further complicated by geographic and climatic variability. It is necessary to develop customized thermal solar solutions that consider local variables, energy demand trends, and technology limitations because what works in one area may not produce equivalent results in another. A promising future trend is the fusion of digitalization and smart technologies. System performance may be improved, energy output can be maximized, and preventive maintenance can be made possible through continuous tracking, predictive analytics, and adaptive control mechanisms. The road to optimization is largely influenced by financial factors. Although the cost of solar technology is declining, early investments are still a problem. To create supportive laws, incentives, and funding programmes that encourage the widespread use of solar thermal and hybrid systems, governments and businesses must work together.

At the intersection of innovation and realistic application is the optimal operation of solar thermal and hybrid energy systems. These systems' development is guided using cutting-edge materials, creative energy storage techniques, hybridization techniques, adaption to various surroundings, integration of digital technology, and responsible financial management. Despite enormous obstacles, there are significant potential advantages in terms of producing sustainable energy and lowering carbon emissions. For these issues to be resolved and the full potential of solar heat and hybrid energy systems to be realized, it is essential that academic institutions, businesses, and policymakers continue to work together on collaborative research, development, and collaboration.

REFERENCES

[1] M. Thirunavukkarasu, Y. Sawle, and H. Lala, "A comprehensive review on optimization of hybrid renewable energy systems using various optimization techniques," *Renew. Sustain. Energy Rev.*, vol. 176, p. 113192, 2023.

[2] X. Yuan et al., "System modelling and optimization of a low temperature local hybrid energy system based on solar energy for a residential district," *Energy Convers. Manag.*, vol. 267, p. 115918, 2022.

[3] B. Willemse, S. Nielsen, J. Pretorius, M. Owen, and A. Rix, "A comparative evaluation of heat dissipation factors for open-rack and floating solar photovoltaic installations," Available SSRN 4547936, 2023. http://dx.doi.org/10.2139/ssrn.4547936

[4] V. P. Singh et al., "Recent developments and advancements in solar air heaters : A detailed review," *Sustainability*, vol. 14, no. 19, pp. 1–57, 2022, https://doi.org/10.3390/su141912149.

[5] K. M. Powell, K. Rashid, K. Ellingwood, J. Tuttle, and B. D. Iverson, "Hybrid concentrated solar thermal power systems: A review," *Renew. Sustain. Energy Rev.*, vol. 80, pp. 215–237, 2017.

[6] C. McGregor, J. Pretorius, A. Attieh, and J. Hoffmann, "Techno-economic assessment of electricity generation from a medium-scale CSP-PV hybrid plant using long-duration storage," Available SSRN 4571684, 2023. https://www.researchgate.net/publication/374447688_Techno-economic_Assessment_of_Electricity_Generation_from_a_Medium-Scale_CSP-PV_Hybrid_Plant_Using_Long-Duration_Storage

[7] V. P. Singh, S. Jain, and A. Kumar, "Establishment of correlations for the thermo-hydraulic parameters due to perforation in a multi-V rib roughened single pass solar air heater," *Exp. Heat Transf.*, vol. 35, no. 5, pp. 1–20, 2022, https://doi.org/10.1080/08916152.2022.2064940.

[8] M. Kıyan, E. Bingöl, M. Melikoğlu, and A. Albostan, "Modelling and simulation of a hybrid solar heating system for greenhouse applications using Matlab/Simulink," *Energy Convers. Manag.*, vol. 72, pp. 147–155, 2013.

[9] V. P. Singh, A. Karn, G. Dwivedi, T. Alam, and A. Kumar, "Experimental assessment of variation in open area ratio on thermohydraulic performance of parallel flow solar air heater," *Arab. J. Sci. Eng.*, vol. 41, no. 12, pp. 1–17, 2022, https://doi.org/10.1007/s13369-022-07525-7.

[10] R. Siddaiah and R. P. Saini, "A review on planning, configurations, modeling and optimization techniques of hybrid renewable energy systems for off grid applications," *Renew. Sustain. Energy Rev.*, vol. 58, pp. 376–396, 2016.

[11] F. Boshoff, S. J. van der Spuy, J. Pretorius, and C. J. Meyer, "Investigation into the predicted performance of a cooling fan for an sCO2 CSP plant," Available SSRN 4528886, 2023. http://dx.doi.org/10.2139/ssrn.4528886

[12] B. Bhandari, S. R. Poudel, K.-T. Lee, and S.-H. Ahn, "Mathematical modeling of hybrid renewable energy system: A review on small hydro-solar-wind power generation," *Int. J. Precis. Eng. Manuf. Technol.*, vol. 1, pp. 157–173, 2014.

[13] V. P. Singh et al., "Heat transfer and friction factor correlations development for double pass solar air heater artificially roughened with perforated multi-V ribs," *Case Stud. Therm. Eng.*, vol. 39, no. September, p. 102461, 2022, https://doi.org/10.1016/j.csite.2022.102461.

[14] A. M. Yusop, R. Mohamed, A. Ayob, and A. Mohamed, "Dynamic modeling and simulation of a thermoelectric-solar hybrid energy system using an inverse dynamic analysis input shaper," *Model. Simul. Eng.*, vol. 2014, p. 22, 2014.

[15] M. Saini, A. Sharma, V. P. Singh, S. Jain, and G. Dwivedi, "Solar thermal receivers – A review," *Lect. Notes Mech. Eng.*, vol. II, pp. 1–25, 2022, https://doi.org/10.1007/978-981-16-8341-1.

[16] V. P. Singh et al., "Mathematical modeling of efficiency evaluation of double-pass parallel flow solar air heater," *Sustainability*, vol. 14, no. 17, p. 10535, 2022, https://doi.org/10.3390/su141710535.

[17] A. T. Azar and N. A. Kamal, *Renewable Energy Systems: Modelling, Optimization and Control*. Academic Press, London, United Kingdom, 2021.

[18] V. P. Singh, S. Jain, and J. M. L. Gupta, "Analysis of the effect of perforation in multi-v rib artificial roughened single pass solar air heater: Part A," *Exp. Heat Transf.*, vol. 36, no. 2, pp. 163–182, 2021, https://doi.org/10.1080/0891 6152.2021.1988761.

[19] M. Assad and M. A. Rosen, *Design and Performance Optimization of Renewable Energy Systems*. Academic Press, London, United Kingdom, 2021.

[20] V. P. Singh, S. Jain, and J. M. L. Gupta, "Performance assessment of double-pass parallel flow solar air heater with perforated multi-V ribs roughness – Part B," *Exp. Heat Transf.*, vol. 35, no. 7, pp. 1059–1076, 2022, https://doi.org/10.1080/08916152.2021.2019147.

[21] Y. Zhou, S. Zheng, and G. Zhang, "Machine learning-based optimal design of a phase change material integrated renewable system with on-site PV, radiative cooling and hybrid ventilations-study of modelling and application in five climatic regions," *Energy*, vol. 192, p. 116608, 2020.

[22] H. S. Jang, K. Y. Bae, H.-S. Park, and D. K. Sung, "Solar power prediction based on satellite images and support vector machine," *IEEE Trans. Sustain. Energy*, vol. 7, no. 3, pp. 1255–1263, 2016.

[23] A. Kumar, V. P. Singh, C. S. Meena, and N. Dutt, *Thermal Energy Systems: Design, Computational Techniques, and Applications*, 1st ed. Boca Raton, FL: CRC Press, 2023.

[24] O. Erdinc and M. Uzunoglu, "A new perspective in optimum sizing of hybrid renewable energy systems: Consideration of component performance degradation issue," *Int. J. Hydrogen Energy*, vol. 37, no. 14, pp. 10479–10488, 2012.

[25] T. Sivasakthivel, V. Verma, R. Tarodiya, C. S. Meena, V. P. Singh, and R. Kumar, "Chapter 11: Analysis of optimum operating parameters for ground source heat pump system for different cases of building heating and cooling mode operations," In *Thermal Energy Systems: Design, Computational Techniques, and Applications*, 1st ed., A. Kumar, N. Dutt, V. Singh, and C. Meena, Eds. Boca Raton, FL: CRC Press, 2023, pp. 183–207.

[26] S. Upadhyay and M. P. Sharma, "A review on configurations, control and sizing methodologies of hybrid energy systems," *Renew. Sustain. Energy Rev.*, vol. 38, pp. 47–63, 2014.

[27] R. Kumar et al., "Experimental and RSM-based process-parameters optimisation for turning operation of EN36B steel," *Materials (Basel)*, vol. 16, no. 1, pp. 1–18, 2023, https://doi.org/10.3390/ma16010339.

[28] J. Assaf and B. Shabani, "Multi-objective sizing optimisation of a solar-thermal system integrated with a solar-hydrogen combined heat and power system, using genetic algorithm," *Energy Convers. Manag.*, vol. 164, pp. 518–532, 2018.

[29] A. Datta, A. Kumar, A. Kumar, A. Kumar, and V. P. Singh, "Advanced materials in biological implants and surgical tools," In *Advanced Materials for Biomedical Applications*, 1st ed., N. D. Ashwani Kumar, Yatika Gori, Avinash Kumar, Chandan Swaroop Meena, Eds. Boca Raton, FL: CRC Press, 2022, pp. 283–298.

[30] J. P. Pretorius, "Optimization and control of a large-scale solar chimney power plant," (Ph.D. Thesis). Univ. Stellenbosch, South Africa, 2007.

[31] M. Chennaif, H. Zahboune, M. Elhafyani, and S. Zouggar, "Electric system cascade extended analysis for optimal sizing of an autonomous hybrid CSP/PV/wind system with battery energy storage system and thermal energy storage," *Energy*, vol. 227, p. 120444, 2021.

[32] R. K. Shubham Srivastava, D. Verma, S. Thusoo, A. Kumar, and V. Pratap Singh, "Nanomanufacturing for energy conversion and storage devices," In *Nanomanufacturing and Nanomaterials Design: Principles and Applications*. Boca Raton, FL: Taylor & Francis; CRC Press, 2022, pp. 165–174.

[33] G. Leonzio, "Solar systems integrated with absorption heat pumps and thermal energy storages: State of art," *Renew. Sustain. Energy Rev.*, vol. 70, pp. 492–505, 2017.

[34] A. G. Olabi, M. Mahmoud, B. Soudan, T. Wilberforce, and M. Ramadan, "Geothermal based hybrid energy systems, toward eco-friendly energy approaches," *Renew. Energy*, vol. 147, pp. 2003–2012, 2020.

[35] N. Dutt, A. Jageshwar, H. Ashwani, K. Mukesh, K. Awasthi, and V. Pratap, "Thermo - hydraulic performance of solar air heater having discrete D-shaped ribs as artificial roughness," *Environ. Sci. Pollut. Res.*, 2023, https://doi.org/10.1007/s11356-023-28247-9.

[36] A. Gupta, R. P. Saini, and M. P. Sharma, "Modelling of hybrid energy system-Part II: Combined dispatch strategies and solution algorithm," *Renew. Energy,* vol. 36, no. 2, pp. 466–473, 2011.

[37] A. Kundu, A. Kumar, N. Dutt, V. P. Singh, and C. S. Meena, "Chapter 7: Modelling and simulation of thermal energy system for design optimization," In *Thermal Energy Systems: Design, Computational Techniques,* and Applications, A. Kumar, N. Dutt, V. Singh, and C. Meena, Eds. Boca Raton, FL: CRC Press, 2023, pp. 103–137.

[38] G. Singh, S. Rathor, A. Karn, and V. P. Singh, "Recent progress in solar dryers using phase changing material: A review," *Int. J. Energy Resour. Appl.,* vol. 2, no. 1, pp. 57–77, 2023, https://doi.org/10.56896/IJERA.2023.2.1.005.

[39] V. P. Singh et al., "Nanomanufacturing and design of high-performance piezoelectric nanogenerator for energy harvesting," In *Nanomanufacturing and Nanomaterials Design: Principles and Applications.* Boca Raton, FL: CRC Press, 2022, pp. 241–272.

[40] K. Chauhan and V. P. Singh, "Proceedings prospect of biomass to bioenergy in India : An overview," *Mater. Today Proc.,* no. xxxx, 2023, https://doi.org/10.1016/j.matpr.2023.01.419.

[41] S. M. Zahraee, M. K. Assadi, and R. Saidur, "Application of artificial intelligence methods for hybrid energy system optimization," *Renew. Sustain. Energy Rev.,* vol. 66, pp. 617–630, 2016.

[42] V. P. Singh, C. S. Meena, A. Kumar, and N. Dutt, "Double pass solar air heater: A review," *Int. J. Energy Resour. Appl.,* vol. 1, no. 2, pp. 22–43, 2022, https://doi.org/10.56896/IJERA.2022.1.2.009.

[43] J. P. Pretorius and D. G. Kröger, "Critical evaluation of solar chimney power plant performance," *Solar Energy,* vol. 80, no. 5, pp. 535–544, 2006. https://doi.org/10.1016/j.solener.2005.04.001.

[44] V. P. Singh and G. Dwivedi, "Technical analysis of a large-scale solar updraft tower power plant," *Energies,* vol. 16, no. 1, p. 103325, 2023, https://doi.org/10.3390/en16010494.

[45] M. Owen, J. Pretorius, B. Buitendag, and C. Nel, "Heat dissipation factors for building-attached Pv modules," Available SSRN 4119337, 2022, https://doi.org/10.2139/ssrn.4119337.

[46] J. D. Garrison, "Optimization of a fixed solar thermal collector," *Sol. Energy,* vol. 23, no. 2, pp. 93–102, 1979.

[47] A. Kundu, A. Kumar, V. P. Singh, C. S. Meena, and N. Dutt, "Chapter 1: Introduction to thermal energy resources and their smart applications," In *Thermal Energy Systems:Design, Computational Techniques, and Applications.* Boca Raton, FL: CRC Press, 2023, pp. 1–15.

[48] J. P. Pretorius and D. G. Kröger, "Basic theory and numerical simulation of large scale solar updraft power plants," In Proceedings of the Second International Conference on Solar Chimney Power Technology, 2010, pp. 45–54.

[49] S. Sinha and S. S. Chandel, "Review of recent trends in optimization techniques for solar photovoltaic-wind based hybrid energy systems," *Renew. Sustain. Energy Rev.,* vol. 50, pp. 755–769, 2015.

[50] M. Herrando, J. Freeman, A. Ramos, I. Zabalza, and C. N. Markides, "Energetic and economic optimisation of a novel hybrid PV-thermal system for domestic combined heating and power," 2017.

[51] V. P. Singh and S. Jain, "Economic analysis of a large scale solar updraft tower power plant," *Sustain. Energy Technol. Assessments*, vol. 58, p. 103325, 2023, https://doi.org/10.1016/j.seta.2023.103325.

Chapter 8

Predicting PV thermal collector performance with apriori algorithm

Sridharan M and Senthil Kumar S
SRM Institute of Science and Technology (Tiruchirappalli Campus)

8.1 INTRODUCTION

The art of finding and extracting patterns in large data sets involving methods from the domain of machine learning, database systems and statistics constitutes data mining (DM). DM techniques and tools allow designers to estimate future trends and make more informed determinations. Machine learning (ML) methods are widely used to cluster, predict, track patterns, and classify data sets. Among the available purposes, prediction and association are closely interrelated (Chen et al. 2020). Prediction is the process through which the results of the given inputs can be obtained. Mathematical and cognitive model-based predictions are widely used by researchers (Xu et al. 2018). The overall accuracy or the performance delivered by the prediction system depends on how trained to predict the results. The main objective of the training is to deliver the relationship between every dataset considered. Before the arrival of ML algorithms, human expertise is used as the only way to convey this to the model in the form of rules or instructions (Sowan et al. 2013). With the arrival of ML algorithms, the option for training rules for a prediction model turns two. Researchers proposed algorithms exclusively to associate data sets considered in the process performance evaluation Apriori, Fp-growth, and Eclat algorithms are widely used by the researchers (Das et al. 2019; Montella 2011; Lee et al. 2019).

Among the available prediction models, cognitive-based ones are preferred by the dominant research community. The advantages of the cognitive model such as flexibility, higher accuracy, and response time differentiated it from the numerical model. From the available literature, fuzzy logic (FL) and neural networks (NN) are the two cognitive models preferred dominantly by the research community. The main difference between FL and NN is observed in terms of its training phase. NN is capable of training itself implicitly from the available data set. However, the training instruction for the FL is to be defined explicitly. Still, FL is more compatible and flexible than the NN models during the design of a control system. This made researchers prefer FL to the NN models during the controller design applications. By providing proper rules, the FL model is capable of predicting

DOI: 10.1201/9781003472629-8

the results of the experiments or processes in line with the NN model. As of now, rules to predict the results are fetched into the model only based on human expertise. This manual mode of rule creation may become tedious while handling the larger volume of the database. Also, the manual mean of rule creation leads to time-consuming, vague, and uncertainties.

To fix this research problem, this study proposed a novel approach that integrates the concepts of association rule creation using the Apriori algorithm (a DM approach) with the performance prediction model. To the best of the author's knowledge, this is the first study to introduce the concepts of association algorithms for improving the prediction performance delivered by the cognitive model. To validate the performance of this proposed concept, the results of this study are compared against the available cognitive model-based literature with the conventional rule framing method.

8.2 MATERIALS AND METHODS

8.2.1 Experimentation

To introduce the concepts of ML and DM techniques in the field of thermal science and engineering applications, this study integrated the concepts of DM with the solar photovoltaic thermal water collector system (S-PV/T-W-C-S). Figure 8.1 presents the schematic view of the S-PV/T-W-C-S rig.. The details of technical specifications, constructions, working principles, uncertainty analysis, technical discussions, etc. are already presented by the corresponding author of this study via his earlier works (Sridharan et al. 2018; Sridharan and Jayaprakash 2020; Sridharan and Prakash 2022; Sridharan 2022).

8.2.1.1 Quantitative analysis

The overall power output from the S-PV/T-W-C-S is calculated using the summation of PV and FPC power output. It is written as Eq. (8.1) by (Sridharan 2020):

$$P_O = P_{PV} + P_{FPC} \tag{8.1}$$

8.2.2 FL expert system

In 1980, Lotfi Zadeh introduced the concept of FL. Later, researchers developed exclusive computer-based framework to integrate the concepts of FL with expert systems. Such integrations of FL and concepts of expert systems are together called as fuzzy logic expert systems.

The detailed description related to the concept of FLES can be inferred from the literature. However, it is a necessity here to present and highlight

Figure 8.1 The real-time S-PV/T-W-C-S experimental setup.

the key concepts of FLES. This will make readers of any stream feel comfortable in understanding the concepts of FLES. Figure 8.2 presents the general architecture of FLES. The datasets from the real time rig are processed through three layers of FLES architecture. They are fuzzification, inference and defuzzification. The process of data conversion from a non-fuzzy to a fuzzy set takes place inside a layer called fuzzification layer. The layer in which human expertise-based data association takes place called inference layer. The process of creating data association with the aid of human expertise includes the following disadvantages. Time consumed to associate data sets is higher; the chance for generating irrelevant rules exists, etc. The process of data conversion from a fuzzy to a non-fuzzy set takes place inside a layer called defuzzification layer. The performance delivered by the FLES depends on a key component called membership function (MF). Generally, MF's are divided into two or more linguistic variables (LV). Some of the commonly used LVs are low, medium, and high. The purpose of using variant LVs supports the expert system to understand the concepts in a more precise manner. At the same time, increment in the LVs leads to the complexion during the data association and inference rule construction.

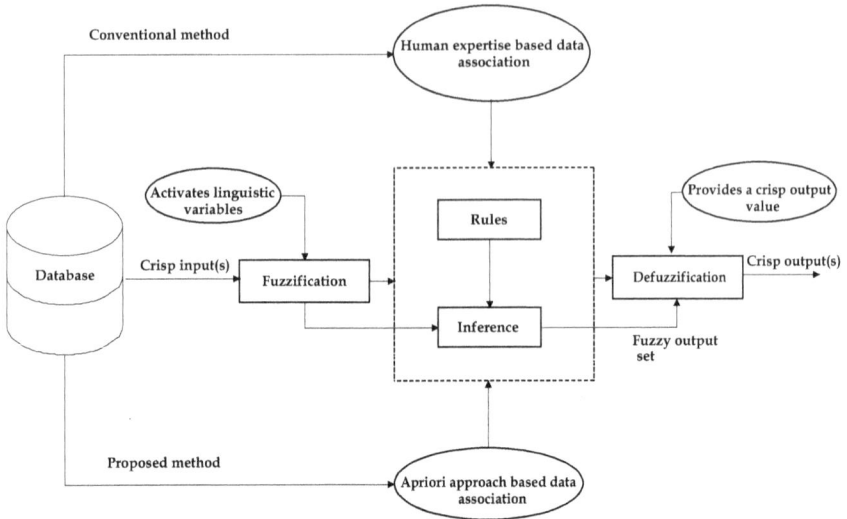

Figure 8.2 Data flow through the C-FLES and N-FLES.

The performance delivered by this conventional FLES depends on the data association rule created by the human expert from that field. At the same time, the process is time-consuming, complex, and subjected to the development of some irrelevant rules. To overcome this, ML-based techniques and algorithms are introduced to create and construct the association rules between the available data sets.

8.2.3 Apriori algorithm

The Apriori algorithm (AA) was first proposed by Agarwal and Srikanth during the year 1994. Dominantly it is used for the creation of association rules, and in mining frequent item sets from the available database. In comparison with the conventional methods, the association analysis using AA does not demand to specify the independent and dependent variables beforehand. Due to its advantages over conventional parametric and non-parametric methods of association rule mining, we employed the AA in this study. The extraction of interesting association rules from this AA depends on three different key indicators namely confidence (C), support (S), and lift (L).

The probabilities of both X and Y occurring together in a data set constitute the term S. It is mathematically given by (Xu et al. 2018) as follows:

$$\text{Supp}(A \rightarrow C) = P(A \cap C) = \frac{\left| A \bigcup C \right|}{|N|} \qquad (8.2)$$

Confidence is defined as the probability of the consequent occurring given that the antecedent is true. It can be calculated as

$$\text{Conf}(A \rightarrow C) = \frac{\text{Supp}\left(A \cup C\right)}{\text{Supp}(A)} = \frac{\left| A \cup C \right|}{\left| A \right|} \tag{8.3}$$

The indicator lift is defined as the probability of the co-occurrence of the antecedent and consequent divided by the probability of their co-occurrence considering both events are independent. It can be calculated as

$$\text{Lift}(A \rightarrow C) = \frac{\text{Conf}(A\ C)}{\text{Supp}(C)} = \frac{\text{Supp}\left(A \cup C\right)}{\text{Supp}(A) \cdot \text{Supp}(C)} \tag{8.4}$$

The indicator lift is used to measure the correlation between the antecedent and the consequent. If the lift value is lower than 1, the interdependence between the antecedent and the consequent is negative. If the lift value is greater than 1, the interdependence is positive. The higher the lift, the greater the interest of the rule. The greater the ratio [lift $(A \Longrightarrow C) > 1$], the stronger the dependency (Xu et al. 2018; Lee et al. 2019; Yu et al. 2019; Das et al. 2019; Chen et al. 2020; Sowan et al. 2013). The detailed discussion relevant to the steps and implementation of AA using the WEKA tool is given by Xu et al. (2018).

8.3 RESULTS AND DISCUSSION

8.3.1 Experimental

The overall power output delivered by the S-PV/T-W-C-S completely depends on the performance delivered by its electrical (PV) and thermal (FPC) components. Equation (8.1) is used to calculate the overall performance delivered by the S-PV/T-W-C-S under observation. Figure 8.3 presents the variations in the overall performance delivered by the S-PV/T-W-C-S with respect to the solar irradiance and time duration. It is clearly inferred that the overall performance variation recorded by the S-PV/T-W-C-S follows the base trend laid by the incident solar irradiance and the nature of electro-thermal units. The kind of variations recorded by the solar irradiance effectively inhibited the modes of heat transfer that takes place between the electrical and thermal units of the S-PV/T-W-C-S under observation.

During the first cycle (10–12 hours) of operation, the variation in the solar irradiance represents an increasing trend. This implies incremental heat transfer processes to happen between the electrical and thermal units. The variations in the overall power output of S-PV/T-W-C-S

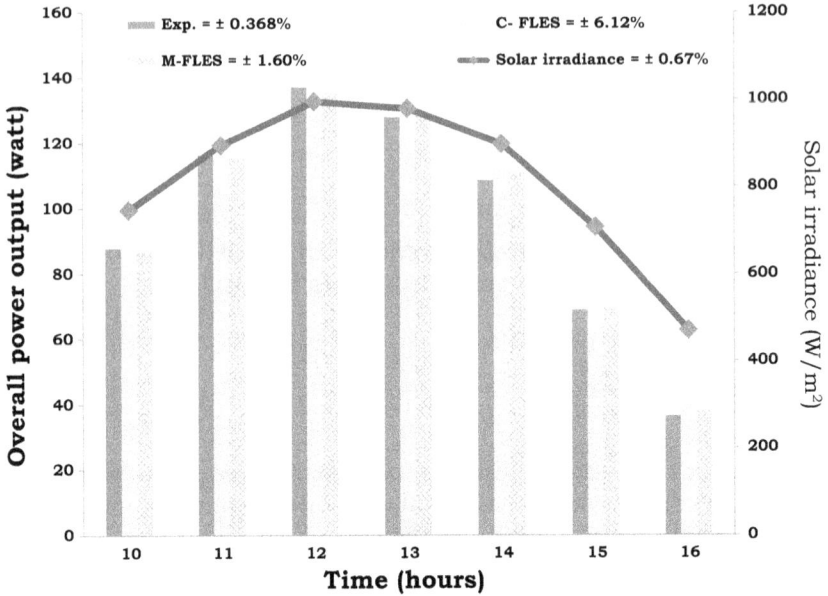

Figure 8.3 Variations in the overall power output predicted by the C-FLES and N-FLES.

range between 88 and 137 W. The value of maximum power output from the S-PV/T-W–C-S is recorded during this cycle of operation. In the second cycle (13–16 hours), the variation in the solar irradiance represents a decreasing trend. This implies decrement in the heat transfer processes happens between the electrical and thermal units. The variations in the overall power output of S-PV/T-W-C-S range between 127.80 and 36 W. The value of minimum power output from the S-PV/T-W-C-S is recorded during this cycle of operation.

8.3.2 FL expert system

A cluster of data sets recorded from the test rig is used to design both the C-FLES and N-FLES. Also, the same data sets are used to develop association rules using the fuzzy and AA. Previous literatures (Sridharan et al. 2018; Sridharan and Jayaprakash 2020; Sridharan and Prakash 2022; Sridharan 2022) presented detailed discussion related to the rules developed based on fuzzy association. The quality and quantity of fuzzy rules developed completely depend on the human expertise. The average time taken to construct fuzzy rules (for the considered data sets $n=7$) is recorded around 3 minutes. Still, the generated rules are subjected to the development of non-interesting rules. This makes the conventional fuzzy approach vague and time-consuming. With the human expertise-based rules, the

FLES proposed by Sridharan et al. (2018) predicted the overall power output with an accuracy of 93.88%. The obtained results are compromising and acceptable. Still, the arrival of ML concepts triggered the research community to attempt the performance prediction accuracy of C-FLES using DM algorithms. Among the available DM algorithms, this study attempts to improve the prediction accuracy of C-FLES using AA.

Through this approach, data sets from the test rigs are initially processed through the input layer of the FL expert system. In a similar fashion to the conventional approach (C-FLES), membership functions and LV are used to convert non-fuzzy (experimental) into fuzzy data sets. Then, by creating association rules by the conventional human approach, this mode aids the AA. This stage of data handling differentiates this proposed approach from the conventional mode. Also, this stage of data handling overcomes the drawbacks of the conventional approach.

At first, data sets are fetched into the AA with reference to the data classifications (LV) considered.. It is followed by the declaration of confidence, support, and lift values based on the requirements of the user end (output layer). From the available research works, the values of confidence and support are considered as 0.90 and 0.14.

By trial and error method, it is clearly inferred that the values of confidence, support and lift are inversely proportional to the number of rules generated. The developed rules clearly defined the relationship between the input and output parameters. The time taken by the AA to generate 36 numbers of rules is recorded as 2 seconds. This mean of generating association rules recorded only the 10% of the time consumed by a human expert. With the maximum values of support, confidence, and lift, the obtained rules are quality and relevant. As a whole, integration of this ML algorithm resulted in the generation of association rules with less time taken than the conventional mean. Also, the integration of AA-based association rule creation with the conventional fuzzy means leads to a better understanding of the transient variations in the experimental process. As a mean of this, the probability of obtaining interesting rules is higher than the conventional approach. The proposed method predicted the overall power output with an accuracy of 98.40%. This overall prediction accuracy recorded by N-FLES is 4.59% higher than the C-FLES.

8.4 CONCLUSIONS

The following are the conclusions inferred from this study:

For the occupied area of $0.4087 \, m^2$, the overall power output delivered by the solar photovoltaic thermal water collector system is around 97.50 W.

For the considered data sets ($n=7$), the average time consumed in creating association rules using human expertise is recorded as 2 minutes. Time taken by this proposed AA to generate the data association rules with minimum support and confidence is around 2 seconds only. Therefore, a significant reduction in time consumed is obtained with the integration of AA.

The overall prediction accuracy of the proposed C-FLES is around 94%. The overall prediction accuracy of the proposed N-FLES is around 98%. This is 4% higher than the C-FLES.

REFERENCES

Chen, J. H., S. C. Hsu, C. L. Chen, H. W. Tai, and T. H. Wu. 2020. "Exploring the Association Rules of Work Activities for Producing Precast Components." *Automation in Construction* 111 (December 2019): 103059. doi:10.1016/j.autcon.2019.103059.

Das, S., A. Dutta, R. Avelar, K. Dixon, X. Sun, and M. Jalayer. 2019. "Supervised Association Rules Mining on Pedestrian Crashes in Urban Areas: Identifying Patterns for Appropriate Countermeasures." *International Journal of Urban Sciences* 23 (1): 30–48. doi:10.1080/12265934.2018.1431146.

Lee, S., Y. Cha, S. Han, and C. Hyun. 2019. "Application of Association Rule Mining and Social Network Analysis for Understanding Causality of Construction Defects." *Sustainability (Switzerland)* 11 (3): 1–13. doi:10.3390/su11030618.

Montella, A. 2011. "Identifying Crash Contributory Factors at Urban Roundabouts and Using Association Rules to Explore Their Relationships to Different Crash Types." *Accident Analysis and Prevention* 43 (4): 1451–1463. doi:10.1016/j.aap.2011.02.023.

Sowan, B., K. Dahal, M. A. Hossain, L. Zhang, and L. Spencer. 2013. "Fuzzy Association Rule Mining Approaches for Enhancing Prediction Performance." *Expert Systems with Applications* 40 (17): 6928–6937. doi:10.1016/j.eswa.2013.06.025.

Sridharan, M. 2023. "Application of Generalized Regression Neural Network in Predicting the Performance of Solar Photovoltaic Thermal Water Collector." *Annals of Data Science* 10: 1–23. doi:10.1007/s40745-020-00273-1.

Sridharan, M. 2022. "Performance Comparison Study on Differently Configured Solar Photovoltaic Thermal Water Collector Systems." *International Journal of Ambient Energy* 43: 8540–8551. doi:10.1080/01430750.2022.2102066.

Sridharan, M., G. Jayaprakash, M. Chandrasekar, P. Vigneshwar, S. Paramaguru, and K. Amarnath. 2018. "Prediction of Solar Photovoltaic/Thermal Collector Power Output Using Fuzzy Logic." *Journal of Solar Energy Engineering, Transactions of the ASME* 140 (6): 0610131–0610136. doi:10.1115/1.4040757.

Sridharan, M. and G. Jayaprakash. 2020. "Verification and Validation of Solar Photovoltaic Thermal Water Collectors Performance Using Fuzzy Logic." *Journal of Verification, Validation and Uncertainty* 4 (December 2019): 1–8. doi:10.1115/1.4045895.

Sridharan, M. and B. Prakash. 2022. "Predicting the Performance of Solar Photovoltaic Thermal Water Collectors Using Hybrid Fuzzy Logic Expert System." *International Journal of Ambient Energy* 43: 8098–8112. doi:10.108 0/01430750.2022.2086913.

Xu, C., J. Bao, C. Wang, and P. Liu. 2018. "Association Rule Analysis of Factors Contributing to Extraordinarily Severe Traffic Crashes in China." *Journal of Safety Research* 67 (2018): 65–75. doi:10.1016/j.jsr.2018.09.013.

Yu, S., Y. Jia, and D. Sun. 2019. "Identifying Factors That Influence the Patterns of Road Crashes Using Association Rules: A Case Study from Wisconsin, United States." *Sustainability (Switzerland)* 11 (7): 1–14. doi:10.3390/su11071925.

Chapter 9

Solar and geothermal-based thermal energy storage for building applications

Madhur Chauhan
G H Raisoni Institute of Engineering & Business Management Jalgaon

Bipasa B. Patra
Padmashri Dr. V. B. Kolte College of Engineering Malkapur

Aarti Rana Chauhan
G H Raisoni Institute of Engineering & Business Management Jalgaon

9.1 INTRODUCTION

The increasing demand for energy and the need to reduce greenhouse gas emissions have spurred the development of sustainable energy solutions for buildings. Solar and geothermal-based thermal energy storage (TES) systems have emerged as promising technologies to address the challenges of energy supply and consumption in the built environment. These innovative systems not only enable the efficient use of renewable energy sources but also facilitate the integration of intermittent renewable energy into building operations [1]. Thermal energy storage (TES) involves capturing excess thermal energy generated during periods of low demand and storing it for later use when demand is higher. This concept aligns perfectly with the intermittent nature of solar and geothermal energy sources, allowing buildings to capture and store energy during periods of abundant solar irradiation or geothermal availability, and utilize it during periods of low or no availability [2]. As a result, these systems not only enhance energy efficiency but also contribute to a more stable and reliable energy supply for buildings [3]. Solar-based TES (Figure 9.1) makes use of solar thermal collectors, such as solar water heaters or concentrated solar power systems, to capture sunlight and convert it into thermal energy [4]. The stored energy can be used for various applications, including space heating, domestic hot water supply, and cooling through absorption chillers. By coupling solar thermal systems with thermal storage, buildings can reduce their dependency on conventional fossil fuels and electricity, leading to a significant reduction in greenhouse gas emissions. Geothermal-based TES leverages the constant and renewable heat emanating from the Earth's subsurface [5].

DOI: 10.1201/9781003472629-9

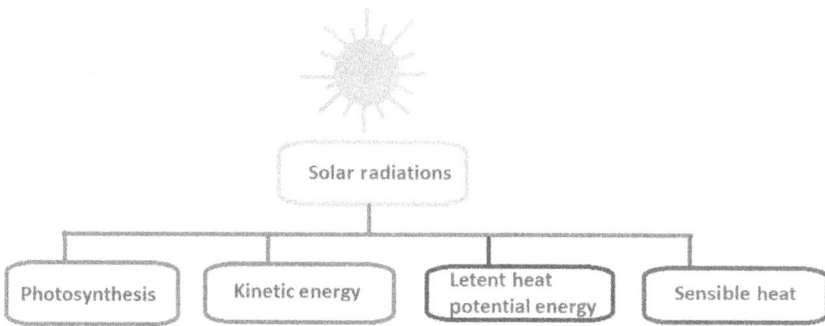

Figure 9.1 Different types of solar radiations.

By incorporating thermal storage technologies into geothermal systems, buildings can optimize their energy use further, enabling them to tap into the stored geothermal energy when required. Numerous studies and research papers have been conducted to evaluate the feasibility and potential benefits of solar and geothermal-based thermal energy storage for buildings [6].

It involves storing excess heat or cold energy during periods of low demand and releasing it when needed to provide heating, cooling, or other thermal services. TES systems offer numerous benefits, including energy cost savings, increased efficiency, and improved grid stability. Sensible heat storage is one type of TES that involves changing the temperature of a material without a phase change. This method is commonly used in applications such as hot water storage for buildings [7]. Latent heat storage, another type of TES, utilizes the phase change of a material to store heat. Phase change materials (PCMs) like paraffin wax or salt hydrates are commonly used in this method [8]. Thermochemical storage, on the other hand, utilizes reversible chemical reactions to store and release heat energy, making use of reactions involving metal hydrides or sorbent materials [9]. Cryogenic energy storage involves the storage of thermal energy by liquefying and storing gases at very low temperatures, such as liquid nitrogen or liquid air [10]. TES finds applications in various sectors. In buildings, TES systems can be integrated into heating, ventilation, and air conditioning (HVAC) systems to reduce energy consumption and peak demand [11]. Industries can benefit from TES by optimizing energy usage in processes like drying, steam generation, or metal casting. TES also plays a crucial role in the integration of renewable energy sources, such as solar or wind, by storing excess energy for later use, thereby addressing the intermittent nature of these sources [12] Additionally, TES can enhance the efficiency and reliability of district heating and cooling networks by balancing supply and demand [13]. The advantages of TES include energy cost savings, improved grid stability, and reduced environmental impact. By storing energy during low-cost or off-peak periods and utilizing it during high-demand or high-cost periods, TES systems contribute to significant energy cost reductions. They also aid in managing the fluctuating nature of renewable energy sources, providing dispatchable energy and enhancing grid stability. Moreover, TES systems promote energy efficiency and the integration of renewable energy, thereby contributing to a reduced environmental footprint and mitigating climate change.

It enables the capture, storage, and subsequent use of excess thermal energy generated by solar power during periods of high solar radiation. TES systems for solar energy storage allow for the decoupling of energy generation from energy use, providing a continuous and reliable energy supply even when sunlight is unavailable. Various storage mediums are employed in solar TES systems, including molten salts, PCMs, and rocks. These mediums absorb and store the thermal energy in the form of sensible heat or latent heat, depending on the specific technology utilized. During periods of low energy demand or limited solar radiation, the stored thermal energy can be retrieved and used to generate electricity, provide heating, or drive cooling systems.

Geothermal energy is a renewable energy source that harnesses the heat stored within the Earth. TES systems play a vital role in maximizing the efficiency and utilization of geothermal energy. TES in geothermal applications involves the capture, storage, and subsequent release of thermal energy to meet fluctuating energy demands. Geothermal TES systems typically utilize underground reservoirs or geological formations as storage mediums. During periods of excess or low-demand energy, geothermal heat pumps or power plants extract heat from the geothermal resource and transfer it to the storage medium. The stored thermal energy can then be retrieved and used when there is a higher demand for heating, cooling, or electricity. Geothermal TES offers several benefits. It allows for the continuous utilization of geothermal energy, even when the demand for heat or electricity is low or when the geothermal resource is not actively produced. TES systems can help balance the supply and demand mismatch by storing excess energy during off-peak periods and releasing it during peak demand, thereby increasing the overall efficiency and reliability of geothermal energy systems [14].

TES plays a crucial role in the utilization and storage of energy derived from hydrogen. Hydrogen, as a clean and versatile energy carrier, can be produced through various methods such as electrolysis, steam reforming, or biomass gasification. TES systems for hydrogen energy storage enable the efficient capture, storage, and subsequent utilization of thermal energy associated with hydrogen production and utilization processes. Hydrogen TES systems utilize thermal energy to store hydrogen or store thermal energy associated with hydrogen production [15]. The thermal energy can subsequently be used for a variety of purposes, such as the production of electricity, industrial activities, or transportation. Reversible chemical processes, such as hydrogenation and dehydrogenation reactions involving metal hydrides, are one method for storing hydrogen thermal energy. Excess thermal energy from the charging phase is used to power the hydrogen absorption process, storing heat energy as hydrogen inside the metal hydrides. The exothermic dehydrogenation reaction used during the discharging phase releases the stored hydrogen, and the resulting thermal energy can be used to generate heat or electricity [16].

9.1.1 Solar energy and TES

Solar energy, while abundant and sustainable, presents challenges of intermittency. This variability, largely contingent upon diurnal cycles and weather conditions, can lead to energy supply disruptions, particularly during evening hours or cloudy days. TES systems can bridge this gap, storing excess solar energy during peak insolation periods. Two primary TES mechanisms are explored: Sensible heat storage, involving mediums like water or rocks, and latent heat storage using PCMs. Recent advancements in PCMs, characterized by their ability to store energy during phase transitions, have shown promise in enhancing the energy density of storage solutions.

Flow Process: Solar Energy Harvesting (Figure 9.2):

- Solar panels or thermal collectors capture solar energy.
- Solar energy is converted into electricity and/or heat.

Solar TES: Materials:

- Molten salts (such as a mixture of sodium nitrate and potassium nitrate)
- PCM (such as paraffin wax or hydrated salts)
- Rocks or gravel (for sensible heat storage)
- Thermal oil

Figure 9.2 Solar panels.

9.1.2 Geothermal energy & TES

Geothermal energy harnesses the Earth's consistent subsurface heat. Though inherently more stable than solar, the integration of TES with geothermal systems can elevate their efficiency. By adjusting the extraction and release of geothermal heat in congruence with building demand profiles, TES ensures that energy wastage is minimized. The potential of Enhanced Geothermal Systems (EGS) combined with advanced TES solutions, particularly in regions with cooler surface temperatures, is discussed.

Flow Process: Geothermal Energy Harvesting (Figure 9.3):

- Geothermal heat pumps extract heat from the ground or water sources.
- Geothermal energy is converted into usable heat or cold.

Geothermal TES: Materials:
Underground reservoirs or geological formations (such as aquifers or rock formations) store thermal energy.

9.1.3 Energy conversion and storage

Electricity generated from solar panels can be used directly or stored in batteries. Excess electricity can be converted into heat through resistive heating or used for other purposes like charging electric vehicles. Heat generated from solar thermal collectors and geothermal sources is stored in a TES system.

TES: Water tanks, subsurface heat storage, PCM, and sensible heat storage systems are a few examples of TES possibilities. When solar or geothermal energy is abundant, extra thermal energy is stored.

Geothermal thermal energy storage

Figure 9.3 Geothermal thermal energy storage.

9.1.4 Energy distribution

The building's heating, cooling, or hot water systems can receive the stored thermal energy when needed. The thermal storage system's stored energy is transferred to the building's heating or cooling loops by heat exchangers and pumps. For space heating, residential hot water supply, or air conditioning, distributed thermal energy is employed. The HVAC systems of the building use the energy. Heat recovery units can be added to the system to increase energy efficiency.

9.1.5 Applications

- **Building Applications:** Buildings, as significant energy consumers, are focal points for sustainability interventions. We delve into multiple building applications:
- **Space Heating & Cooling:** With over 50% of a building's energy needs dedicated to climate control, the role of TES becomes paramount. By storing solar or geothermal heat, TES systems can offset peak demand, reduce grid reliance, and ensure consistent thermal comfort.
- **Domestic Hot Water (DHW):** Integrating TES with solar thermal collectors or geothermal heat exchangers, buildings can achieve year-round hot water supply.
- **Integration with Building Design:** Emerging architectural trends, such as Building-Integrated Photovoltaics (BIPV) and passive solar design, underscore the harmonization of renewable energy and storage. TES systems, when seamlessly integrated into building aesthetics and functionality, can transform structures into self-sufficient energy entities.

9.2 SOLAR TES METHODS

Here are four commonly used types of solar TES:

9.2.1 Sensible heat storage

Sensible heat storage involves storing thermal energy by raising the temperature of a material. It is commonly used with molten salts or rocks/gravel as the storage medium. During the charging phase, excess thermal energy from the solar collector is transferred to the storage medium, raising its temperature. This stored heat can be retrieved later when needed by transferring it back to a heat exchanger for various applications, such as electricity generation or space heating [17].

9.2.2 Latent heat storage

The ability of materials to alter their phase allows for the storage and release of thermal energy through latent heat storage. PCMs are substances that may transform from a solid to a liquid or from a liquid to a gas while absorbing or releasing a significant amount of energy. By melting and hardening, PCMs like paraffin wax or hydrated salts are employed to store thermal energy. The PCM is melted by solar thermal energy during the charging phase, and the stored energy is released as the solid PCM returns to its solid state during the discharging phase.

9.2.3 Thermochemical storage

Reversible chemical processes are used in thermochemical storage to store and release thermal energy. Chemical compounds, like metal oxides or carbonates, are transformed between states during these reactions. Thermal energy causes an endothermic process that stores energy as chemical potential during the charging phase. The reversible cycle is finished when heat is required when the energy is released through an exothermic process [18].

9.2.4 Thermo-electrochemical storage

Thermo-electrochemical storage employs redox reactions in an electrolyte solution to store and release thermal energy. During the charging phase, the electrolyte is reduced at the cathode, storing energy, and oxidized at the anode. The stored energy can be released by reversing the reactions. Thermo-electrochemical storage has the advantage of high energy density and potentially long-term storage capabilities [19].

9.2.5 Solar energy and TES for buildings application

- **Efficiency and Reliability:** Solar energy, when combined with TES, ensures that buildings can access stored heat during non-solar hours, leading to more consistent energy supply and higher energy security.
- **Economic Benefits:** While the initial investment for integrating TES with solar energy systems can be high, the payback period can be shortened with increasing energy prices and incentives for renewable installations.
- **Grid Support:** TES can store excess solar energy during peak production times, which can be released during peak demand periods, reducing the strain on energy grids.
- **Versatility:** Solar-based TES systems can be designed for both high-temperature applications (like Concentrated Solar Power) and low-temperature applications (domestic hot water and space heating) [20–25].

9.3 GEO TES METHODS

Geothermal energy systems can incorporate various types of TES to enhance their efficiency and flexibility. Here are three types of geothermal TES methods commonly employed:

9.3.1 Aquifer TES (ATES)

ATES involves the use of underground aquifers as TES reservoirs. During the charging phase, excess thermal energy from geothermal sources or other sources is transferred to the aquifer by injecting water or a heat transfer fluid into the ground. The heat is stored in the aquifer by raising its temperature. During the discharging phase, the stored thermal energy is extracted by pumping the heated water or fluid back to the surface for direct use in heating applications or to generate electricity [26–30].

9.3.2 Borehole thermal energy storage (BTES)

BTES involves the use of vertical boreholes for thermal energy storage. In this method, a closed-loop system with heat exchangers is installed in the boreholes. During the charging phase, excess thermal energy is transferred to the ground via a heat transfer fluid circulated through the heat exchangers, thereby increasing the temperature of the surrounding soil or rock. During the discharging phase, the stored thermal energy is retrieved by circulating the fluid through the heat exchangers to extract the heat for various applications.

9.3.3 Rock thermal energy storage (RTES)

RTES utilizes the thermal properties of rocks or gravel to store thermal energy. In this method, excess thermal energy is transferred to the rock or gravel during the charging phase, either by circulating a heat transfer fluid through rock-filled boreholes or by heating the rocks directly. The high thermal capacity of the rocks allows for efficient storage of heat. During the discharging phase, the stored thermal energy is extracted by circulating a heat transfer fluid through the rock-filled boreholes or by using the heated rocks for various applications [22].

9.3.4 Geothermal energy and TES for buildings application [31–37]

- **Consistent Energy Supply:** Geothermal energy is already a consistent energy source, but when integrated with TES, the efficiency and energy delivery rates can be further optimized.

- **Environmental Benefits**: Geothermal energy has a low carbon foot-print. Coupling it with TES can further reduce emissions associated with peak energy demand.
- **Economic Savings**: In areas with high geothermal potential, using geothermal energy with TES can lead to substantial long-term savings, considering the stability of geothermal energy prices compared to fossil fuels.
- **Extended Lifespan**: TES can reduce the strain on geothermal systems by managing demand. This can potentially extend the lifespan of the geothermal equipment.

9.4 HYDROGEN ENERGY SYSTEMS

HES can incorporate various types of TES methods to enable efficient storage and utilization of thermal energy. Here are three types of hydrogen TES methods commonly used:

9.4.1 Metal hydride storage

Metal hydride storage involves the use of certain metal alloys that can reversibly absorb and release hydrogen through a chemical reaction. During the charging phase, excess thermal energy is used to drive the hydrogen absorption process, causing the metal hydride to store thermal energy in the form of hydrogen. During the discharging phase, the stored hydrogen is released by supplying heat, resulting in the release of thermal energy. Metal hydrides offer the advantage of high hydrogen storage densities and can be used for both small-scale and large-scale applications [23].

9.4.2 Thermochemical storage

Thermochemical storage utilizes reversible chemical reactions to store and release thermal energy. In this method, thermal energy is used to initiate an endothermic reaction that results in the formation of a chemical compound with high energy content. During the discharging phase, the stored energy is released through an exothermic reaction, regenerating the initial reactants. Thermochemical storage offers the advantage of high energy density and potentially long-term storage capabilities [24].

9.4.3 Liquid organic hydrogen carriers (LOHC)

LOHC systems involve the storage of hydrogen in the form of liquid organic compounds that can reversibly bind and release hydrogen. During the charging phase, excess thermal energy is used to drive the hydrogenation process, resulting in the formation of a liquid organic compound loaded with hydrogen. The stored hydrogen can be released during the discharging phase by supplying heat, allowing the regenerated liquid organic compound to be reused for hydrogen storage. LOHC systems offer advantages such as high hydrogen density and ease of transport [25].

9.5 CHALLENGES & OPPORTUNITIES

Despite the evident synergies of solar, geothermal, and TES, challenges persist. Initial investment costs, space requirements for storage systems, and technological complexities can be deterrents. However, the plummeting costs of solar technologies, advancements in compact TES solutions, and increasing governmental incentives are fostering a conducive environment for widespread adoption. The nexus of solar and geothermal energy with TES presents a promising avenue for sustainable building applications. By ensuring consistent energy availability, optimizing system efficiencies, and championing innovative building designs, this integration paves the way for a resilient, sustainable, and energy-efficient urban future [38–44].

9.6 CONCLUSIONS

Sustainability: Both solar and geothermal energy are sustainable energy sources. When combined with TES, their potential to provide clean energy for building applications is magnified.

Resilience and Security: TES systems enhance energy security, allowing buildings to maintain essential functions during power outages or periods of low energy generation.

Design Flexibility: Modern TES systems can be incorporated into building designs in various ways, including underground storage, water tanks, or PCMs, offering architects and engineers flexibility.

Future Development: Continued research and innovation in TES materials and technologies, such as PCMs and thermochemical storage, will further enhance the benefits and reduce the costs of integrating TES with renewable energy sources.

In conclusion, integrating solar and geothermal energy with thermal energy storage for building applications presents a promising pathway towards a more sustainable, reliable, and economically viable energy future. As technology advances and the importance of reducing greenhouse gas emissions becomes even more paramount, these integrated systems will likely play an increasingly significant role in global energy strategies [45–49].

9.7 FUTURE SCOPE

The future scope of integrating solar and geothermal energy with TES for building applications is promising. Given the increasing urgency to mitigate climate change, reduce dependence on fossil fuels, and transition towards sustainable energy solutions, here's an exploration of potential future developments [50–52]:

- **Advancements in Storage Materials:** Research is ongoing to develop advanced TES materials like PCMs and Thermochemical Storage. These materials can store and release heat at specific temperatures, enhancing the efficiency of energy storage and release.
- **Integration with Smart Grids:** Future TES systems in buildings will likely be integrated with smart grids. This means buildings can not only consume energy but also store and supply excess energy back to the grid, turning them into active grid participants.
- **Modular and Scalable Designs:** As urban areas continue to expand, modular and scalable TES systems that can easily integrate with both new and existing infrastructures will become crucial.
- **Hybrid Systems:** Integrating solar and geothermal energy sources into a single, cohesive system can maximize efficiency. We might see designs where solar energy is used primarily during sunny days, while geothermal sources act as a backup or supplementary source, ensuring a continuous energy supply.
- **Building-Integrated Photovoltaics (BIPV):** Future buildings might not only have solar panels on roofs but integrated into the very structure (walls, windows, facades). Combined with TES, such designs can maximize energy capture and storage.
- **District Heating and Cooling:** On a larger scale, TES might be used in district heating and cooling systems. An entire neighbourhood or district could rely on a central TES system that gets its energy from solar or geothermal sources.
- **Enhanced Geothermal Systems (EGS):** Future geothermal systems might harness energy from deeper within the Earth. Combined with TES, these systems could offer even more significant energy outputs.
- **Cost Reduction and Financial Incentives:** As technologies mature and economies of scale come into play, the costs associated with implementing these systems will decrease. Additionally, governments might provide more incentives to encourage the adoption of such sustainable technologies.
- **Interdisciplinary Research:** The intersection of materials science, architecture, environmental science, and engineering will lead to holistic solutions that consider multiple facets of the energy challenge.

- **Regulations and Standards:** Future building codes might mandate a certain percentage of renewable energy usage, where TES systems become a standard feature rather than an optional one.
- **Education and Training:** As these technologies become mainstream, there will be a greater emphasis on training professionals in the field of sustainable building design and energy storage solutions.
- In summary, the future of solar and geothermal-based TES for building applications is bright. As technological, economic, and societal landscapes evolve, these sustainable energy solutions will become increasingly central to our efforts to create resilient, sustainable, and energy-efficient urban spaces [53–59].

REFERENCES

[1] Olabi, A. G., Shehata, N., Maghrabie, H. M., Heikal, L. A., Abdelkareem, M. A., Rahman, S. M., Shah, S. K., & Sayed, E. T. (2022). Progress in solar thermal systems and their role in achieving the sustainable development goals. *Energies*, 15(24), 9501.

[2] Dincer, I., & Rosen, M. A. (2013). *Thermal Energy Storage: Systems and Applications* (2nd ed.). John Wiley & Sons, Hoboken, NJ[ISBN: 978-1-118-22996-6].

[3] Akbarzadeh, A., & Rowe, A. (2018). Review of solar thermal storage systems with heat transfer fluids in concentrating solar power plants. *Renewable and Sustainable Energy Reviews*, 82(3), 2584–2617. doi:10.1016/j.rser.2017.10.043.

[4] Cui, B., Ma, Z., Li, P., & Yan, Z. (2020). Recent progress in geothermal energy storage: A review. *Applied Energy*, 261, 114320. doi:10.1016/j.apenergy.2019.114320.

[5] Zondag, H. A., De Boer, R., Van Helden, W. G. J., Van Zolingen, R. J. C., & Van Steenhoven, A. A. (2003). The performance of a low-cost solar water heater with integrated storage. *Solar Energy*, 74(3), 253–269. doi:10.1016/S0038-092X(03)00041-3.

[6] Allouhi, A., Kousksou, T., Belarbi, R., Jamil, A., Mourad, Y., & Zeraouli, Y. (2017). Solar thermal energy storage in a two-tank system: A review. *Renewable and Sustainable Energy Reviews*, 75, 1367–1377. doi:10.1016/j.rser.2016.11.050.

[7] Cabeza, L. F., et al. (2017). Thermal energy storage in buildings: A state-of-the-art review. *Applied Energy*, 182, 813–845.

[8] Zalba, B., et al. (2003). Review on thermal energy storage with phase change: materials, heat transfer analysis and applications. *Applied Thermal Engineering*, 23(3), 251–283.

[9] Dincer, I., & Rosen, M. A. (2012). *Thermal Energy Storage: Systems and Applications* (2nd ed.). John Wiley & Sons, Hoboken, NJ.

[10] Gómez-Villarejo, R., et al. (2019). Cryogenic energy storage for renewable sources integration: A review. *Renewable and Sustainable Energy Reviews*, 105, 295–312.

[11] Wang, L., et al. (2020). Review of thermochemical heat storage systems: From material selection to system integration. *Renewable and Sustainable Energy Reviews*, 130, 109961.

[12] Sharma, A., Tyagi, V. V., Chen, C. R., & Buddhi, D. (2009). Review on thermal energy storage with phase change materials and applications. *Renewable and Sustainable Energy Reviews*, 13(2), 318–345.

[13] Lund, J. W. (2015). *Geothermal Energy Systems: Exploration, Development, and Utilization.* Academic Press, Cambridge, MA.

[14] Acuna, J. A., Munoz, F. D., & Uihlein, A. (2016). Thermal energy storage in deep geothermal systems: A review. *Renewable and Sustainable Energy Reviews*, 61, 127–137.

[15] Moreno, E., Sánchez, L., Daza, L., Sánchez-Aguilar, R., & Romero, M. (2019). Thermal energy storage technologies for concentrated solar power plants. *Energies*, 12(11), 2201.

[16] Sarker, M. S. I., & Keum, J. K. (2019). Recent advances in thermal energy storage systems for sustainable energy: A review. *Energies*, 12(9), 1699.

[17] Kalogirou, S. A. (2009). *Solar Energy Engineering: Processes and Systems.* Academic Press, Cambridge, MA.

[18] Hasnain, S. M. (1998). Review on sustainable thermal energy storage technologies, Part I: Heat storage materials and techniques. *Energy Conversion and Management*, 39(11), 1127–1138.

[19] Tucker, D., Gorensek, M., & Webb, L. (2011). Thermochemical energy storage system. U.S. Patent No. 7,902,529.

[20] Felsmann, C., & Rybach, L. (2014). Underground thermal energy storage – A review of definitions, opportunities and challenges. *Renewable and Sustainable Energy Reviews*, 30, 613–635.

[21] Meier, A., Gruber, S., & Bayer, P. (2013). Borehole thermal energy storage – A review on experiments and numerical simulations. *Renewable and Sustainable Energy Reviews*, 26, 207–225.

[22] Yang, H., Wang, J., Wang, H., & Wang, Y. (2017). Thermal energy storage in porous media for concentrated solar power plants – A review. *Renewable and Sustainable Energy Reviews*, 79, 1102–1115.

[23] Züttel, A., Wenger, P., Sudan, P., Mauron, P., Emmenegger, C., & Schlapbach, L. (2002). Hydrogen storage in metal hydrides. *Advanced Materials*, 14(21), 1619–1623.

[24] Piatkowski, N., Adelhelm, P., Borchardt, L., Bormann, H., Dieterle, L., Fischer, P., & Schlemmer, M. (2018). Thermochemical energy storage: An overview of the current status and potential developments. *Energies*, 11(7), 1601.

[25] Preuster, P., Papp, C., Wasserscheid, P., & Libuda, J. (2017). Liquid organic hydrogen carriers (LOHCs): Toward a hydrogen infrastructure based on sustainable fuels. *Accounts of Chemical Research*, 50(1), 74–85.

[26] Jradi, M., Veje, C., & Jørgensen, B. N. (2017). Performance analysis of a soil-based thermal energy storage system using solar-driven air-source heat pump for Danish buildings sector. *Applied Thermal Engineering*, 114, 360–373.

[27] Koukou, M. K., Dogkas, G., Vrachopoulos, M. G., Konstantaras, J., Pagkalos, C., Stathopoulos, V. N., Pandis, P. K., Lymperis, K., Coelho, L., & Rebola, A. (2020). Experimental assessment of a full scale prototype thermal energy storage tank using paraffin for space heating application. *International Journal of Thermofluids*, 1, 100003.

[28] Chavan, S., Rudrapati, R., & Manickam, S. (2022). A comprehensive review on current advances of thermal energy storage and its applications. *Alexandria Engineering Journal*, 61(7), 5455–5463.

[29] Patra, B., Nema, P., Khan, M. Z., & Khan, O. (2023). Optimization of solar energy using MPPT techniques and industry 4.0 modelling. *Sustainable Operations and Computers*, 4, 22–28.

[30] Patra, B., & Nema, P. (2022). Comparative design analysis of modified solar based 15 level multi level inverter for power quality improvement. *Journal of Algebraic Statistics*, 13(3), 1084–1095.

[31] Patra, B., & Nema, P. (2021). Analysis of solar integrated multilevel inverter for smart grid power filters. In *International Conference on Advances in Electrical, Computing, Communication and Sustainable Technologies*, pp. 1–5.

[32] Alva, G., Lin, Y., & Fang, G. (2018). An overview of thermal energy storage systems. *Energy*, 144, 341–378.

[33] Leonzio, G. (2017). Solar systems integrated with absorption heat pumps and thermal energy storages: State of art. *Renewable and Sustainable Energy Reviews*, 70, 492–505.

[34] Hassan, F., Jamil, F., Hussain, A., Ali, H. M., Janjua, M. M., Khushnood, S., Farhan, M., Altaf, K., Said, Z., & Li, C. (2022). Recent advancements in latent heat phase change materials and their applications for thermal energy storage and buildings: A state of the art review. *Sustainable Energy Technologies and Assessments*, 49, 101646.

[35] Yang, T., Liu, W., Kramer, G. J., & Sun, Q. (2021). Seasonal thermal energy storage: A techno-economic literature review. *Renewable and Sustainable Energy Reviews*, 139, 110732.

[36] Miró, L., Oró, E., Boer, D., Cabeza, L. F. (2015). Embodied energy in thermal energy storage (TES) systems for high temperature applications. *Applied Energy*, 137, 793–799.

[37] Lin, Y., Jia, Y., Alva, G., & Fang, G. (2018). Review on thermal conductivity enhancement, thermal properties and applications of phase change materials in thermal energy storage. *Renewable and Sustainable Energy Reviews*, 82, 2730–2742.

[38] Meena, C. S., Prajapati, A. N., Kumar, A., & Kumar, M. (2022). Utilization of solar energy for water heating application to improve building energy efficiency: An experimental study. *Buildings*, 12, 2166. doi:10.3390/buildings12122166.

[39] Meena, C. S., Kumar, A., Jain, S., Ur Rehman, A., & Mishra, S. (2022). Innovation in green building sector for sustainable future. *Energies*, 15, 6631. doi:10.3390/en15186631.

[40] Dutt, N., Hedau, A. J., Kumar, A., Awasthi, M. K., Singh, V. P., & Dwivedi, G. (2023). Thermo-hydraulic performance of solar air heater having discrete D-shaped ribs as artificial roughness. *Environmental Science and Pollution Research*. doi:10.1007/s11356-023-28247-9.

[41] Kushwaha, P. K., Sharma, N. K., Kumar, A., & Meena, C. S. (2023). Recent advancements in augmentation of solar water heaters using nanocomposites with PCM: Past, present, and future. *Buildings*, 13, 79. doi:10.3390/buildings13010079.

[42] Singh, V. P., Jain, S., Karn, A., & Kumar, A. (2022). Experimental assessment of variation in open area ratio on thermohydraulic performance of parallel flow solar air heater. *Arabian Journal for Science and Engineering*, 48, 11695–11711. doi:10.1007/s13369-022-07525-7.

[43] Singh, V. P., Jain, S., Kumar, A., Mishra, S., & Sharma, N. K. (2022). Heat transfer and friction factor correlations development for double pass solar air heater artificially roughened with perforated multi-V ribs. *Case Studies in Thermal Engineering*, 39, 102461, ISSN 2214-157X. doi:10.1016/j.csite.2022.102461.

[44] Singh, V. P., Jain, S., Karn, A., Kumar, A., Dwivedi, G., & Meena, C. S. (2022). Recent developments and advancements in solar air heaters: A detailed review. *Sustainability*, 14, 12149. doi:10.3390/su141912149.

[45] Singh, V. P., Jain, S., Karn, A., Kumar, A., Dwivedi, G., Meena, C. S., & Cozzolino, R. (2022). Mathematical modeling of efficiency evaluation of double-pass parallel flow solar air heater. *Sustainability*, 14, 10535. doi:10.3390/su141710535.

[46] Singh, V. P., Jain, S., & Kumar, A. (2022). Establishment of correlations for the thermo-hydraulic parameters due to perforation in a multi-V rib roughened single pass solar air heater. *Experimental Heat Transfer*, 35(5), 1–20. doi:10.1080/08916152.2022.2064940.

[47] Saxena, A., Prajapati, A. N., Pant, G., Meena, C. S., Kumar, A., & Singh, V. P. (2023). Water consumption optimization of hybrid heat pump water heating system. In: Shukla, A. K., Sharma, B. P., Arabkoohsar, A., Kumar, P. (eds) *Recent Advances in Mechanical Engineering. FLAME 2022. Lecture Notes in Mechanical Engineering*. Springer, Singapore, pp. 721–732. doi:10.1007/978-981-99-1894-2_61.

[48] Kaygusuz, K. (2011). Prospect of concentrating solar power in Turkey: The sustainable future. *Renewable and Sustainable Energy Reviews*, 15(1): 808–814.

[49] Verma, V., Meena, C. S., Thangavel, S., Kumar, A., Choudhary, T., & Dwivedi, G. (2023). Ground and solar assisted heat pump systems for space heating and cooling applications in the northern region of India – A study on energy and CO2 saving potential. *Sustainable Energy Technologies and Assessments*, 59, 103405, ISSN 2213-1388. doi:10.1016/j.seta.2023.103405.

[50] Awasthi, M. K., Dutt, N., Kumar, A., & Kumar, S. (2023). Electrohydrodynamic capillary instability of Rivlin-Ericksen viscoelastic fluid film with mass and heat transfer. *Heat Transfer*, 53(1), 115–133. doi:10.1002/htj.22944.

[51] Meena, C. S., Kumar, A., Roy, S., Cannavale, A., & Ghosh, A. (2022). Review on boiling heat transfer enhancement techniques. *Energies*, 15, 5759. doi:10.3390/en15155759.

[52] Kundu, A., Kumar, A., Dutt, N., Meena, C. S., & Singh, V. P. (2023). Introduction to thermal energy resources and their smart applications. In: Kumar, A., Singh, V. P., Meena, C. S., Dutt, N. (eds) *Thermal Energy Systems: Design, Computational Techniques and Applications.* CRC Press, Boca Raton, FL, pp. 1–15; Chapter 1. doi:10.1201/9781003395768-1.

[53] Dewangan, A. K., Moinuddin, S. Q., Cheepu, M., Sajjan, S. K., & Kumar, A. (2023). Thermal energy storage: Opportunities, challenges and future scope. In: Kumar, A., Singh, V. P., Meena, C. S., Dutt, N. (eds) *Thermal Energy Systems: Design, Computational Techniques and Applications.* CRC Press, Boca Raton, FL, pp. 17–28; Chapter 2. doi:10.1201/9781003395768-2.

[54] Kundu, A., Kumar, A., Dutt, N., Singh, V. P., & Meena, C. S. (2023). Modelling and simulation of thermal energy system for design optimization. In: Kumar, A., Singh, V. P., Meena, C. S., Dutt, N. (eds) *Thermal Energy Systems: Design, Computational Techniques and Applications.* CRC Press, Boca Raton, FL, pp. 103–140; Chapter 7. doi:10.1201/9781003395768-7.

[55] Pant, G., Meena, C. S., Saxena, A., Kumar, A., Singh, V. P., & Dutt, N. (2023). Study the temperature variation in alternate coils of insulated condenser cum storage tank: Experimental study. In: Sikarwar, B. S., Sharma, S. K., Jain, A., Singh, K. M. (eds) *Advances in Fluid and Thermal Engineering. FLAME 2022. Lecture Notes in Mechanical Engineering.* Springer, Singapore, pp. 627–638. doi:10.1007/978-981-99-2382-3_52.

[56] Kumar, A., Singh, V. P., Meena, C. S., & Dutt, N. (2023). *Thermal Energy Systems: Design, Computational Techniques and Applications.* Boca Raton, FL: Taylor & Francis (CRC Press). doi:10.1201/9781003395768.

[57] Singh, S., Kumar, A., Behura, S. K., & Verma, K. (2022). Challenges and opportunities in nanomanufacturing. In: Singh, S., Behura, S. K., Kumar, A., Verma, K. (eds) *Nanomanufacturing and Nanomaterials Design: Principles and Applications.* CRC Press, Boca Raton, FL, pp. 17–30; Chapter 02. doi:10.1201/9781003220602-2.

[58] Srivastava, S., Verma, D., Thusoo, S., Kumar, A., Singh, V. P., Kumar, R. (2022). Nanomanufacturing for energy conversion and storage devices. In: Singh, S., Behura, S. K., Kumar, A., Verma, K. (eds) *Nanomanufacturing and Nanomaterials Design: Principles and Applications.* CRC Press, Boca Raton, FL, pp. 165–173; Chapter 10. doi:10.1201/9781003220602-10.

[59] Singh, V. P., Dwivedi, A., Karn, A., Kumar, A., Singh, S., Srivastava, S., & Srivastava, K. (2022). Nanomanufacturing and design of high-performance piezoelectric nanogenerator for energy harvesting. In: Singh, S., Behura, S. K., Kumar, A., Verma, K. (eds) *Nanomanufacturing and Nanomaterials Design: Principles and Applications.* CRC Press, Boca Raton, FL, pp. 241–242; Chapter 15. doi:10.1201/9781003220602-15.

Chapter 10

Study and analysis of low-level geothermal heat pumps in Oman

*Morteza Khashehchi, Sivasakthivel Thangavel,
Pooyan Rahmanivahid, Milad Heidari,
and Taleb Moazzeni*
Global College of Engineering and Technology (GCET) Muscat

10.1 INTRODUCTION

The urgent global need for sustainable energy sources to combat climate change, enhance energy security, and support sustainable development is evident. Oman, historically reliant on fossil fuels, is highly motivated to diversify its energy sources and reduce carbon emissions. The geological features of Oman, including hot springs and geothermal anomalies, suggest untapped subsurface heat reservoirs. This geological advantage aligns with Oman's commitment to environmental preservation. Geothermal heat pump technology, known for its efficiency and adaptability, is of interest in Oman's hot climate. This study in Oman aims to address knowledge gaps and assess the feasibility of geothermal technology, contributing to a more sustainable and resilient energy future [1–3] (Figure 10.1).

The motivation behind exploring low-level geothermal heat pumps in Oman stems from a convergence of environmental, economic, and technological imperatives. Oman's commitment to mitigating climate change and reducing its carbon footprint drives the search for sustainable energy alternatives. The unique geological characteristics of the region, marked by geothermal indicators like hot springs and anomalies, provide a compelling incentive to tap into the Earth's heat reservoirs. As Oman seeks to diversify its energy sources and bolster energy security, geothermal heat pumps offer a promising solution by harnessing the Earth's natural thermal energy for heating and cooling purposes [4].

10.1.1 Objectives of the study

The objectives of the study on low-level geothermal heat pumps in Oman encompass a comprehensive and systematic exploration of various dimensions, driven by the ambition to facilitate informed decision-making and contribute to the sustainable energy trajectory of the country. Foremost,

DOI: 10.1201/9781003472629-10

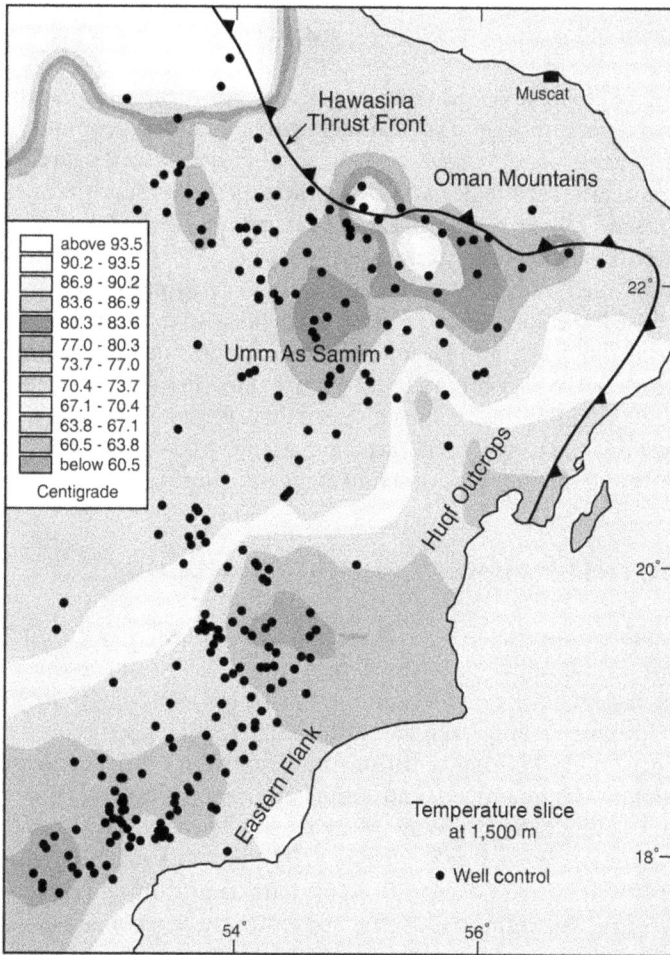

Figure 10.1 Geothermal resources in Oman (1,500 m).

the study aims to assess the intricate interplay between Oman's geological composition and climatic conditions, determining the viability and suitability of implementing low-level geothermal heat pump systems. This involves mapping geothermal potential, identifying areas with favorable subsurface heat reservoirs, and evaluating the practicality of capturing and utilizing this indigenous source of renewable energy. Furthermore, the study seeks to conduct a rigorous technical analysis to decipher the operational efficiency, energy performance, and potential benefits of adopting low-level geothermal heat pumps within Oman's diverse energy landscape. By quantifying energy savings, greenhouse gas emissions reductions, and the system's overall thermal performance, the research aims to provide valuable insights into the technology's practicality and alignment with Oman's energy needs and aspirations.

Addressing environmental concerns is another focal point of the study. By assessing potential environmental impacts, such as groundwater contamination, land use changes, and noise pollution, the research aims to propose mitigation strategies and best practices that ensure the sustainable deployment of low-level geothermal heat pumps without compromising Oman's natural ecosystems and water resources.

Ultimately, the study's overarching objective is to culminate in a set of evidence-based recommendations and guidelines that equip policymakers, industry stakeholders, and investors with the insights needed to strategically incorporate low-level geothermal heat pumps into Oman's sustainable energy roadmap. By aligning technological feasibility, economic viability, environmental sustainability, and social acceptance, the study aspires to contribute significantly to Oman's journey toward a greener, more resilient energy future.

10.2 GEOTHERMAL HEAT PUMP TECHNOLOGY

Geothermal heat pump (GHP) technology, also known as ground-source heat pump technology, presents a remarkable solution at the intersection of energy efficiency, sustainability, and environmental stewardship [5]. This innovative approach capitalizes on the Earth's intrinsic thermal energy to provide heating, cooling, and hot water for a diverse range of residential, commercial, and industrial applications (Figure 10.2). Through a sophisticated process that leverages the natural temperature gradient beneath the Earth's surface, GHP systems offer an alternative to conventional heating, ventilation, and air conditioning (HVAC) methods, delivering substantial benefits for both energy consumers and the environment [6].

10.2.1 Operating principles

The fundamental principle underlying GHP technology is the transfer of heat between the ground and a building's interior. In the heating mode, the system extracts heat from the Earth, even in colder weather, and transfers it to the indoor space. Conversely, during the cooling mode, the system removes heat from the building and deposits it into the Earth [7]. This versatile and reversible operation is achieved through a network of underground pipes (Figure 10.3) or boreholes containing a heat exchange fluid, which facilitates the transfer of heat to or from the ground.

The heart of GHP systems lies within the heat pump unit, a sophisticated assembly comprising a compressor, condenser, evaporator, and expansion valve. This unit orchestrates the circulation and manipulation of refrigerant

Heat pumps

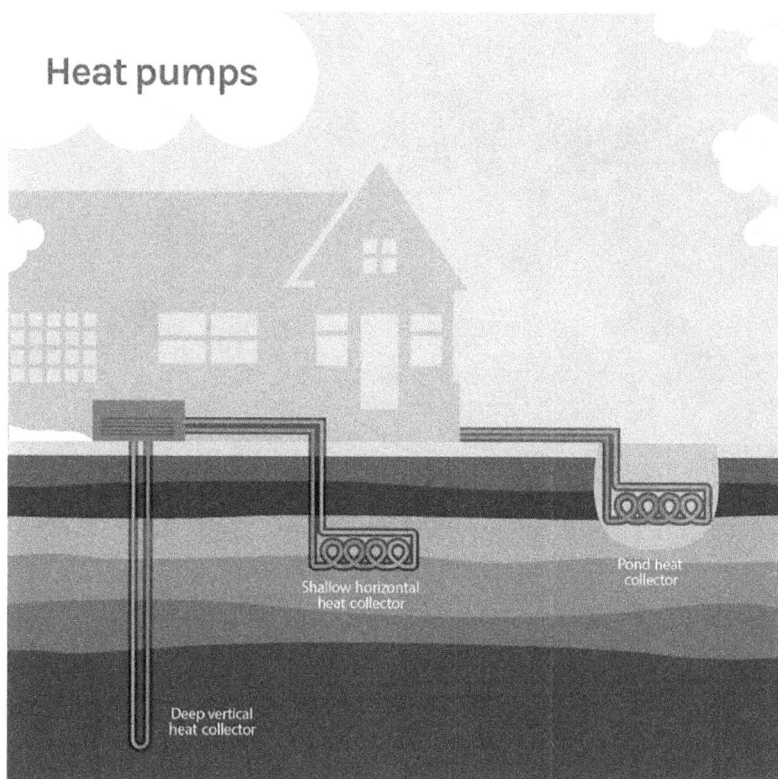

Figure 10.2 Energy production from geothermal energy.

gas, enabling it to change states from a liquid to a gas and vice versa. In the heating mode, the refrigerant absorbs heat from the heat exchanger loop, evaporates into a gas, and then releases the captured heat indoors upon condensation. In contrast, during the cooling mode, the refrigerant extracts heat from the indoor environment evaporates and subsequently deposits the heat into the Earth upon condensation. This dynamic process ensures a consistent and comfortable indoor climate throughout the year.

Integral to GHP operations is the heat exchanger loop, which can take the form of closed-loop or open-loop configurations. Closed-loop systems circulate a heat exchange fluid, often a mixture of water and antifreeze, through underground pipes or boreholes. Open-loop systems, on the other hand, employ groundwater for heat exchange before returning it to the environment [8]. Both configurations leverage the Earth's stable temperatures to facilitate efficient heat transfer, serving as a bridge between the Earth and the building.

The environmental and energy benefits resulting from these operating principles are profound. By tapping into the Earth's heat, GHP systems

Figure 10.3 Geothermal heat pumps – e.g., Carolina country.

achieve commendable energy efficiency ratios (EER) and coefficient of performance (COP) values, leading to reduced energy consumption, lower operational costs, and diminished greenhouse gas emissions [9]. Ultimately, the intricate interplay of natural thermal properties and advanced technology within GHP operating principles underpins their role as a sustainable and forward-looking solution for indoor climate control, aligning harmoniously with the broader goals of energy efficiency and environmental responsibility [10,11].

A typical GHP system comprises several essential components, each playing a vital role in its seamless operation [12]:

Heat Exchanger Loop: This loop, also known as the ground loop, is responsible for facilitating the transfer of heat between the building and the Earth's subsurface. The heat exchange fluid circulates within this loop, carrying heat to or from the ground.

Heat Pump Unit: The heart of the GHP system, the heat pump unit, houses critical components including a compressor, condenser, evaporator, and expansion valve. These components collaborate to manipulate the refrigerant's temperature and pressure, facilitating efficient heat transfer.

Distribution System: The conditioned air, heated or cooled by the GHP system, is distributed throughout the building using traditional ductwork, radiant floor systems, or other distribution methods, ensuring uniform comfort.

Control System: Advanced control systems oversee the operation of the heat pump, regulating temperature settings and optimizing performance based on the building's heating and cooling requirements.

Ground Heat Exchanger: This component consists of underground pipes or boreholes, which act as conduits for heat exchange between the Earth and the heat pump system.

10.2.2 Advantages of GHP technology

The advantages of GHP technology are multifaceted and impactful [13]:

Exceptional Energy Efficiency: GHP systems exhibit commendable energy efficiency due to the stable temperatures below the Earth's surface. This results in high EER and COP values, translating to reduced energy consumption and operational costs.

Environmental Stewardship: GHP systems contribute to a greener environment by producing minimal direct emissions. Compared to conventional HVAC systems, they significantly reduce greenhouse gas emissions and contribute to improved air quality.

Longevity and Reliability: The underground components of GHP systems have longer lifespans than conventional HVAC systems, translating to lower maintenance requirements and enhanced system reliability.

Significant Energy Savings: The consistent energy performance of GHP systems can lead to substantial energy savings over the system's lifespan, offsetting the initial installation costs.

Adaptability and Versatility: GHP systems are adaptable to various building types, ranging from residential houses to commercial complexes. Furthermore, they can be integrated with existing HVAC systems to enhance overall efficiency.

While GHP technology offers a plethora of advantages, proper system design, installation, and maintenance are crucial to ensure optimal performance. Whereas initial installation costs may be higher compared to traditional HVAC systems, the long-term energy savings often outweigh this investment.

Thermal Efficiency and Performance of GHP

The pursuit of sustainable and efficient heating and cooling solutions has propelled GHP systems to the forefront of innovative energy technologies. At the heart of GHP systems lies their remarkable thermal efficiency and performance, which stems from the intricate interplay of scientific principles, advanced engineering, and a deep understanding of heat transfer dynamics. These systems harness the Earth's stable temperatures, unlocking a realm of possibilities that optimize energy utilization, minimize environmental impact, and redefine our approach to indoor climate control [14].

GHP systems exemplify thermal efficiency by exploiting the Earth's consistent subsurface temperatures. In heating mode, the systems draw heat from the Earth and transfer it indoors, even during frigid winters. Conversely, in cooling mode, the systems remove heat from indoor spaces and dissipate it into the cooler Earth, providing comfort during scorching summers. This unique ability to transfer heat from where it's least needed to where it's most desired underscores the resourcefulness of GHP systems [15].

The intricate components within GHP systems orchestrate this heat transfer dance with finesse. The ground heat exchanger, often in the form of horizontal or vertical loops, serves as the bridge connecting the Earth's heat reservoir to the refrigerant fluid. This fluid, propelled by a compressor, undergoes phase changes as it absorbs and releases heat. The result is a system that operates with remarkable efficiency, capitalizing on the energy-saving potential of heat exchange [16].

Thermal efficiency is further enhanced by the utilization of variable-speed compressors, which adapt the system's operation based on real-time heating or cooling demands [17]. By modulating their speed, these compressors ensure that the system operates at its optimal capacity, minimizing energy wastage and maximizing performance. The integration of sophisticated electronic controls and sensors furthers this efficiency, ensuring that the system adapts to changing conditions and fine-tuning its operation for optimal energy transfer.

10.2.3 Types of GHP systems

The realm of GHP technology encompasses a diverse array of system types, each meticulously designed to cater to specific applications, geological considerations, and environmental factors. These distinct variations in GHP systems exemplify the nuanced approach required to effectively harness the Earth's renewable thermal energy for optimal heating, cooling, and hot water provision. The delineation of these types underscores the multifaceted nature of GHP technology and its capacity to adapt to a wide range of settings, climates, and building requirements.

10.2.3.1 Closed-loop systems

One prominent category within the spectrum of GHP systems is the "closed-loop" configuration. In this system, a heat exchange fluid, often a mixture of water and antifreeze, circulates within a closed circuit of pipes buried underground or vertical boreholes drilled into the Earth [18]. These pipes or boreholes serve as conduits through which heat is transferred to or from the Earth. The closed-loop design offers several subtypes, each characterized by distinct layouts and installation methods:

- Horizontal Closed-Loop System: In this configuration, pipes are laid in shallow trenches excavated horizontally beneath the Earth's surface [19]. This design is particularly suitable for regions with ample land availability but limited vertical space. The horizontal layout allows for efficient heat exchange while utilizing the expansive surface area of the trenches.

- **Vertical Closed-Loop System:** Contrasting with the horizontal system, the vertical closed-loop design involves drilling boreholes to significant depths [20]. This configuration is ideal for areas with space constraints or challenging soil conditions. The vertical arrangement maximizes the depth of heat exchange, accessing the stable temperatures of the Earth at greater depths.
- **Pond/Lake Closed-Loop System:** Leveraging nearby water bodies, the pond/lake closed-loop system utilizes pipes submerged underwater for heat transfer. This approach capitalizes on the consistent thermal properties of water, allowing efficient heat exchange between the Earth and the system [21]. This type of system is particularly applicable when bodies of water are present in proximity to the building.

10.2.3.2 Open-loop systems

In contrast to closed-loop systems, "open-loop" GHP systems directly employ groundwater for heat exchange. These systems extract water from an aquifer using a well, which then passes through a heat exchanger to transfer heat to or from the Earth. The water is subsequently discharged back into a surface water body or reinjected into the ground. Open-loop systems offer remarkable efficiency due to the relatively constant temperature of groundwater, making them highly effective for heating and cooling purposes [22].

10.2.3.3 Hybrid systems

In some instances, hybrid GHP systems combine elements of both closed-loop and open-loop designs, capitalizing on the strengths of each approach [23]. These hybrid configurations offer flexibility and customization, allowing system designers to optimize performance based on the specific characteristics of the site and its surroundings.

The selection of a specific GHP system type is influenced by a myriad of factors, including available land, geological conditions, water availability, local regulations, and the building's heating and cooling demands. Each type caters to different site-specific requirements, reflecting the adaptability and versatility of GHP technology. However, the intricate taxonomy of GHP systems underscores the diverse range of solutions that this technology offers for sustainable climate control. Whether through closed-loop, open-loop, or hybrid configurations, GHP systems exemplify innovation in utilizing the Earth's renewable thermal energy, highlighting the complex interplay between technology and nature to achieve energy-efficient and environmentally conscious heating and cooling solutions.

10.3 COMPARISON WITH OTHER HEATING AND COOLING TECHNOLOGIES

In the context of heating and cooling technologies, the innovative prowess of GHP systems emerges as a game-changer, setting them apart from conventional alternatives. When juxtaposed with traditional methods like fossil fuel-based heating and air conditioning, GHP systems offer a paradigm shift towards energy efficiency, environmental responsibility, and long-term sustainability. Unlike combustion-based systems that rely on finite resources and emit pollutants, GHPs harness the Earth's inexhaustible thermal energy, resulting in reduced carbon emissions and improved air quality. Moreover, the efficiency ratios of GHP systems often surpass those of conventional HVAC systems, translating to significant energy savings and reduced operational costs for end-users [24–26]. This distinction becomes even more pronounced in colder climates, where GHPs' constant access to geothermal warmth renders them superior to air-source heat pumps that struggle in extreme conditions. Additionally, GHPs minimize the vulnerability of heating and cooling systems to fuel price fluctuations, enhancing stability and financial predictability for users. When compared to electric resistance heating, which is notably less efficient and environmentally friendly, GHPs stand out as an exemplar of energy conservation. As renewable energy initiatives gain prominence, GHP technology's versatility and ability to provide both heating and cooling further cement its superiority over single-purpose systems. Ultimately, the comparison underscores GHP systems' transformative potential, positioning them as a cornerstone of sustainable energy solutions that align with the urgent global imperative to reduce energy consumption, combat climate change, and foster a greener future.

10.4 GEOLOGICAL AND CLIMATIC CONTEXT OF OMAN

The geological and climatic context of Oman is marked by unique features that hold significant implications for the exploration and utilization of GHP technology. Situated in the southeastern corner of the Arabian Peninsula, Oman's geological composition showcases diverse formations, including sedimentary rocks, limestone, and igneous intrusions. This geological diversity presents opportunities for GHP systems, as certain rock formations can host geothermal reservoirs suitable for heat exchange.

Additionally, the presence of hot springs and geothermal anomalies signifies the potential for subsurface heat reservoirs, a vital factor in assessing the viability of GHP technology in the region. Climatically, Oman experiences extreme temperatures, with scorching summers and mild winters. This climate accentuates the demand for efficient cooling solutions, making GHPs a compelling option due to their ability to provide cooling alongside heating. Moreover, Oman's geographical positioning results in varying microclimates, which should be considered when determining the best locations for GHP installations. The geological richness and climatic diversity of Oman offer a promising foundation for the successful implementation of GHP systems, signaling the potential to harness the Earth's thermal energy to address the country's heating and cooling needs sustainably.

10.4.1 Geological features and heat sources

The Earth's geological features and the intricate dance of its heat sources weave a complex tapestry that underlies the potential for harnessing geothermal energy, a renewable and sustainable resource. Geological formations, sculpted by millions of years of tectonic activity, volcanic processes, and the forces of nature, hold the key to unlocking the Earth's inner warmth (Figure 10.4). These formations, from fault lines to permeable rocks, serve

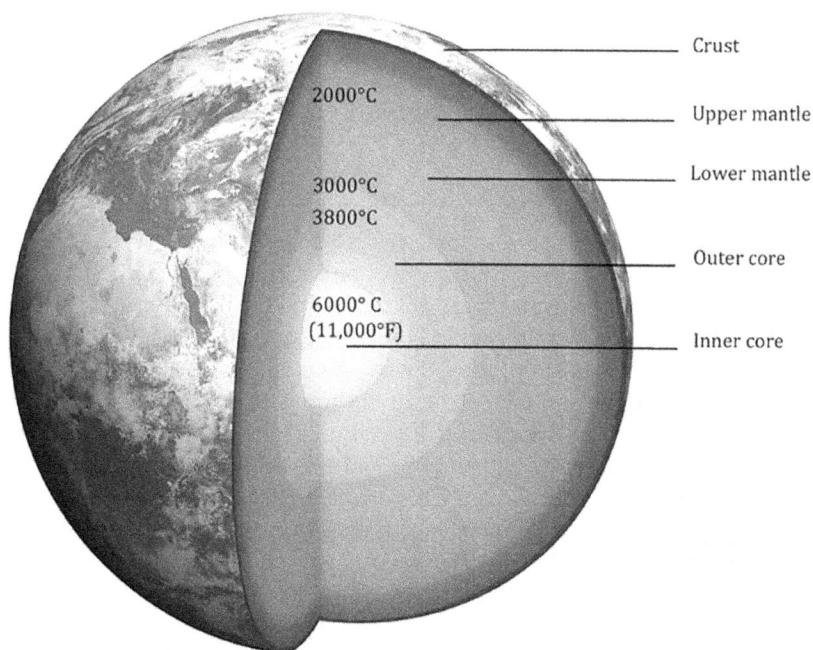

Figure 10.4 Taking the temperature of Earth's core | Discover Magazine.

as conduits through which the heat from within the Earth's core makes its way to the surface [27–29]. Volcanic regions, shaped by molten magma chambers and dynamic plate movements, often possess geological features like geysers and hot springs, showcasing the direct influence of Earth's internal heat on the landscape.

At the heart of this geothermal energy potential are the diverse heat sources that fuel its existence. Primarily, the Earth's core, a fiery realm of molten iron and nickel, emanates residual heat from the planet's tumultuous formation. This ancient energy, trapped within the Earth's core, continually escapes through conduction, convection, and radiation, generating heat gradients that reach the surface. Additionally, the Earth's crust harbors radioactive elements like uranium, thorium, and potassium, remnants of the planet's formation, which decay over time, emitting heat in the process. This radiant energy weaves through the Earth's layers, permeating the rocks, the waters, and the very fabric of the planet.

10.4.2 Climatic considerations and energy demand

The climatic conditions of a region play a profound role in shaping its energy demand landscape and the strategies adopted to meet these demands sustainably. Oman, situated in the southeastern part of the Arabian Peninsula, possesses a unique and challenging climate characterized by arid desert conditions. For example, Figure 10.5 shows data tables and charts of monthly and yearly climate conditions in Ibra, Oman [30]. This climatic context has significant implications for energy consumption patterns, particularly with regard to cooling and heating needs, as well as the exploration of innovative solutions like GHP technology.

Oman's climate is defined by its scorching summers and relatively mild winters. During the summer months, temperatures can soar to well over 40°C, creating an urgent need for effective cooling systems. The intense heat necessitates air conditioning and cooling solutions for residential, commercial, and industrial spaces to maintain comfortable indoor environments. This demand for cooling is further exacerbated by the country's rapid urbanization and economic growth, which have led to increased construction of buildings and infrastructure [30].

However, Oman's climatic demands for cooling align with the peak electricity consumption period. The very need for cooling during the hottest times of the year coincides with the highest electricity demand, stressing the energy grid and challenging the country's power supply infrastructure. Conventional air conditioning systems, which rely heavily on electricity, contribute significantly to the strain on the grid during these peak periods. The resultant spikes in electricity consumption not only pose challenges to the grid's stability but also raise concerns about energy security, sustainability, and environmental impact.

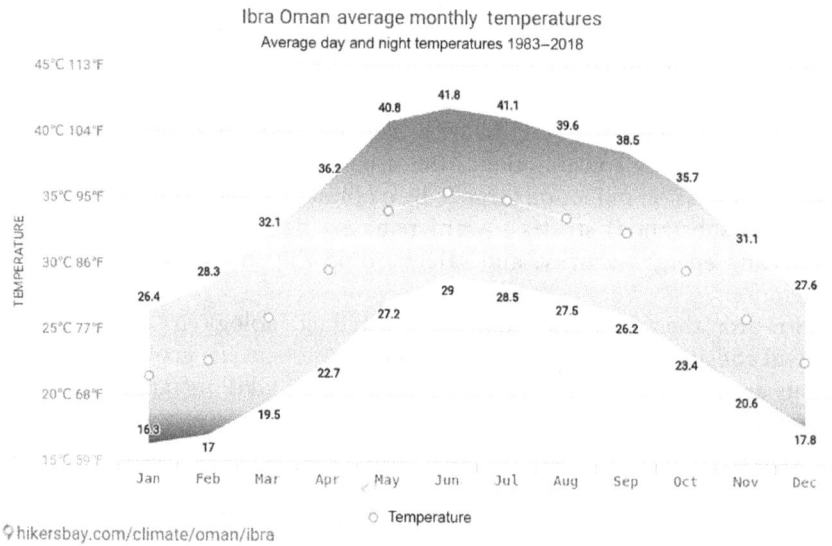

Figure 10.5 Data tables and charts of monthly and yearly climate conditions in Ibra Oman.

In this context, GHP technology emerges as a transformative solution that addresses Oman's climatic demands and energy challenges. GHPs, also known as ground-source heat pumps, are an innovative and sustainable technology that harnesses the Earth's natural thermal energy for both heating and cooling purposes. They operate on the principle that the Earth maintains a relatively constant temperature below the surface, regardless of external weather conditions. This inherent stability offers a unique advantage for providing efficient climate control.

GHP systems consist of heat exchange loops buried underground, typically in a horizontal, vertical, or pond/lake configuration. These loops contain a heat exchange fluid, often a mixture of water and antifreeze, which transfers heat between the building's interior and the Earth. In the cooling mode, GHPs extract heat from the building and release it into the relatively cooler ground, effectively cooling the indoor space. During the heating mode, the system reverses the process, extracting heat from the Earth and distributing it indoors for heating.

The adoption of GHP technology in Oman has the potential to alleviate the strain on the electricity grid during peak cooling periods. By leveraging the Earth's thermal energy, GHPs offer an efficient alternative to traditional air conditioning systems, which often contribute to surges in electricity consumption. GHPs can reduce the electricity load during peak periods, leading to a more stable and resilient energy grid. Additionally, GHPs' ability to provide both heating and cooling solutions makes them a versatile choice for Oman's climate, where temperature fluctuations between seasons can be considerable.

The integration of GHPs into Oman's energy mix aligns with the country's broader goals of energy diversification, sustainability, and environmental responsibility. As global efforts to combat climate change intensify, transitioning away from fossil fuel-intensive cooling and heating methods becomes imperative. GHPs, with their minimal carbon footprint and reduced electricity consumption, offer a pathway for Oman to contribute to global emission reduction targets. By embracing GHP technology, Oman can take substantial strides toward reducing its reliance on fossil fuels, enhancing energy security, and safeguarding the environment for future generations.

However, the widespread adoption of GHP technology in Oman requires careful consideration and strategic planning. Site-specific geological assessments are essential to determine the suitability of GHP installations in different regions of the country. Geological surveys can identify areas with the potential for geothermal reservoirs that can serve as efficient heat exchange sources for GHP systems. Factors such as soil composition, thermal conductivity, and depth to the water table are crucial determinants of a site's geothermal potential.

10.5 CASE STUDIES AND APPLICATIONS

10.5.1 GHPs in residential buildings

As the world gravitates toward more sustainable and energy-efficient solutions, the integration of GHPs into residential buildings emerges as a game-changer in redefining modern living. These innovative systems, rooted in harnessing the Earth's thermal energy, offer a compelling fusion of comfort, efficiency, and environmental responsibility [16,31–34]. The adoption of GHPs in residential settings not only revolutionizes the way we experience indoor climate control but also showcases their potential to mitigate energy consumption, reduce carbon emissions, and create a sustainable blueprint for the future (Figure 10.6).

At the core of the GHP's impact on residential buildings lies its ability to provide consistent heating and cooling, enhancing occupants' comfort while drastically reducing energy consumption. In cold winters, GHPs efficiently draw heat from the Earth, ensuring cozy interiors even during freezing temperatures. In scorching summers, these systems seamlessly remove excess heat, creating a refreshing oasis. The result is a steady and comfortable indoor environment, free from the temperature fluctuations commonly experienced with conventional heating and cooling methods.

GHPs go beyond comfort; they epitomize energy efficiency and environmental consciousness. By utilizing the Earth's consistent subsurface temperatures, GHPs dramatically reduce the need for electricity to power traditional HVAC systems. This translates into substantial energy savings

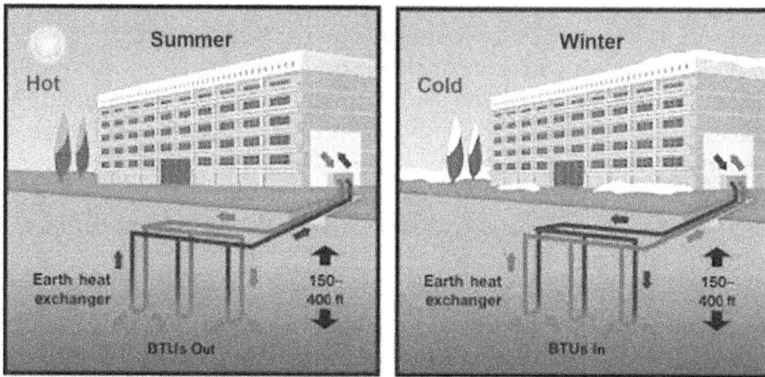

Figure 10.6 Geothermal and ground-source heat pumps in residential buildings.

and a corresponding reduction in carbon emissions. The replacement of fossil fuel-dependent systems with GHPs represents a significant stride toward achieving sustainability goals, contributing to local and global efforts to combat climate change.

Furthermore, GHPs align seamlessly with residential renewable energy integration. By harnessing the Earth's natural heat reservoirs, GHPs can be paired with solar panels to create energy-efficient hybrid systems. The synergy between these technologies ensures a consistent supply of electricity and heating/cooling energy while minimizing the reliance on conventional power sources. This integration showcases the holistic nature of sustainable residential living, where energy generation and consumption harmonize with nature's inherent resources.

10.5.2 GHPs in commercial and industrial settings

In the dynamic landscape of commercial and industrial sectors, where energy consumption and environmental impact are of paramount concern, GHPs) emerge as transformative solutions that marry operational efficiency with sustainable practices. These innovative systems, rooted in the utilization of the Earth's thermal energy, have the potential to reshape the way businesses approach heating, cooling, and energy management [24,35–37]. The integration of GHPs in commercial and industrial settings not only optimizes operational costs and enhances comfort but also sets a new benchmark for corporate environmental responsibility.

The energy efficiency inherent in GHPs directly translates to substantial cost savings for businesses. By utilizing the Earth's natural heat reservoirs, GHPs drastically reduce electricity consumption, often one of the most significant operational expenses. This reduction in energy demand not only results in lower utility bills but also mitigates peak load requirements, thus preventing potential energy grid strain. Additionally, the minimized energy consumption leads to a correspondingly reduced carbon footprint, aligning with corporate sustainability goals and environmental regulations.

GHPs' integration with renewable energy sources further amplifies their impact in commercial and industrial contexts. By coupling GHPs with solar panels or other renewable technologies, businesses can create comprehensive energy solutions that not only meet their heating and cooling needs but also generate electricity sustainably. This integration showcases a holistic approach to energy management, where efficiency, sustainability, and operational continuity coalesce.

The financial benefits of GHPs extend beyond energy savings. Many governments and municipalities offer incentives and grants for adopting energy-efficient technologies. Businesses that embrace GHPs can leverage these incentives to offset installation costs and accelerate their return on investment. Additionally, the enhanced environmental credentials of GHP-equipped facilities may confer a competitive advantage in today's environmentally conscious market, appealing to environmentally-minded customers and stakeholders.

10.5.3 Integration with existing HVAC systems

In HVAC, the integration of GHP systems with existing infrastructure stands as a strategic pathway to elevate energy efficiency, bolster comfort, and pave the way for sustainable energy practices [38,39]. This dynamic integration unites the innovative potential of GHPs with the operational familiarity of conventional HVAC systems, unveiling a synergy that optimizes energy utilization, reduces environmental impact, and enhances the overall indoor environment.

GHP integration with existing HVAC systems capitalizes on the strengths of both technologies, creating a holistic approach that meets the multifaceted demands of modern buildings. GHPs excel in maintaining stable indoor temperatures, tapping into the Earth's thermal reservoirs to deliver efficient heating and cooling. Conventional HVAC systems, on the other hand, often handle ventilation and air distribution adeptly. By integrating the two, businesses and homeowners can create a harmonious blend of technologies that capitalize on each system's strengths, resulting in an optimal indoor climate.

One of the primary advantages of this integration lies in its potential to enhance energy efficiency. GHPs are inherently energy-efficient due to their utilization of the Earth's consistent temperatures, reducing the electricity required for heating and cooling. By supplementing existing HVAC systems with GHPs, the load on conventional equipment is lightened, allowing it to operate more efficiently. The result is a reduction in overall energy consumption, contributing to lower utility bills and reduced carbon emissions.

The integration process often involves retrofitting GHP components into the existing HVAC framework, such as adding ground heat exchangers, compressors, and controls. The compatibility between GHPs and conventional systems simplifies the retrofitting process, minimizing disruption and facilitating a smooth transition. This adaptability extends to various building types, from residential to commercial and industrial, ensuring that the benefits of GHPs can be harnessed across diverse environments.

10.6 CHALLENGES AND CONSIDERATIONS

10.6.1 Geological and hydrogeological constraints

The integration of GHP systems presents a transformative opportunity for sustainable energy solutions, yet the journey is not without its challenges. Geological and hydrogeological constraints emerge as critical factors that demand careful consideration and strategic planning in the implementation of GHP systems. The success of these systems hinges on understanding the intricate interplay between geological formations, hydrogeological conditions, and system design, in order to overcome limitations and harness the Earth's thermal potential effectively [36].

Geological constraints encompass a range of factors, including soil composition, rock types, and land structure. These elements influence the design of the ground heat exchangers, which are integral to GHP systems. The choice between horizontal and vertical ground loops is influenced by available land area, soil conductivity, and geology. Unsuitable geological conditions, such as rocky terrain or shallow bedrock, can complicate installation and increase costs. Comprehensive geological surveys and subsurface investigations are crucial to identifying challenges early and devising solutions that optimize system performance.

Hydrogeological constraints, related to groundwater availability and quality, also impact GHP implementation. Groundwater serves as a heat source or heat sink in GHP systems, and its temperature and flow rate can affect system efficiency. High groundwater flow rates can impact heat transfer rates and alter subsurface temperatures over time, potentially affecting system performance. Water quality considerations are vital, as corrosive water can damage heat exchanger components. Balancing the sustainable use of groundwater resources with system requirements necessitates a thorough understanding of local hydrogeological conditions and regulatory frameworks [39].

10.7 POLICY RECOMMENDATIONS FOR GEOTHERMAL ENERGY DEVELOPMENT

In the pursuit of a more sustainable energy landscape, policy frameworks play an instrumental role in shaping the trajectory of technology adoption, innovation, and environmental stewardship. As geothermal energy emerges as a potent player in the sustainable energy portfolio, a set of comprehensive policy recommendations can lay the groundwork for the effective development, integration, and utilization of geothermal resources. These recommendations, rooted in a commitment to energy security, environmental conservation, and economic prosperity, offer a roadmap to unlock the full potential of geothermal energy and pave the way toward a greener and more resilient future.

- Financial Incentives and Support Mechanisms
- Renewable Energy Standards and Targets
- Grid Integration and Energy Storage
- Research and Development Funding
- Education and Workforce Development
- Public Awareness and Education
- International Collaboration and Best Practices Sharing

In fact, policy recommendations for geothermal energy development form a roadmap toward a more sustainable and resilient energy future. By prioritizing financial incentives, renewable energy targets, grid integration, research funding, education, and international collaboration, governments can drive the widespread adoption of geothermal energy. These recommendations underscore the transformative potential of geothermal resources in mitigating climate change, enhancing energy security, and fostering economic growth while paving the way for a greener, more sustainable, and inclusive energy landscape.

10.8 CONCLUSION

The exploration into the GHP systems has unearthed a trove of key findings that illuminate their significance as transformative energy solutions. Through an in-depth analysis of technological principles, environmental implications, climatic considerations, and policy frameworks, a mosaic of insights has emerged, highlighting the multifaceted benefits and challenges that define GHP systems' role in the sustainable energy landscape.

At the core of GHP technology lies its remarkable energy efficiency, driven by the utilization of the Earth's consistent thermal reservoirs. The operating principles of GHP systems, including heat transfer and refrigeration cycles, underscore their capacity to provide both heating and cooling, all while minimizing energy consumption and carbon emissions. The classification of GHP systems into closed-loop and open-loop configurations offers design flexibility to suit diverse geological and hydrogeological conditions, expanding their applicability across regions.

The environmental advantages of GHP systems resonate as a resounding testament to their sustainable essence. By tapping into renewable geothermal resources, GHPs exhibit a lower environmental impact than conventional heating and cooling methods, contributing to reduced greenhouse gas emissions and a diminished reliance on fossil fuels. This alignment with climate goals and sustainable development objectives underscores GHP systems' potential to bridge the gap between comfort and environmental consciousness.

Climatic considerations specific to Oman have revealed the nation's vast geothermal potential, rooted in the geological characteristics of the region. The exploration of Oman's climatic context underscores the viability of GHP systems in mitigating energy demand for cooling, enhancing energy security, and curbing carbon emissions. This geographical insight fuels a vision for the integration of GHP systems within Oman's energy portfolio, highlighting the nation's ability to harness its natural resources for a sustainable future.

However, the journey toward GHP adoption is not without its challenges. The intricate interplay between geological and hydrogeological conditions underscores the importance of comprehensive site assessments, efficient ground heat exchanger design, and adept system integration. Technical complexities, ranging from system design and controls to maintenance and troubleshooting, demand a skilled workforce capable of navigating the intricacies of GHP systems.

Policy recommendations serve as guiding lights, unveiling a roadmap for the effective development and integration of geothermal energy. These recommendations underscore the role of financial incentives, renewable energy standards, grid integration strategies, and international collaborations in fostering GHP adoption. Public awareness and education emerge as catalysts for widespread acceptance, shedding light on GHP systems' benefits, dispelling misconceptions, and fostering a culture of sustainability.

10.9 IMPLICATIONS FOR SUSTAINABLE ENERGY DEVELOPMENT IN OMAN

The potential of GHP systems carries profound implications for the trajectory of sustainable energy development in Oman. As a nation characterized by its unique geological and climatic context, Oman stands poised to harness the transformative power of GHPs to reshape its energy landscape, enhance energy security, and advance its sustainability goals. The integration of GHP systems into Oman's energy portfolio holds promise as a catalyst for change, ushering in a new era of energy efficiency, environmental stewardship, and economic growth.

Oman's climatic considerations, particularly its high demand for cooling in response to the region's scorching temperatures, align seamlessly with GHP technology's ability to provide efficient cooling while minimizing electricity consumption. By tapping into the Earth's thermal reservoirs,

GHPs can alleviate peak electricity demand, reducing strain on the grid during periods of high air conditioning use. This has the potential to mitigate energy supply challenges, enhance grid stability, and decrease reliance on conventional fossil fuel-based cooling methods, ultimately contributing to a more sustainable energy mix.

The adoption of GHP systems in Oman carries implications beyond energy efficiency. By reducing carbon emissions and energy consumption, GHPs align with Oman's commitment to mitigate climate change and meet its international climate targets. The integration of GHP systems also presents opportunities for job creation, skills development, and technological innovation. A skilled workforce capable of designing, installing, and maintaining GHP systems can emerge, fostering local expertise and contributing to economic growth.

REFERENCES

[1] Tariq Umar, Geothermal energy resources in Oman, *Proceedings of the Institution of Civil Engineers – Energy*, Volume 171, Issue 1, 2018, 37–43.

[2] Steven Griffiths, A review and assessment of energy policy in the Middle East and North Africa region, *Energy Policy*, Volume 102, 2017, 249–269.

[3] Trent Jacobs, Journal of Petroleum Technology, SLB Evaluating Economic Potential of Oman's Geothermal Prospects, 2022. Available at: https://jpt.spe.org/slb-evaluating-economic-potential-of-omans-geothermal-prospects.

[4] Authority for Electricity Regulation, Oman (AERO), Study on Renewable Energy Resources, AERO, Muscat, Oman, 2008. Available at: https://regulationbodyofknowledge.org/wpcontent/uploads/2013/09/AuthorityforElectricityRegulation_Oman_Study_on.pdf.

[5] Pooya Farzanehkhameneh, M. Soltani, Farshad Moradi Kashkooli, Masoud Ziabasharhagh, Optimization and energy-economic assessment of a geothermal heat pump system, *Renewable and Sustainable Energy Reviews*, Volume 133, 2020, 110282.

[6] Chao Zeng, Yanping Yuan, Fariborz Haghighat, Karthik Panchabikesan, Xiaoling Cao, Li Yang, Ziyu Leng, Thermo-economic analysis of geothermal heat pump system integrated with multi-modular water-phase change material tanks for underground space cooling applications, *Journal of Energy Storage*, Volume 45, 2022, 103726.

[7] Ladislaus Rybach, Shallow systems: Geothermal heat pumps, Editor(s): Ali Sayigh, *Comprehensive Renewable Energy*, Elsevier, Cambridge, MA, 2012, pp. 189–207.

[8] Stuart J. Self, Bale V. Reddy, Marc A. Rosen, Geothermal heat pump systems: Status review and comparison with other heating options, *Applied Energy*, Volume 101, 2013, 341–348.

[9] Ladislaus Rybach, Walter J. Eugster, Sustainability aspects of geothermal heat pump operation, with experience from Switzerland, *Geothermics*, Volume 39, Issue 4, 2010, 365–369.

[10] Jean-Louis Comlan Fannou, Louis Lamarche, Clarence Semassou Guy, Kajl Stanislaw, Antoine Vianou, Cooling mode experimental analysis of a direct expansion geothermal heat pump with a control depending on the discharge pressure, *Applied Thermal Engineering*, Volume 160, 2019, 113955.

[11] Giovanni Russo, Alexandros S. Anifantis, Giuseppe Verdiani, Giacomo Scarascia Mugnozza, Environmental analysis of geothermal heat pump and LPG greenhouse heating systems, *Biosystems Engineering*, Volume 127, 2014, 11–23.

[12] Maryam Karami, Shahram Delfani, Jafar Esmaeelian, Chapter 7 - Recent progress in geothermal heat pumps, Editor(s): Mejdi Jeguirim, *Recent Advances in Renewable Energy Technologies*, Academic Press, Cambridge, MA, 2022, pp. 287–320.

[13] Lili Wang, Xinyu Huang, Masoud Babaei, Zhengguang Liu, Xiaohu Yang, Jinyue Yan, Full-scale utilization of geothermal energy: A high-efficiency CO_2 hybrid cogeneration system with low-temperature waste heat, *Journal of Cleaner Production*, Volume 403, 2023, 136866.

[14] Oguz Arslan, Asli Ergenekon Arslan, Taqiy Eddine Boukelia, Modelling and optimization of domestic thermal energy storage based heat pump system for geothermal district heating, *Energy and Buildings*, Volume 282, 2023, 112792.

[15] Oguz Arslan, Asli Ergenekon Arslan, Irfan Kurtbas, Exergoeconomic and exergoenvironmental based multi-criteria optimization of a new geothermal district heating system integrated with thermal energy storage driven heat pump, *Journal of Building Engineering*, Volume 73, 2023, 106733.

[16] Vittoria Battaglia, Laura Vanoli, Clara Verde, Perumal Nithiarasu, Justin R. Searle, Dynamic modelling of geothermal heat pump system coupled with positive-energy building, *Energy*, Volume 284, 2023, 128557.

[17] Dong Zhang, Su Xinyue, Liu Pengfei, Hui Bo, Hou Gang, Liu Chunyang, Xu Baorui, Zhang Tao, An Zhoujian, Experiment study on startup characteristics and operation performance of PV/T solar assisted heat pump water heater system driven by direct current variable frequency compressor, *Solar Energy*, Volume 263, 2023, 111771.

[18] Vikram Soni, Alireza Darzi, Hannah McPhee, Sepehr Saber, Mohammad Zargartalebi, Jason Riordon, Adnan Ozden, Michael Holmes, Vlad Zatonski, Matthew Toews, David Sinton, Performance analysis of phase change slurries for closed-loop geothermal system, *Renewable Energy*, Volume 216, 2023, 119044.

[19] Ali Ghavidel, Robert Gracie, Maurice B. Dusseault, Design parameters impacting electricity generation from horizontal multilateral closed-loop geothermal systems in Hot Dry Rock, *Geothermics*, Volume 105, 2022, 102469.

[20] Dmytro Rudakov, Oleksandr Inkin, Evaluation of vertical closed loop system performance by modeling heat transfer in geothermal probes, *Geothermics*, Volume 106, 2022, 102567.

[21] Guide to Geothermal Heat Pumps, US Department of Energy, EERE Information Center, 2011. Available at: https://www.energy.gov/sites/prod/files/guide_to_geothermal_heat_pumps.pdf.

[22] Gaosheng Wang, Xianzhi Song, Chao Yu, Yu Shi, Guofeng Song, Fuqiang Xu, Jiayan Ji, Zihao Song, Heat extraction study of a novel hydrothermal open-loop geothermal system in a multi-lateral horizontal well, *Energy*, Volume 242, 2022, 122527.

[23] Willian M. Duarte, Tiago F. Paulino, Sinthya G. Tavares, Antônio A.T. Maia, Luiz Machado, Feasibility of solar-geothermal hybrid source heat pump for producing domestic hot water in hot climates, *International Journal of Refrigeration*, Volume 124, 2021, 184–196.

[24] Sofía Vargas-Payera, Cecilia Ibarra, Nicolás Hurtado, Social and cultural aspects in the adoption of geothermal heat pump systems to replace wood-burning heaters in educational spaces: The Chilean Patagonian case, *Journal of South American Earth Sciences*, Volume 128, 2023, 104426.

[25] Spiros Gkousis, Gwenny Thomassen, Kris Welkenhuysen, Tine Compernolle, Dynamic life cycle assessment of geothermal heat production from medium enthalpy hydrothermal resources, *Applied Energy*, Volume 328, 2022, 120176.

[26] Mehri Akbari Kordlar, Florian. Heberle, and Dieter Brüggemann, Thermo-economic evaluation of a tri-generation system driven by geothermal brine to cover flexible heating and cooling demand, *Geothermics*, Volume 110, 2023, 102678.

[27] Diego Viesi, Antonio Galgaro, Giorgia Dalla Santa, Eloisa Di Sipio, Tomas Garbari, Paola Visintainer, Alberto Zanetti, Raffaele Sassi, Luigi Crema, Combining geological surveys, sizing tools and 3D multiphysics in designing a low temperature district heating with integrated ground source heat pumps, *Geothermics*, Volume 101, 2022, 102381.

[28] Wenjing Lin, Guiling Wang, Haonan Gan, Shengsheng Zhang, Zhen Zhao, Gaofan Yue, Xiting Long, Heat source model for Enhanced Geothermal Systems (EGS) under different geological conditions in China, *Gondwana Research*, Volume 122, 2023, 243–259.

[29] Isaack Kanda, Yasuhiro Fujimitsu, Jun Nishijima, Geological structures controlling the placement and geometry of heat sources within the Menengai geothermal field, Kenya as evidenced by gravity study, *Geothermics*, Volume 79, 2019, 67–81.

[30] Climate Conditions in Ibra, Practical travel advice agency, 2014. Available at: https://hikersbay.com/climate-conditions/oman/ibra/climate-conditions-in-ibra.html?lang=en.

[31] Hyeonsoo Kim, Lars Junghans, Economic feasibility of achieving net-zero emission building (NZEB) by applying solar and geothermal energy sources to heat pump systems: A case in the United States residential sector, *Journal of Cleaner Production*, Volume 416, 2023, 137822,

[32] Alessandro Franco, Fabio Fantozzi, Experimental analysis of a self consumption strategy for residential building: The integration of PV system and geothermal heat pump, *Renewable Energy*, Volume 86, 2016, 1075–1085.

[33] Onder Ozgener, Use of solar assisted geothermal heat pump and small wind turbine systems for heating agricultural and residential buildings, *Energy*, Volume 35, Issue 1, 2010, 262–268.

[34] Beom-Jun Kim, Su-Young Jo, Jae-Weon Jeong, Comparative analysis of sizing procedure for cooling and water heating cascade heat pump applied to a residential building, *Case Studies in Thermal Engineering*, Volume 43, 2023, 102775.

[35] Jeffrey Molavi, James McDaniel, A review of the benefits of geothermal heat pump systems in retail buildings, *Procedia Engineering*, Volume 145, 2016, 1135–1143.

[36] Adela Ramos-Escudero, Isabel C. Gil-García, Maria. Socorro García-Cascales, Angel Molina-Garcia, Energy, economic and environmental GIS-based analysis of shallow geothermal potential in urban areas – A Spanish case example, *Sustainable Cities and Society*, Volume 75, 2021, 103267.

[37] John W. Lund, Aniko N. Toth, Direct utilization of geothermal energy 2020 worldwide review, *Geothermics*, Volume 90, 2021, 101915.

[38] Irfan Ahmad Gondal, Prospects of shallow geothermal systems in HVAC for NZEB, *Energy and Built Environment*, Volume 2, Issue 4, 2021, 425–435.

[39] S Sergio Díaz de Garayo, Alvaro Martínez, David Astrain, Annual energy performance of a thermoelectric heat pump combined with a heat recovery unit to HVAC one passive house dwelling, *Applied Thermal Engineering*, Volume 204, 2022, 117832.

Chapter 11

Green hydrogen production

Methods, designs and applications

Mohit Nayal, Abhishek Kr. Sharma,
Siddharth Jain, and Varun Pratap Singh
University of Petroleum and Energy Studies Dehradun

11.1 INTRODUCTION

Due to rising living standards, population expansion, and industrial development in developing nations, there is an increase in worldwide demand for energy and consumption. By 2030, worldwide energy consumption is expected to have increased by 50%, according to the International Energy Agency. Fossil fuels account for more than 95% of this demand, but they also contribute significantly to environmental pollution and global warming because of the large amount of greenhouse gas emissions. Therefore, the creation of new green energy solutions is crucial for the advancement of sustainability. A rise in the global temperature below 1.5°C is prohibited under the Paris Agreement, which was ratified by 196 countries at COP 21. For ecological sustainability and to reduce carbon emissions, several nations have also adopted ecologically friendly energy regulations. To reduce the release of greenhouse gases from 709.1 million tonnes in 2017 to 536 million tonnes by 2030, South Korea released the 2nd Climate Change Strategy Master Plan in 2019 [1].

Around the world, research institutes and organisations are creating cutting-edge technologies that use renewable resources to produce green energy and fuels. The most prevalent element in the universe, hydrogen, is a viable source of clean, renewable energy. However, it is not easily accessible directly on Earth and is only found in forms of water, fossil fuels, and biomass that have been chemically mixed. To inexpensively and effectively isolate hydrogen from naturally existing chemicals is the major problem [2].

11.2 HYDROGEN PRODUCTION

Hydrogen is a critical element in the quest for sustainable energy solutions, and its production methods have significant environmental implications. Approximately 87 million tonnes of hydrogen are produced annually, but the sources and processes behind this production vary in terms of environmental impact.

DOI: 10.1201/9781003472629-11

The majority of hydrogen production relies on steam reforming natural gas, yielding blue hydrogen, which generates a colossal 830 million tonnes of CO_2 per year. Although efforts are made to capture and store some of this CO_2 underground, a substantial portion is still released into the atmosphere. Grey hydrogen, derived from non-renewable fossil fuels like coal or natural gas through steam reforming or auto-thermal reforming, also emits CO_2 directly into the atmosphere. Brown hydrogen, the most common form of hydrogen, is produced by gasifying hydrocarbon-rich fuels, emitting 10–12 tonnes of CO_2 per tonne of hydrogen. The more environmentally friendly option is green hydrogen, generated via water electrolysis, which splits water into hydrogen and oxygen using electricity. It has no carbon emissions during production [3].

Tables 11.1–11.3 provide valuable insights into the different hydrogen colour, their associated technologies, costs, and carbon emissions. These data-driven tables help in understanding the environmental and economic impacts of hydrogen production methods, highlighting the importance of transitioning towards cleaner and more sustainable practices, particularly green hydrogen, in our pursuit of a low-carbon energy future.

Green hydrogen, produced through water electrolysis powered by renewable sources such as solar and wind, offers a compelling solution to combat environmental pollution and reduce greenhouse gas emissions. Water electrolysis, depicted in Figure 11.1, efficiently splits water into hydrogen and oxygen while harnessing clean energy. This approach also addresses energy supply-demand imbalances, enabling surplus renewable electricity to be stored as hydrogen. Beyond its role as a primary energy source, hydrogen serves as a versatile energy carrier, powering fuel cell vehicles and bolstering the industrial sector. It finds applications in chemical and petrochemical industries, facilitating ammonia and synthetic fuel production. Notably, hydrogen storage devices exhibit superior efficiency compared to batteries [5].

Table 11.1 Hydrogen colour shades and their technology, cost, and CO_2 emissions [4]

Hydrogen colour	Technology	Source	Products	CO_2 emission	Cost ($ kg/H_2)
Green hydrogen	Electrolysis	Water	$H_2 + O_2$	Minimal	3.6–5.8
Blue hydrogen	Reforming + carbon capture	Nature gas	$H_2 + CO_2$ (captured 80%–90%)	Low	1.5–2.9
Brown hydrogen	Gasification	Brown coal (lignite)	$H_2 + CO_2$	High	1.2–2.1
Black hydrogen	Gasification	Black coal (bituminous)	$H_2 + CO_2$	High	1.2–2.1
Grey hydrogen	Reforming	Natural gas	$H_2 + CO_2$ (released)	Medium	1–2.1

Table 11.2 Technical characteristics of typical water electrolysis technology [7]

	Alkaline	PEM	Solid oxide
Anode reaction	$2OH^- \rightarrow H_2O + \frac{1}{2}O_2 + 2e^-$	$H_2O \rightarrow 2H^+ + \frac{1}{2}O_2 + 2e^-$	$O^{2-} \rightarrow \frac{1}{2}O_2 + 2e^-$
Cathode reaction	$2H_2O + 2e^- \rightarrow H_2 + 2OH^-$	$2H^+ + 2e^- \rightarrow H_2$	$H_2O + 2e^- \rightarrow H_2 + O^{2-}$
Overall cell	$H_2O \rightarrow H_2 + \frac{1}{2}O_2$	$2H_2O \rightarrow H_2 + \frac{1}{2}O_2$	$H_2O \rightarrow H_2 + \frac{1}{2}O_2$
Electrolyte	KOH/NaOH (5 M)	Solid-polymer-electrolyte (PFSA)	Yttria-stabilized-zirconia (YSZ)
Separator	Asbestos/zirfon/Ni	Nafion	Solid electrolyte YSZ
Electrode/catalyst (hydrogen side)	Nickel-coated perforated stainless steel	Iridium oxide	Ni/YSZ
Electrode/catalyst (oxygen side)	Nickel-coated-perforated stainless steel	Platinum carbon	Perovskites (LSCF, LSM) (La, Sr, Co, FE) (La, Sr, Mn)
Gas diffusion layer	Nickel mesh	Titanium mesh/carbon cloth	Nickel mesh/foam
Bipolar plates	Stainless steel/nickel-coated stainless steel	Platinum/gold-coated titanium or titanium	Cobalt coated stainless steel
Nominal-current density	0.20–0.80 A/cm²	1.0–2.0 A/cm²	0.30–1.0 A/cm²
Voltage-range (limits)	1.40–3.0V	1.40–2.50V	1.0–1.5V
Operating temperature	70°C–90°C	50°C–80°C	700°C–850°C
Cell-pressure	<30.0 bar	<70.0 bar	1.0 bar
H2 purity	99.500%–99.9998%	99.900%–99.9999%	99.900%
Efficiency	50.0%–78.0%	50.0%–83.0%	89.0% (laboratory)
Lifetime (stack)	60,000 hours	50,000–80,000 hours	20,000 hours
Development-status	Mature	Commercialized	R & D
Electrode-area	10,000–30,000 cm²	1500 cm²	200 cm²
Capital costs (stack) minimum 1.0 MW	$270/kW	$400/kW	>$2000/kW
Capital costs (stack) minimum 10.0 MW	$500–1000/kW	$700–1400/kW	Unknown

Table 11.3 Advantage and disadvantage of typical water electrolysis technology [22]

Technology	Advantages	Disadvantages
Alkaline electrolysis	• Well-established technology • For industry application • Nobel metal-free electrocatalyst • Relativity low cost • Long-term stability	• Limited current densities • Crossover by the gases • High concentrated (5 M KOH) liquid electrolyte
PEM electrolysis	• Commercialized technology • Operates higher current densities • High purity of the gases • Compact system design • Quick response	• Cost of the cell components • Noble metal electrocatalyst • Acidic electrolyte
Solid oxide electrolysis	• Excellent efficiency (80%) • High working temperature	• Limited stability • Under development

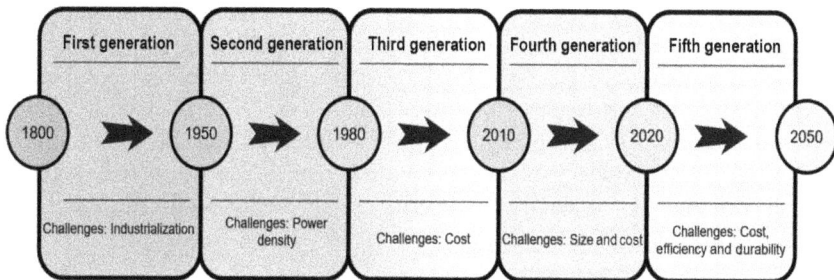

Figure 11.1 Generation of water electrolysis development.

11.3 ELECTROLYSIS: MECHANISMS, TECHNOLOGIES, AND EFFICIENCY ENHANCEMENT

Water electrolysis, a clean electrochemical process, is harnessed to produce green hydrogen by splitting water into hydrogen and oxygen at ambient temperature through the interaction of H_2O with electricity and heat. However, due to economic limitations, only 4% of global hydrogen production is achieved through this method, with the majority being a by-product of the chlor-alkali industry. Notably, the fertilizer sector boasts the largest electrolysis unit. This technology has evolved through approximately five generations, marked by innovations, challenges, and changing significance. Four key types of water electrolysis technologies have emerged: Alkaline, PEM (Proton Exchange Membrane), and solid oxide electrolysis. Each offers unique features, advantages, and drawbacks while adhering to the same fundamental principles, contributing to the dynamic landscape of green hydrogen production [6].

11.3.1 Alkaline electrolysis

Alkaline water electrolysis has a rich history dating back to 1789 when Troostwijk and Diemann introduced this technique for industrial hydrogen production. Over 400 units have been installed and operated worldwide, demonstrating its commercial viability. This method utilizes a concentrated alkaline solution (5 M KOH/NaOH) and operates within a relatively lower temperature range (30°C–80°C). Nickel-coated stainless steel electrodes and asbestos/ZrO_2-based diaphragms serve as separators, and hydroxyl ions (OH) are the ionic charge carriers. The porous diaphragm structure facilitates the electrochemical reaction by allowing the flow of KOH/NaOH and water. Alkaline water electrolysis exhibits an investment cost ranging from USD 500 to 1000 per kW, with a lifespan of approximately 90,000 hours [8]. However, it faces challenges related to low current densities due to moderate OH mobility and the use of corrosive KOH electrolytes. Sensitivity to ambient CO_2 and the subsequent formation of K_2CO_3 salt decrease the quantity of hydroxyl ions and ionic conductivity, resulting in reduced hydrogen generation. K_2CO_3 salt also restricts the anode gas diffusion layer's pores, hindering ion transferability across the diaphragm. As a consequence, alkaline water electrolysis yields gases of lower quality (99.9%), primarily hydrogen and oxygen. Key components in this technology include diaphragms/separators, current collectors (gas diffusion layers), separator plates (bipolar plates), and end plates, each contributing to its operational efficiency and challenges [9] (Figure 11.2).

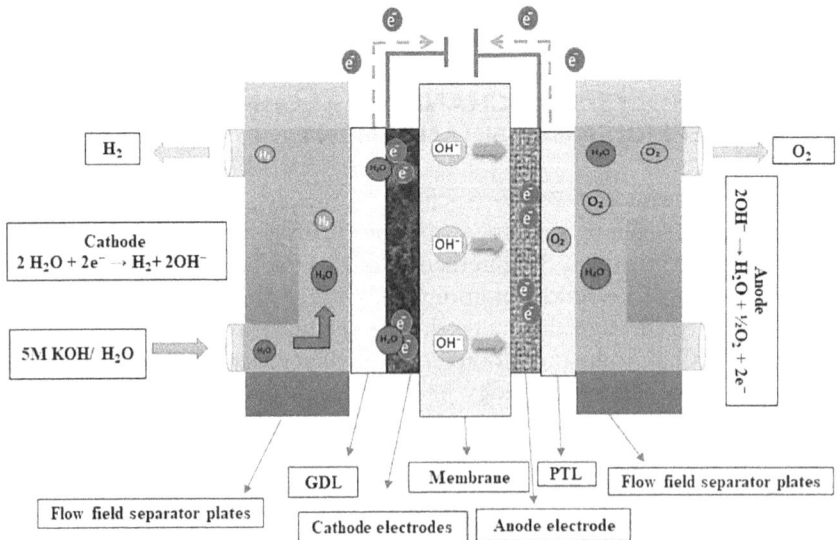

Figure 11.2 Schematic illustration of alkaline water electrolysis working principle.

Alkaline water electrolysis, a well-established industrial technique, is pivotal in the pursuit of cost-effective and efficient hydrogen production. Researchers and organizations are continuously working to enhance its performance and lower production costs, with several noteworthy developments in recent years. These advancements underscore the potential of alkaline water electrolysis in reducing hydrogen production costs and enhancing efficiency. Future research should continue to focus on novel electrode materials, separators, and the integration of this technology with renewable energy sources to further propel the hydrogen economy [10].

11.3.2 PEM electrolysis

PEM water electrolysis, introduced by General Electric Co. in 1966, effectively addresses the limitations of alkaline water electrolysis. This method utilizes a sulfonated polymer membrane as an electrolyte, permitting the passage of H^+ ions and deionized (DI) water through the proton-conducting membrane, serving as ionic charge carriers. This approach yields exceptionally pure gases (99.999% purity) of hydrogen and oxygen, operates at lower temperatures (30°C–80°C), and achieves higher current densities. The superior kinetics of the hydrogen evolution reaction in PEM water electrolysis, attributed to the highly active Pt electrode surface and lower electrolyte pH, contribute to its efficiency compared to alkaline water electrolysis. PEM water electrolysis boasts safety advantages stemming from the absence of caustic electrolytes and a smaller footprint [11]. Numerous global water electrolyzer manufacturers have developed large-scale PEM water electrolyzer for diverse industrial and transportation applications. These systems offer claimed stability of 60,000 hours with minimal performance degradation and an ultimate goal of achieving 100,000 hours of stability. Nonetheless, a significant challenge associated with PEM water electrolysis remains the high cost of components such as electrode materials, current collectors, and bipolar plates [12].

Key components in PEM water electrolysis include Nafion as the commonly used membrane, recognized for its high proton conductivity, current density, mechanical strength, and chemical stability. Noble metal-based electrocatalysts, such as IrO_2 for the Oxygen Evolution Reaction (OER) and carbon-supported Pt for the Hydrogen Evolution Reaction (HER), are employed in the anode and cathode electrode materials. Porous titanium/titanium mesh and carbon cloth serve as anode and cathode gas diffusion layers, while different flow field designed bipolar plates are used as separators and endplates, with straight parallel flow field designed separator plates demonstrating superior performance in PEM water electrolyzer [13] (Figure 11.3).

PEM water electrolysis technology continues to play a significant role in industrial and transportation applications, with large-scale producers offering capacities up to megawatt (MW) levels. However, the cost of hydrogen

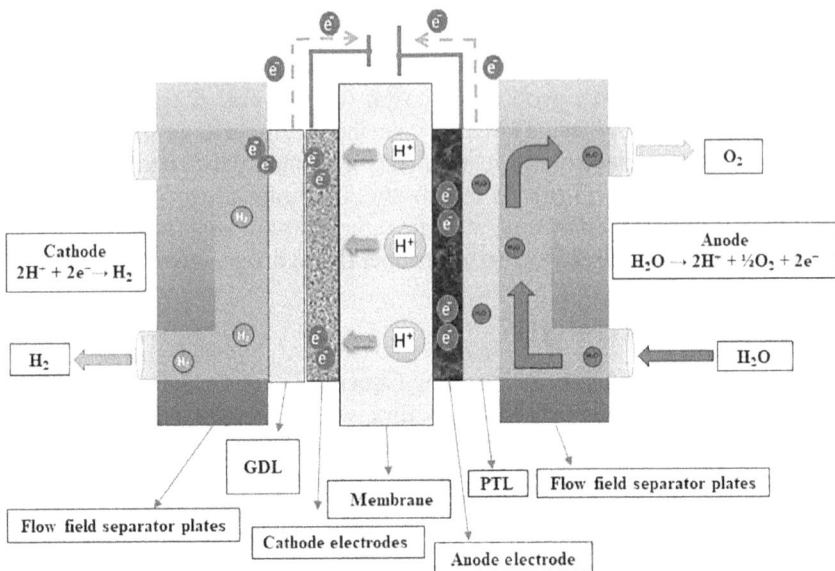

Figure 11.3 Schematic view of PEM water electrolysis working principle.

generation remains a concern, currently ranging between \$700 and 1400 per kilowatt-hour (kW/H_2). Addressing this issue necessitates overcoming various challenges, such as reducing or substituting platinum group metals with more cost-effective transition metals, eliminating expensive coatings on bipolar plates, and developing new bipolar plates using economical materials [14].

Researchers from both academic and commercial institutions are actively engaged in innovative efforts, with a specific focus on cost reduction. For instance, some researchers have devised integrated electrode designs featuring platinum nanowires on ultrathin titanium gas diffusion layers for hydrogen evolution reactions in PEM water electrolysis. The resulting ultrathin titanium integrated PtNW electrode (PtNW/Ti) demonstrated a reduced over potential of 63 mV at a higher current density of $100 mA/cm^2$. Additionally, as a cathode electrode in a PEM water electrolyzer, it exhibited a lower cell voltage of 1.643 V at 1 A/cm^2 and a remarkable efficiency of 90.08%, requiring significantly fewer catalyst loadings compared to typical Membrane Electrode Assemblies (MEAs) [15].

In 2020, a new titanium-doped molybdenum phosphide (Ti-MoP) electrocatalyst for the hydrogen evolution process in an acidic electrolyte was introduced. The Ti-MoP electrocatalyst outperformed undoped MoP, with a mere 81.5 mV overpotential at $10 mA/cm^2$. Another development in 2021 involved a platinum-free cobalt phosphide (Co-P) electrode for the hydrogen evolution process, demonstrating superior electrochemical performance with an overpotential of 143.85 mV at a current density of $10 mA/cm^2$.

Continuing in 2021, a graphene-encapsulated noble metal-free NiMo alloy was identified as an acid-stable HER catalyst. During a 10,000-cycle potential cycling test, the N-doped graphene (4–8 layers) enclosed NiMo cathode catalyst exhibited exceptional stability. Moreover, in the same year, a transition metal phosphide catalyst supported on carbon black (FeP/CB) was proposed as an alternative to platinum-based catalysts for HER. The synthesized 50% FeP/CB catalyst demonstrated improved electrocatalytic activity, with an overpotential of 252 mV at a current density of 10 mA/cm^2, comparable to the 220 mV of the platinum catalyst [16].

In the realm of the OER, a 2021 study introduced an electrode with an iridium film adorned with WOx Nano rods (Ir@WOxNRs). This electrode displayed outstanding electrochemical performance and stability, with a reduced noble metal loading of 0.14 mg/cm^2 during a 1030-hour stability test at a constant current density of 0.5 mA/cm^2. In addition, researchers in 2021 synthesized nitrogen-deficient graphite carbon nitride (N-CN) supported iridium oxide (IrO$_2$/N-CN) to enhance the OER activity of IrO$_2$ in PEM water electrolysis. This IrO$_2$/N-CN electrocatalyst surpassed bare IrO$_2$ in terms of mass activity and specific activity and exhibited satisfactory stability during a 300-hour durability test at a constant current density of 1.6 A/cm^2 [17].

Further developments included the creation of a high-performance IrOx/W-TiO$_2$ electrocatalyst for noble metal reduction in PEM water electrolysis, which outperformed undoped IrO$_2$ in terms of Ir mass activity and stability. Researchers also added magnetic Fe$_3$O$_4$ to IrO$_2$ to improve bubble separation during OER. Furthermore, a core-shell nanostructured catalyst of Ru@Ir-O with tensile stresses and oxygen insertion in the Ir shell demonstrated 78-fold greater mass activity than commercial IrO$_2$ at 1.55 V in 0.5 M H$_2$SO$_4$ electrolyte. Finally, a low-Ir core-shell OER electrocatalyst with an intermetallic IrGa (IrGa-IMC) core and a partly oxidized Ir (IrOx) shell achieved a decreased over potential and Tafel slope. Moreover, a straightforward electrodeposition strategy was utilized in 2022 to create a WO$_3$ nano-array electrode with a heterogeneous IrRu coating, exhibiting remarkable electrocatalytic performance and stability in acidic conditions. The Ir$_{3.2}$Ru1.6@WO$_3$ electrode displayed a competitive over potential of 245 mV at 10 mA/cm^2, surpassing commercial IrO$_2$. Single-cell PEM water electrolysis tests revealed an extraordinarily high current density of 4.5 A/cm^2 at a cell voltage of 2.13 V and reduced Ir loading of 115 g/cm^2. Durability investigations extended over 500 hours at a constant current density of 0.5 mA/cm^2, further confirming the promising performance of this technology [18].

11.3.3 Solid oxide electrolysis

The solid oxide water electrolysis cell (SOEC) is an electrochemical conversion device that was originally developed in the 1970s by General Electric, the Brookhaven National Laboratory, and the German company Dornier in the USA. Operating at high temperatures, the SOEC reduces power consumption and enhances energy efficiency, resulting in a significant reduction in hydrogen production costs. Key advantages of the SOEC include its high operating temperature, favourable thermodynamics and reaction kinetics, and seamless thermal integration into downstream chemical synthesis processes. Moreover, it offers impressive conversion efficiency and eliminates the need for electrocatalysts made of noble metals. However, the commercialization of SOEC has been hindered by limited long-term stability, with reported stability for yttria-stabilized zirconia thin electrolytes being only 20,000 hours, which remains a challenge in advancing this technology. Efforts are ongoing to improve the durability and longevity of SOEC systems for wider adoption in the hydrogen production sector [19].

11.3.3.1 Working principle of solid oxide water electrolysis

Solid oxide water electrolysis (SOEC) typically produces green hydrogen and oxygen at elevated temperatures using steam as the source of water. Initially, at the cathode, water molecules are reduced into hydrogen (H_2) and oxide ions (O_{2-}) with the addition of two electrons. The remaining oxide ions and the hydrogen generated are then transported through an ion exchange membrane to the anode. At the anode, oxide ions are further reduced to form oxygen, releasing electrons that travel through an external circuit back to the cathode due to the cathode's positive attraction. SOECs comprise two porous electrodes (anode and cathode) and a dense ceramic electrolyte with the capability to conduct oxide ions. The most common electrolyte used is yttria-stabilized zirconia (YSZ), renowned for its high ionic conductivity, and excellent chemical and thermal stability. State-of-the-art hydrogen (cathode) electrode materials include a ceramic metal composition of YSZ and nickel (Ni-YSZ), which serves as a non-noble metal catalyst with high electronic conductivity. These advancements enable efficient and sustainable hydrogen production in SOEC systems [20] (Figure 11.4).

Solid oxide water electrolysis (SOEC) technology is undergoing active research and development with an eye toward commercialization. One notable trend is the use of non-noble metal electrocatalysts at higher operating temperatures, which enhances energy efficiency. Researchers have been focusing on addressing challenges related to long-term stability, hydrogen production costs, and overall cost reduction. Recent developments include the creation of cobalt-free perovskite materials, hybrid solid oxide electrolysis cells, and the utilization of Mn-doped Ruddlesden-Popper oxide $La_{1.5}Sr_{0.5}NiO_{4+}$ (LSNMx) and layered perovskite-based rare earth nickelates, along with composite oxide materials for hydrogen electrodes. Notably, the Pr_2NiO_{4+} (PNO) electrode exhibited improved electrochemical performance, though it experienced accelerated degradation at higher current densities. To mitigate this issue, researchers have successfully enhanced

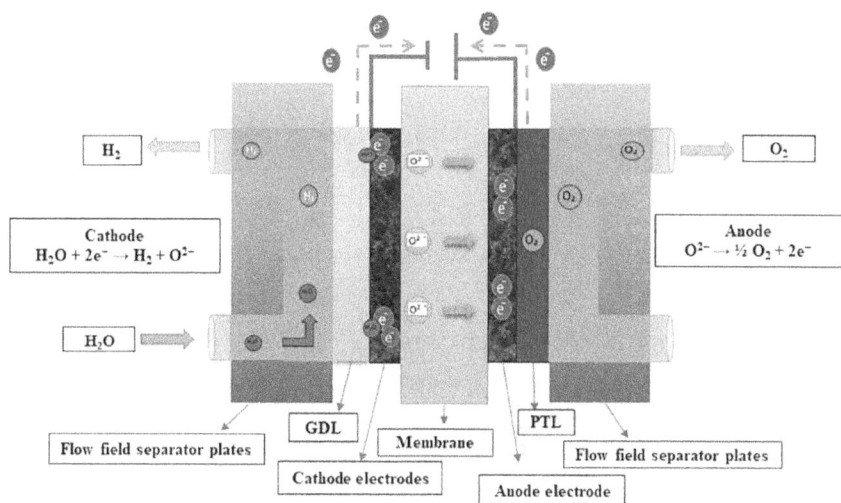

Figure 11.4 Schematic view of solid oxide water electrolysis working principle.

electrochemical efficiency and stability at higher current densities by replacing cobalt in the modified $Pr_2Ni_{0.8}Co_{0.2}O_{4+}$ electrode. These advancements underscore the ongoing progress in SOEC technology [21].

In conclusion, solid oxide water electrolysis technology has the potential to produce hydrogen and deliver high energy efficiency. To accomplish these objectives, though, issues like long-term stability, cost-cutting, and the usage of sources of renewable energy must be addressed [23].

11.4 BIOMASS CONVERSION FOR GREEN HYDROGEN PRODUCTION

Biomass, a renewable primary energy source, comprises energy crops, crop residues, forest wood, grass, industrial and forest residues, animal and municipal waste, and various organic materials. Plants utilize photosynthesis to store solar energy in chemical bonds, forming biomass. While using biomass for energy production releases CO_2, the emissions are roughly equivalent to what the organisms absorbed during their lifetimes. Hydrogen production from biomass involves two methods: thermochemical and biological processes. Biological processes offer higher energy efficiency and environmental friendliness but yield lower hydrogen rates and molar yields (mol H_2/mol feedstock) depending on the feedstock. Thermochemical processes like gasification, on the other hand, are more economically and environmentally advantageous due to their faster rates and larger stoichiometric hydrogen output [24].

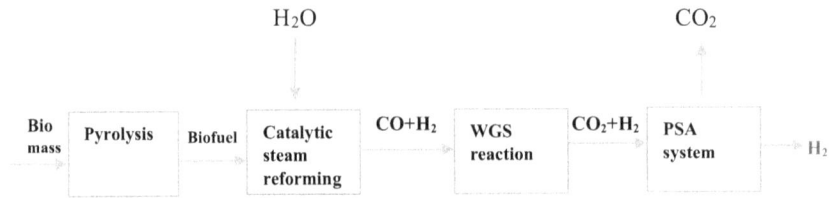

Figure 11.5 Flow diagram of the biomass pyrolysis process.

11.4.1 Thermochemical processes

Biomass conversion into hydrogen and hydrogen-rich gases remains pivotal for sustainable development, primarily through advanced thermochemical processes. Pyrolysis and gasification are key methods that yield gaseous by-products, such as CH_4 and CO, and offer pathways for further hydrogen production via steam reforming and the Water-Gas Shift (WGS) reaction [25]. In contrast, combustion and liquefaction are less ideal due to lower hydrogen generation and the production of undesirable by-products [26] (Figure 11.5).

Through biomass heating at temperatures between 650 and 800 K and pressures of 0.1–0.5 MPa, this process results in the production of liquid oils, solid charcoal, and gaseous compounds. Hydrocarbon gases, including methane, can be steam reformed, and the WGS reaction is utilized to enhance hydrogen production. Following the conversion of CO to CO_2 and H_2, pure hydrogen is obtained [27]. The gasification of biomass in various media, including air, oxygen, or steam, results in the production of gaseous fuel (syngas). Reactor types and operating conditions are determined by the gasifying agent's flow rate and velocity, typically ranging from air to 33 bar and temperatures between 500°C and 1400°C. The yield of hydrogen from biomass pyrolysis is influenced by factors such as feedstock type, catalyst choice, temperature, and residence time [28]. Steam gasification, with an overall efficiency of up to 52%, surpasses fast pyrolysis in terms of hydrogen production rates. For a plant with an expected daily hydrogen output of 139,700 kg and biomass costs ranging from $46 to $80 per dry tonne, the projected hydrogen production cost is estimated at $1.77–$2.05/kg, aligning with the conventional route of biomass gasification, steam reforming, and Pressure Swing Adsorption (PSA) purification. These advancements demonstrate the potential for cost-effective and sustainable hydrogen production from biomass [29].

11.4.2 Biological routes

The increased emphasis on sustainable development and waste reduction has led to an increase in research into biological hydrogen generation. The majority of biological activities function at room temperature and pressure,

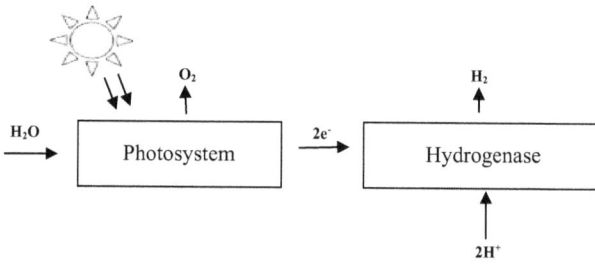

Figure 11.6 Schematic of the direct bio-photolysis process.

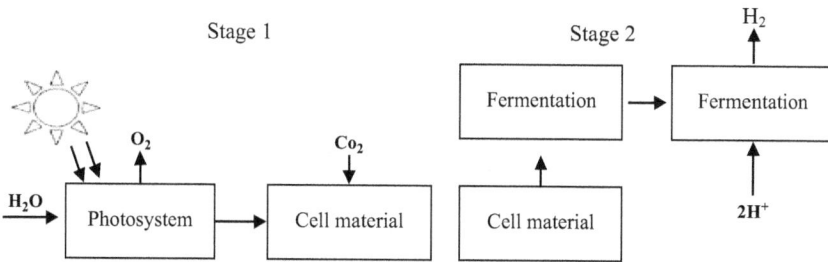

Figure 11.7 Schematic of the indirect bio-photolysis process.

which uses renewable energy sources and makes them less energy-intensive. By utilizing a variety of waste products as feedstock, they also help with trash recycling. Direct and indirect bio-photolysis, photo and dark fermentations, and multi-stage or sequential dark and photo-fermentation are important biological processes utilized to produce hydrogen gas [30] (Figures 11.6 and 11.7).

In the latest advancements, green and blue-green algae have been harnessed for their ability to perform direct and indirect bio-photolysis, facilitating the breakdown of water molecules into hydrogen ions and oxygen. Green algae utilize photosynthesis to initiate this process, generating hydrogen ions and oxygen. The hydrogenase enzyme then transforms these components into hydrogen gas, a phenomenon referred to as direct bio-photolysis. To maintain an oxygen-sensitive environment, it is crucial to regulate oxygen concentrations below 0.1%. Researchers have identified mutants derived from microalgae with reduced pigment concentrations, lower chlorophyll levels, and heightened oxygen tolerance, resulting in increased hydrogen production. These findings contribute to the ongoing efforts to optimize algae-based hydrogen generation [31] (Figure 11.8).

Recent progress in bio-hydrogen generation technologies involves cyanobacteria or blue-green algae synthesizing both hydrogenase and nitrogenase enzymes, enabling indirect bio-photolysis for hydrogen production

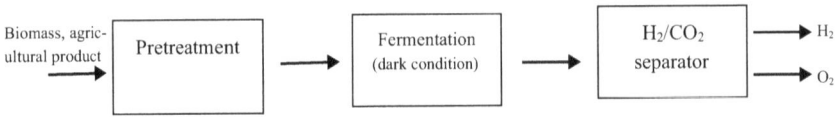

Figure 11.8 Schematic of the dark fermentation process.

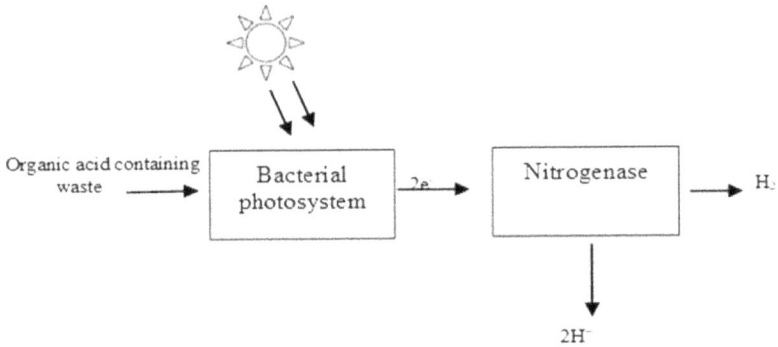

Figure 11.9 Schematic of the photo-fermentation process.

from water. While still in the conceptual phase, the anticipated manufacturing cost is approximately $10/GJ or $1.42/kg of H_2. This approach utilizes water as a renewable resource and mitigates CO_2 emissions, presenting a cost-effective and sustainable solution. However, limitations include modest hydrogen production potential, the necessity for a large surface area for light capture, and limited waste utilization [32].

Biochemical fermentation processes are used to convert organic feedstocks into alcohols, acetone, and hydrogen. These processes are appealing for bio-hydrogen production because they utilize waste materials, offer cost-effective energy production, and address waste treatment. Dark fermentation, accounting for over 80% of total end products, involves anaerobic bacteria working on carbohydrate-rich substrates under anoxic conditions. While glucose is the preferred feedstock, its cost and limited availability have led to the use of cellulose and starch-containing materials as alternatives [32] (Figure 11.9).

Since the amount of hydrogen generated relies on pH, keeping a pH between 5 and 6 is necessary for optimal production. Dark fermentation, on the other hand, is a straightforward procedure that doesn't need a lot of space because it doesn't rely on light sources. It is possible to constantly create hydrogen from a variety of potentially useful substrates, such as garbage and rubbish, day and night [33].

Another biochemical activity that utilizes solar energy and organic acids in nitrogen-deficient environments is photo-fermentation. Some photosynthetic

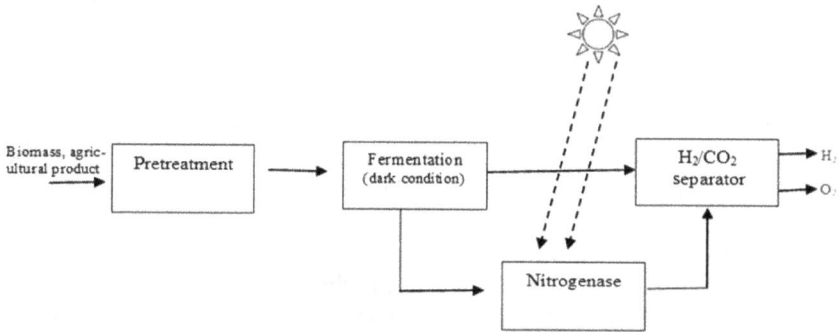

Figure 11.10 Diagram the progression of the dark and photo-fermentation processes.

bacteria may transform organic acids to generate hydrogen and carbon dioxide, but the yield and production rate of H_2 are stimulated by increasing light intensity. The main obstacles to dark fermentation's competition, however, are low solar energy conversion rates, the need for complex anaerobic photo-bioreactors covering wide regions, and the scarcity of organic acids [34].

The latest advancements in hydrogen production technology involve hybrid systems that combine anaerobic and photosynthetic bacteria. These systems achieve higher hydrogen production yields with reduced light energy requirements. This approach is often referred to as sequential dark/photo-fermentation, indicating a two-stage process where dark and photo-fermentation are combined. Key factors influencing hydrogen (H_2) production include temperature, pH levels, and the presence of organic acids. This integrated approach holds promise for enhancing the efficiency and sustainability of hydrogen production processes [35] (Figure 11.10).

11.5 SOLAR-DRIVEN HYDROGEN PRODUCTION METHODS

Due to the typical water electrolysis system's dependence on non-renewable fossil fuel sources for electricity, it has been thought that solar-driven water electrolysis is a potential method for producing green hydrogen [36].

11.5.1 Photovoltaic-electrolysis systems

A thermoelectric electrolysis system (PV-EC) is a device that uses a photovoltaic (PV) cell to generate electricity, which is then used to split water molecules into hydrogen and oxygen gases. A PV cell converts sunlight into direct current (DC) to feed an electrochemical cell. An electrochemical cell uses a direct current to drive a water-splitting reaction to produce hydrogen and oxygen gases at the electrodes [37].

PV cells in PV-EC systems are typically made of silicon, but other materials such as gallium arsenide and copper indium gallium selenide (CIGS) can also be used. Solar cells are typically around 20% efficient, but can be even higher with newer technologies. An electrochemical cell in a PV-EC system is a device that uses electricity to split water molecules into hydrogen and oxygen gases [38,39]. A cell typically consists of two electrodes, a cathode and an anode, immersed in an electrolyte solution. Hydrogen gas is evolved at the cathode and oxygen gas at the anode. Electrochemical cells are typically around 80% efficient, but can be higher with certain materials and designs. A pump in the PV-EC system circulates water between the PV cells and the electrochemical cells. This ensures that the water is always fresh and free of contaminants. The purity of water used in PV-EC systems is important as contaminants can affect the efficiency of the system. The water should be free of dissolved solids, organics, and bacteria [40].

PV-EC system is a promising technology for hydrogen production from solar power. They are relatively efficient, scalable, and modular. It is also a clean and renewable source of hydrogen production. However, PV-EC systems are still relatively expensive and require high water purity. Continued research and development are expected to improve the efficiency and cost-effectiveness of PV-EC systems, making them a more viable option for hydrogen production [41].

11.5.2 Photo-electrochemical cells

A photo-electro-chemical cell (PEC) is a device that uses light to split water molecules into hydrogen and oxygen gases. A PEC consists of two electrodes, a photoanode and a cathode, immersed in an electrolyte solution. A photoanode is made of a semiconductor material that absorbs sunlight and creates an electric field that drives the water-splitting reaction. Cathodes are usually made of metals that catalyze the reduction of oxygen. The efficiency of PEC is determined by the efficiencies of the photoanode, electrolyte, and cathode. Photoanode efficiencies are typically around 10%, but can be higher with certain materials and designs. Electrolyte efficiencies are typically around 80%, but can be lower depending on the material. Cathode efficiency is typically around 90%, but can be lower depending on the material. The cost of PEC is determined by the cost of the photoanode, electrolyte, cathode, and other components [42]. Photoanode cost is the most important factor and is expected to drop as the technology matures. The purity of water used in PEC is important as contaminants can affect the efficiency of the system. The water should be free of dissolved solids, organics, and bacteria.

Photo-electrochemical (PEC) technology shows promise for hydrogen production using solar power, offering advantages such as relative efficiency, scalability, and modular design for renewable hydrogen generation by harnessing sunlight to split water into hydrogen and oxygen gases with up to a 40% efficiency rate. Nevertheless, challenges include its relatively high cost, the need for high-purity water, sensitivity to water contaminants affecting efficiency, and potential material degradation over time. To make PEC systems more efficient, cost-effective, and durable for large-scale deployment, ongoing research and development efforts are crucial, as they work towards unlocking the full potential of PEC in solar-powered hydrogen production [43].

11.6 SYSTEM DESIGN CONSIDERATIONS FOR GREEN HYDROGEN PRODUCTION

The design of a green hydrogen production system requires meticulous consideration of various factors to ensure its environmental efficiency and long-term sustainability. The primary objective is to reduce carbon emissions, making the use of clean energy sources critical. Hydropower, wind power, and solar energy are preferred choices due to their low environmental impact. These energy sources facilitate the efficient conversion of water into oxygen and hydrogen using technologies like electrolysis. Efficiency, durability, and scalability of electrolyzer are key factors in system design. Additionally, distribution and storage are of equal importance. Hydrogen, with its low energy density, must be safely and cost-effectively stored using methods such as compressed liquid or gaseous storage. Compatibility with existing infrastructure, including pipelines, storage facilities, and refuelling stations, is crucial for seamless integration into energy and transportation systems.

Lifecycle analyses are essential to assess the environmental impact and identify areas for improvement within the system. Economic viability is also a critical consideration, weighing long-term benefits against initial capital and ongoing operational costs. Increasing electrolyzer utilization through demand-side control and synergistic connections can enhance economic viability. The impact of policy and regulatory frameworks is significant, with supportive regulations, incentives, and carbon pricing mechanisms accelerating implementation and attracting capital. International collaboration fosters technology transfers and joint research initiatives, contributing to system design by enabling the sharing of diverse experiences and expertise [44].

11.6.1 Process integration and optimization

Process integration and optimization are vital engineering approaches that systematically enhance the efficiency, sustainability, and overall effectiveness of industrial processes. These methods involve in-depth analysis, redesign, and reconfiguration to minimize resource consumption, waste generation, and energy usage, all while increasing productivity and profitability. Process integration delves into the complex interrelationships among different process units or stages, aiming to leverage synergies for improved overall performance. It primarily focuses on identifying opportunities for the exchange of heat, mass, and energy within a manufacturing system. By carefully balancing the heat and material requirements of various process units, energy losses are reduced, potentially through techniques like energy recovery and combined heat and power generation.

Process optimization concentrates on streamlining specific process components for optimal efficiency. This can entail selecting ideal operating conditions, developing superior catalysts or materials, and enhancing reaction kinetics while considering the economic viability of changes. Engineers rely on data analysis tools, simulation software, and mathematical modelling during the integration and optimization processes, aiding in understanding the intricate relationships between various process parameters and making informed decisions for system reconfiguration or improvement. These methods find application across diverse industries, from manufacturing and petrochemicals to energy generation. For example, they can lead to the development of more energy-efficient refineries and reduced greenhouse gas emissions in the oil and gas sector. In the food industry, process integration can result in decreased energy consumption and waste generation, aligning with sustainability goals. An essential advantage of process integration and optimization is their holistic perspective, which not only benefits ecological and sustainable operations but also enhances economic performance. By identifying areas of waste and inefficiency, companies can make targeted investments in technologies and procedures that yield long-term benefits [45].

11.6.2 Techno-economic analysis

A technical project or process can be assessed for its viability, economic potential, and other factors using a complete assessment methodology called a "techno-economic analysis" (TEA). To give information on the technical efficiency, cost structure, and prospective profitability of a venture, it integrates technical and economic aspects. The technical aspect of a TEA entails a careful analysis of the technology's engineering, design, and operating components. This entails assessing elements including production capacity, material needs, energy usage, and process efficiency. Technical factors aid in defining the fundamental criteria that affect the project's overall performance. On the financial side, TEA explores the costs related to the equipment or process. It entails predicting both the operational costs, such as labour, maintenance, services, and raw materials, as well as the capital costs needed for infrastructure, equipment, and facilities. Additionally, based on variables like pricing of goods, market demand, and prospective sales volume, revenue predictions are created. Technical and economic data are combined to create TEA, which offers a thorough perspective of the full project lifespan [46]. It assists in locating important cost factors, potential bottlenecks, and possibilities for improvement. Additionally, TEA can be used to assess the effects of various design decisions, compare various technological solutions, and calculate the point of break-even or return on investment. In the decision-making processes for numerous companies and sectors, TEA can be quite important. For instance, in the field of renewable energy, TEA assists in determining whether solar, wind, or bioenergy projects are financially feasible by taking installation costs, energy production, and possible revenue streams into account [47]. TEA can be used in chemical engineering to evaluate the commercial viability of novel manufacturing techniques or the creation of novel materials [48].

11.7 APPLICATIONS OF GREEN HYDROGEN: ENERGY STORAGE AND CLEAN FUEL SYNTHESIS

Green hydrogen, which is generated from sources that are renewable and water electrolysis, has important applications in storing energy and clean fuel production. Green hydrogen, as a versatile energy carrier, addresses the intermittent nature of energy from renewable sources by keeping excess energy during peak generation periods and releasing it when needed. This allows for grid stabilization and promotes the incorporation of greater proportions of renewable into the energy mix. Green hydrogen can also be used as a clean fuel in a variety of industries, including transportation and industrial. It can power fuel cell electric vehicles, which have larger ranges and faster refilling than battery electric vehicles. Green hydrogen replaces fossil fuels in businesses such as refining and ammonia production, lowering emissions and promoting environmental goals. Its ability to be turned into synthetic fuels such as e-fuels opens the door to decarbonizing aircraft, shipping, and heavy-duty transportation. As efforts to address climate change grow, green hydrogen's numerous applications position it as a critical component of the transition to a low-carbon energy future [49].

11.7.1 Hydrogen energy storage

Hydrogen storage of energy has emerged as a possible answer to the issues faced by renewable energy sources' intermittent nature and the need for system stability. This novel strategy entails transforming excess energy from renewable sources, such as wind or solar electricity, into hydrogen via a process known as electrolysis. Electrolysis divides molecules of water into hydrogen and oxygen during periods of extra electrical output, providing "green hydrogen" when fuelled by renewable sources. The hydrogen can then be stored in a variety of forms, such as compressed gas, liquid, or chemical compounds. One of the key benefits of the storage of hydrogen energy is its capacity to store enormous amounts of energy over long periods of time. Unlike battery packs, which are best suited for short-term storage, the storage of hydrogen can be scaled up to satisfy seasonal or long-term energy demands. As a result, it is an excellent contender for bridging the gap between energy output and consumption, delivering a consistent and steady power supply. The adaptability of hydrogen increases its utility as an energy storage medium. It has a wide range of energy applications, including power generation, heat creation, and usage as a clean fuel in transportation and industrial activities. Depending on the use, stored hydrogen can be turned into power or heat via fuel cells or combustion. There are some issues with storing hydrogen energy that must be solved. Methods for creating, storing, and converting hydrogen that are efficient and cost-effective are still being developed. Because of its tiny amount of energy per unit volume, hydrogen storage and transportation require careful study, demanding innovative storage systems [50].

11.7.2 Synthetic fuel production

Synthetic fuel production, also known as synfuel or synthesized hydrocarbon gasoline production, is a novel technique that tries to produce liquid or gaseous fuels artificially. Synthetic fuels, as opposed to conventional fuels, are created by combining various raw materials, such as carbon dioxide (CO_2) and hydrogen, which are often obtained from water electrolysis or natural gas, using advanced chemical processes such as Fischer-Tropsch synthesis or methanol-to-gasoline conversion. One of the primary objectives for synthetic fuel generation is to address the issues of global warming and energy security. Carbon-neutral or carbon-negative synthetic fuel generation has the potential to reduce overall emissions of carbon in the transportation sector, which contributes significantly to greenhouse gas emissions. Synthetic fuels can be created with minimal net greenhouse gas emissions by absorbing CO_2 from industrial operations or straight from the atmosphere and using sources of clean energy to power the synthesis. This closed carbon cycle can disentangle fuel usage from carbon emissions. Multiple phases are involved in the creation of synthetic fuels, including feedstock preparation, gasification or reforming of steam methane to extract hydrogen, and the actual fuel synthesis process. To facilitate chemical transformations, these processes necessitate sophisticated catalysts and reactors. The resulting synthetic fuels are compatible with existing combustion engines, turbines, and other combustion-based systems, providing a potential transitional solution as renewable energy technologies advance [51].

11.8 GREEN HYDROGEN IN INDUSTRIAL AND TRANSPORTATION SECTORS

Green hydrogen is being touted as a possible answer to the dual issues of decarbonizing the manufacturing and transportation sectors. Green hydrogen presents a unique potential to cut emissions of greenhouse gases while satisfying the energy needs of these critical industries as the world seeks greener and more renewable sources of energy. Green hydrogen has the potential to transform operations that are currently dependent on fossil fuels in the industrial sector. Steel, cement, chemicals, and refining are among the greatest sources of world carbon emissions. In these processes, green hydrogen can serve as a clean substitute for natural gas and coal. Renewable energy sources such as solar, wind, and hydroelectric power may generate hydrogen by breaking the molecules of water into hydrogen and oxygen via electrolysis. When fuelled by renewable energy, this process emits no greenhouse gases, making it an important tool in decarbonizing industrial activity. These sectors may retain productivity while substantially reducing their environmental impact by incorporating green hydrogen [52].

Similarly, the incorporation of green hydrogen has the potential to assist the transportation sector. While electric cars (EVs) have gained popularity, certain means of transportation, such as long-distance trucks, ships, and airplanes, suffer limitations due to battery technological limitations. In these applications, hydrogen that is green can serve as a clean and efficient fuel. When compared to standard battery-powered vehicles, hydrogen-powered fuel cell electric vehicles (FCEVs) have larger ranges and faster refuelling periods. Furthermore, hydrogen fuel cell technology can be used in heavy-duty transportation modes where electrification is difficult. Hydrogen-powered vehicles and buses are currently being evaluated and deployed in several places, demonstrating green hydrogen's potential to transform the transportation industry [53].

Several hurdles must be overcome before green hydrogen may be widely adopted in these sectors. The cost for manufacturing is a major consideration. While the cost of renewable energy has decreased, the overall cost of green hydrogen remains greater than that of hydrogen produced from fossil sources. Governments and companies must work together to design laws and incentives that encourage investment in sustainable hydrogen infrastructure while lowering manufacturing costs. Another crucial issue is infrastructure development [54]. Hydrogen transportation and distribution necessitate the installation of specific pipelines, storage facilities, and refuelling stations. Establishing this infrastructure necessitates significant investments and meticulous planning. Governments and commercial players must collaborate to build a network that allows for the efficient and secure distribution of gas across industrial centres and transportation routes [4].

11.8.1 Hydrogen-fuelled transportation solutions

Hydrogen-powered modes of transportation are gaining popularity as a possible option to reduce the transportation sector's carbon impact. With growing worries about global warming and the need to transition away from fossil fuels, hydrogen has emerged as a versatile and clean energy carrier capable of powering a wide range of modes of transportation, including automobiles, buses, trains, trucks, ships, and airplanes. The primary benefit of hydrogen is its ability to produce no emissions from the tailpipe because the only consequence of its combustion is water vapour. This makes hydrogen especially appealing for tackling urban air quality issues and contributing to worldwide efforts to reduce greenhouse gas emissions. FCEVs are one major application of hydrogen-fuelled mobility. Fuel cells in these vehicles turn hydrogen gas into electricity, which then drives the vehicle's electric motor. The complete process is highly efficient, emitting just water vapour and heat and so avoiding dangerous pollutants and greenhouse gas emissions. FCEVs have various advantages over battery electric vehicles, such longer driving ranges and faster refilling times. Furthermore, hydrogen can be created using renewable sources and the process of electrolysis resulting in "green hydrogen" that improves the environmental advantages of FCEVs even further [55].

Hydrogen fuel cell buses have proven their ability to provide long driving ranges, quick refilling, and low emissions, making them appropriate for routes with stringent operational needs. As part of their environmentally friendly transportation policies, many cities throughout the world are exploring or installing hydrogen-powered buses. Aside from road transportation, hydrogen has the potential to decarbonize heavy-duty applications including trucks, railroads, and maritime vessels. Because of the weight and infrastructure for charging needs, long-haul transportation, for example, frequently meets issues with battery electric cars. Hydrogen-powered trucks have longer ranged and shorter refuelling stops, making them a viable freight transportation option. Fuel cells made from hydrogen have been proposed for ship propulsion in the marine sector, providing a path to emissions-free shipping and lowering the negative environmental effects of maritime trade.

11.9 CONCLUSION

In conclusion, this chapter has provided a thorough examination of green hydrogen production methods, designs, and their wide-ranging applications. The importance of green hydrogen in the transition towards a sustainable energy future cannot be overstated, and the knowledge shared within these pages will undoubtedly serve as a valuable resource for researchers, engineers, and policymakers. The various hydrogen production methods were discussed with a keen focus on the sustainable pathways leading to green hydrogen production. The electrolysis, biomass conversion, and solar-driven processes shed light on the myriad possibilities for generating carbon-neutral hydrogen are also discussed. The subsections on alkaline electrolysis, PEM electrolysis, and solid oxide electrolysis presented a comprehensive overview of the mechanisms, technologies, and efficiency enhancements in electrolysis, emphasizing their potential to revolutionize the hydrogen industry.

Biomass conversion for green hydrogen production explored the exciting frontiers of thermochemical and biological routes, offering an eco-friendly alternative to traditional hydrogen production methods. Solar-driven hydrogen production methods, specifically photovoltaic-electrolysis systems and photo-electrochemical cells, demonstrated the immense promise of harnessing renewable energy sources to generate clean hydrogen.

System design considerations were discussed in detail, addressing the essential engineering aspects of green hydrogen production. This understanding of system design will be crucial in optimizing efficiency and cost-effectiveness in real-world applications.

The applications of green hydrogen were also emphasized, with a particular focus on energy storage and clean fuel synthesis. The ability to store excess renewable energy as green hydrogen and use it to produce clean fuels opens up a world of opportunities for sustainable energy management.

Furthermore, the integration of green hydrogen in industrial and transportation sectors was explored, highlighting its role in decarbonizing key sectors and reducing greenhouse gas emissions. The chapter has underscored that green hydrogen is not just a potential solution but a necessary one for the future of clean energy and environmental preservation. In summary, this chapter has presented a comprehensive understanding of green hydrogen, from its production methods to its wide-ranging applications. With its potential to revolutionize the energy landscape, green hydrogen stands as a vital component of the transition to a more sustainable and eco-friendly future.

REFERENCES

[1] N. Höhne et al., "The Paris Agreement: Resolving the inconsistency between global goals and national contributions," In *Climate Policy after the 2015 Paris Climate Conference*, France, 2021, pp. 30–46.

[2] O. Bičáková and P. Straka, "Production of hydrogen from renewable resources and its effectiveness," *Int. J. Hydrogen Energy*, vol. 37, no. 16, pp. 11563–11578, 2012.

[3] L. Kaiwen, Y. Bin, and Z. Tao, "Economic analysis of hydrogen production from steam reforming process: A literature review," *Energy Sources, Part B Econ. Planning, Policy*, vol. 13, no. 2, pp. 109–115, 2018. doi:10.1080/1556 7249.2017.1387619.

[4] S. Shiva Kumar and H. Lim, "An overview of water electrolysis technologies for green hydrogen production," *Energy Rep.*, vol. 8, pp. 13793–13813, 2022. doi: 10.1016/j.egyr.2022.10.127.

[5] C. B. Agaton, K. I. T. Batac, and E. M. Reyes Jr., "Prospects and challenges for green hydrogen production and utilization in the Philippines," *Int. J. Hydrogen Energy*, vol. 47, no. 41, pp. 17859–17870, 2022. doi:10.1016/j.ijhydene.2022.04.101.

[6] M. Wang, Z. Wang, X. Gong, and Z. Guo, "The intensification technologies to water electrolysis for hydrogen production – A review," *Renew. Sustain. Energy Rev.*, vol. 29, pp. 573–588, 2014. doi:10.1016/j.rser.2013.08.090.

[7] S. A. Grigoriev, V. N. Fateev, D. G. Bessarabov, and P. Millet, "Current status, research trends, and challenges in water electrolysis science and technology," *Int. J. Hydrogen Energy*, vol. 45, no. 49, pp. 26036–26058, 2020. doi:10.1016/j.ijhydene.2020.03.109.

[8] A. R. Singh, S. K. Singh, and V. P. Singh, "Process parameters optimization of carbon nano tube based catalytic transesterification of algal oil," *Mater. Today Proc.*, no. xxxx, 2023. doi:10.1016/j.matpr.2023.01.418.

[9] K. Zeng and D. Zhang, "Recent progress in alkaline water electrolysis for hydrogen production and applications," *Prog. Energy Combust. Sci.*, vol. 36, no. 3, pp. 307–326, 2010. doi:10.1016/j.pecs.2009.11.002.

[10] N. Guillet and P. Millet, "Alkaline water electrolysis," A. Godula-Jopek, Ed., In *Hydrogen Production*, 2015, pp. 117–166. https://onlinelibrary.wiley.com/doi/10.1002/9783527676507.ch4

[11] F. Barbir and T. Gómez, "Efficiency and economics of proton exchange membrane (PEM) fuel cells," *Int. J. Hydrogen Energy*, vol. 21, no. 10, pp. 891–901, 1996. doi:10.1016/0360-3199(96)00030-4.

[12] N. Dutt, A. Jageshwar, H. Ashwani, K. Mukesh, K. Awasthi, and V. Pratap, "Thermo - hydraulic performance of solar air heater having discrete D - shaped ribs as artificial roughness," *Environ. Sci. Pollut. Res.*, 2023. doi:10.1007/s11356-023-28247-9.

[13] M. Carmo, D. L. Fritz, J. Mergel, and D. Stolten, "A comprehensive review on PEM water electrolysis," *Int. J. Hydrogen Energy*, vol. 38, no. 12, pp. 4901–4934, 2013. doi:10.1016/j.ijhydene.2013.01.151.

[14] M. Langemann, D. L. Fritz, M. Müller, and D. Stolten, "Validation and characterization of suitable materials for bipolar plates in PEM water electrolysis," *Int. J. Hydrogen Energy*, vol. 40, no. 35, pp. 11385–11391, 2015. doi:10.1016/j.ijhydene.2015.04.155.

[15] Z. Xie et al., "Ultrathin platinum nanowire based electrodes for high-efficiency hydrogen generation in practical electrolyzer cells," *Chem. Eng. J.*, vol. 410, p. 128333, 2021.

[16] I. Jang et al., "Electron-deficient titanium single-atom electrocatalyst for stable and efficient hydrogen production," *Nano Energy*, vol. 78, p. 105151, 2020.

[17] K. Zhu, X. Zhu, and W. Yang, "Application of in situ techniques for the characterization of NiFe-based oxygen evolution reaction (OER) electrocatalysts," *Angew. Chemie Int. Ed.*, vol. 58, no. 5, pp. 1252–1265, 2019. doi:10.1002/anie.201802923.

[18] B. Thokchom, P. Qiu, M. Cui, B. Park, A. B. Pandit, and J. Khim, "Magnetic Pd@ Fe3O4 composite nanostructure as recoverable catalyst for sonoelectro-hybrid degradation of Ibuprofen," *Ultrason. Sonochem.*, vol. 34, pp. 262–272, 2017.

[19] S. D. Ebbesen and M. Mogensen, "Electrolysis of carbon dioxide in solid oxide electrolysis cells," *J. Power Sources*, vol. 193, no. 1, pp. 349–358, 2009.

[20] A. Hauch et al., "Recent advances in solid oxide cell technology for electrolysis," *Science*, vol. 370, no. 6513, p. eaba6118, 2020. doi:10.1126/science.aba6118.

[21] M. Ni, M. K. H. Leung, and D. Y. C. Leung, "Technological development of hydrogen production by solid oxide electrolyzer cell (SOEC)," *Int. J. Hydrogen Energy*, vol. 33, no. 9, pp. 2337–2354, 2008. doi:10.1016/j.ijhydene.2008.02.048.

[22] B. P. Chaplin, "Chapter 17 – Advantages, disadvantages, and future challenges of the use of electrochemical technologies for water and wastewater treatment," In *Electrochemical Water and Wastewater Treatment*, C. A. Martínez-Huitle, M. A. Rodrigo, and O. B. T.- E. W. and W. T. Scialdone, Eds. Oxford, Waltham, MA: Butterworth-Heinemann, 2018, pp. 451–494. https://doi.org/10.1016/B978-0-12-813160-2.00017-1

[23] V. P. Singh, S. Jain, and A. Kumar, "Establishment of correlations for the thermo-hydraulic parameters due to perforation in a multi-V rib roughened single pass solar air heater," *Exp. Heat Transf.*, vol. 35, no. 5, pp. 1–20, 2022. doi:10.1080/08916152.2022.2064940.

[24] S. E. Hosseini and M. A. Wahid, "Hydrogen production from renewable and sustainable energy resources: Promising green energy carrier for clean development," *Renew. Sustain. Energy Rev.*, vol. 57, pp. 850–866, 2016. doi:10.1016/j.rser.2015.12.112.

[25] V. P. Singh and S. Jain, "Economic analysis of a large scale solar updraft tower power plant," *Sustain. Energy Technol. Assessments*, vol. 58, p. 103325, 2023. doi:10.1016/j.seta.2023.103325.

[26] V. P. Singh et al., "Recent developments and advancements in solar air heaters: A detailed review," *Sustainability*, vol. 14, no. 19, pp. 1–57, 2022. doi:10.3390/su141912149.

[27] M. Saini, A. Sharma, V. P. Singh, S. Jain, and G. Dwivedi, "Solar thermal receivers – A review," *Lect. Notes Mech. Eng.*, vol. II, pp. 1–25, 2022. doi:10.1007/978-981-16-8341-1.

[28] V. P. Singh, A. Karn, G. Dwivedi, T. Alam, and A. Kumar, "Experimental assessment of variation in open area ratio on thermohydraulic performance of parallel flow solar air heater," *Arab. J. Sci. Eng.*, vol. 41, no. 12, pp. 1–17, 2022. doi:10.1007/s13369-022-07525-7.

[29] S. Jain, N. Kumar, V. P. Singh, S. Mishra, and N. K. Sharma, "Transesterification of algae oil and little amount of waste cooking oil blend at low temperature in the presence of NaOH," *Energies*, vol. 16, no. 1, pp. 1–12, 2023.doi:10.3390/en16031293.

[30] V. P. Singh and G. Dwivedi, "Technical analysis of a large-scale solar updraft tower power plant," *Energies*, vol. 16, no. 1, p. 103325, 2023. doi:10.3390/en16010494.

[31] K. Chauhan and V. P. Singh, "Proceedings prospect of biomass to bioenergy in India: An overview," *Mater. Today Proc.*, no. xxxx, 2023. doi:10.1016/j.matpr.2023.01.419.

[32] A. Kundu, A. Kumar, V. P. Singh, C. S. Meena, and N. Dutt, "Chapter 1 – Introduction to thermal energy resources and their smart applications," In *Thermal Energy Systems:Design, Computational Techniques, and Applications.* Boca Raton, FL: CRC Press, 2023, pp. 1–15.

[33] A. Kundu, A. Kumar, N. Dutt, V. P. Singh, and C. S. Meena, "Chapter 7 – Modelling and simulation of thermal energy system for design optimization," In *Thermal Energy Systems: Design, Computational Techniques*, and Applications, A. Kumar, N. Dutt, V. Singh, and C. Meena, Eds. Boca Raton, FL: CRC Press, 2023, pp. 103–137.

[34] T. Sivasakthivel, V. Verma, R. Tarodiya, C. S. Meena, V. P. Singh, and R. Kumar, "Chapter 11 – Analysis of optimum operating parameters for ground source heat pump system for different cases of building heating and cooling mode operations," In *Thermal Energy Systems: Design, Computational Techniques, and Applications*, 1st ed., A. Kumar, N. Dutt, V. Singh, and C. Meena, Eds. Boca Raton, FL: CRC Press, 2023, pp. 183–207.

[35] G. S. Rathore, S. R, A. N, A. Karn, and V. P. Singh, "Recent progress in solar dryers using phase changing material : A review," *Int. J. Energy Resour. Appl.*, vol. 2, no. 1, pp. 57–77, 2023. doi:10.56896/IJERA.2023.2.1.005.

[36] R. K. S. Srivastava, D. Verma, S. Thusoo, A. Kumar, and V. P. Singh, "Nanomanufacturing for energy conversion and storage devices," In *Nanomanufacturing and Nanomaterials Design: Principles and Applications.* Boca Raton, FL: CRC Press, 2022, pp. 165–174.

[37] A. Datta, A. Kumar, A. Kumar, A. Kumar, and V. P. Singh, "Advanced materials in biological implants and surgical tools," In *Advanced Materials for Biomedical Applications*, 1st ed., N. D. A. Kumar, Y. Gori, A. Kumar, and C. S. Meena, Eds. Boca Raton, FL: CRC Press, 2022, pp. 283–298.

[38] Y. G. A. Kumar, A. K. S. Gangwar, A. Kumar, C. S. Meena, V. P. Singh, N. Dutt, and A. Prasad, "Biomedical study of femur bone fracture and healing," In *Advanced Materials for Biomedical Applications.* Boca Raton, FL: CRC Press, 2022, pp. 283–298.

[39] V. P. Singh et al., "Nanomanufacturing and design of high-performance piezoelectric nanogenerator for energy harvesting," In *Nanomanufacturing and Nanomaterials Design: Principles and Applications.* Boca Raton, FL: Taylor & Francis;CRC Press, 2022, pp. 241–272. https://doi.org/10.1201/9781003220602

[40] C. McGregor, J. Pretorius, A. Attieh, and J. Hoffmann, "Techno-economic assessment of electricity generation from a medium-scale CSP-PV hybrid plant using long-duration storage," Available SSRN 4571684, 2023. https://www.researchgate.net/publication/374447688_Techno-economic_Assessment_of_Electricity_Generation_from_a_Medium-Scale_CSP-PV_Hybrid_Plant_Using_Long-Duration_Storage

[41] R. Kumar et al., "Experimental and RSM-based process-parameters optimisation for turning operation of EN36B steel," *Materials (Basel)*, vol. 16, no. 1, pp. 1–18, 2023. doi:10.3390/ma16010339.

[42] A. M. M. I. Qureshy, M. Ahmed, and I. Dincer, "Performance assessment study of photo-electro-chemical water-splitting reactor designs for hydrogen production," *Int. J. Hydrogen Energy*, vol. 44, no. 18, pp. 9237–9247, 2019. doi:10.1016/j.ijhydene.2019.01.280.

[43] I. Dincer and C. Acar, "Innovation in hydrogen production," *Int. J. Hydrogen Energy*, vol. 42, no. 22, pp. 14843–14864, 2017. doi:10.1016/j.ijhydene.2017.04.107.

[44] T. Hai et al., "Optimal design and transient simulation next to environmental consideration of net-zero energy buildings with green hydrogen production and energy storage system," *Fuel*, vol. 336, 2023. doi:10.1016/j.fuel.2022.127126.

[45] E. Facchinetti, M. Gassner, M. D'Amelio, F. Marechal, and D. Favrat, "Process integration and optimization of a solid oxide fuel cell – Gas turbine hybrid cycle fueled with hydrothermally gasified waste biomass," *Energy*, vol. 41, no. 1, pp. 408–419, 2012. doi:10.1016/j.energy.2012.02.059.

[46] J. P. Pretorius, "Optimization and control of a large-scale solar chimney power plant," (Ph.D. Thesis), Univ. Stellenbosch, South Africa, 2007.

[47] J. P. Pretorius and J. A. Erasmus, "Effect of tip vortex reduction on air-cooled condenser axial flow fan performance: An experimental investigation," *J. Turbomach.*, vol. 144, no. 3, 2021. doi:10.1115/1.4052234.

[48] Y. S. M. Camacho, S. Bensaid, G. Piras, M. Antonini, and D. Fino, "Techno-economic analysis of green hydrogen production from biogas autothermal reforming," *Clean Technol. Environ. Policy*, vol. 19, no. 5, pp. 1437–1447, 2017. doi:10.1007/s10098-017-1341-1.

[49] C. Tarhan and M. A. Çil, "A study on hydrogen, the clean energy of the future: Hydrogen storage methods," *J. Energy Storage*, vol. 40, p. 102676, 2021. doi:10.1016/j.est.2021.102676.

[50] F. Zhang, P. Zhao, M. Niu, and J. Maddy, "The survey of key technologies in hydrogen energy storage," *Int. J. Hydrogen Energy*, vol. 41, no. 33, pp. 14535–14552, 2016. doi:10.1016/j.ijhydene.2016.05.293.

[51] T. H. Maugh, "Hydrogen: Synthetic fuel of the future," *Science (80-.)*, vol. 178, no. 4063, pp. 849–852, 1972.

[52] T. R. Ayodele and J. L. Munda, "Potential and economic viability of green hydrogen production by water electrolysis using wind energy resources in South Africa," *Int. J. Hydrogen Energy*, vol. 44, no. 33, pp. 17669–17687, 2019. doi:10.1016/j.ijhydene.2019.05.077.

[53] A. V. Abad and P. E. Dodds, "Green hydrogen characterisation initiatives: Definitions, standards, guarantees of origin, and challenges," *Energy Policy*, vol. 138, p. 111300, 2020. doi:10.1016/j.enpol.2020.111300.

[54] J. Pretorius and D. Kröger, "Incorporating vegetation under the collector," *R D Journal, South African Inst. Mech. Eng.*, vol. 24, no. 1, pp. 3–11, 2008.

[55] M. Balat, "Potential importance of hydrogen as a future solution to environmental and transportation problems," *Int. J. Hydrogen Energy*, vol. 33, no. 15, pp. 4013–4029, 2008, doi: 10.1016/j.ijhydene.2008.05.047.

Chapter 12

Comparative assessment of hydrogen production methods from renewable energy

A review

Abdul Quaiyoom, Nikhil Dev, and Bhupender Singh
J.C. Bose University of Science and Technology, YMCA

Ashok Kumar Yadav
Raj Kumar Goel Institute of Technology

12.1 INTRODUCTION

A steady rise in energy consumption during the 20th and early 21st centuries has been caused by global population growth and rising living standards. Figure 12.1 depicts the world's total primary energy supply by fuels in 2016. A rise in energy production capacity will be required as a result. It may be possible to meet the world's energy needs while reducing and eliminating greenhouse gas emissions by locating more reliable,

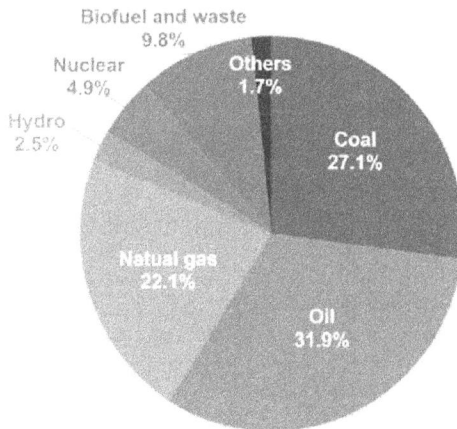

Figure 12.1 The world's total primary energy supply by fuels in 2016. (Source: IEA, International Energy Agency.)

DOI: 10.1201/9781003472629-12

sustainable, and diverse energy sources. Hydrogen offers a lot of advantages over other fuels; as a result, it might be used to lower pollution and reliance on foreign oil.

Hydrogen, the universe's most abundant, simplest, and lightest elemental constituent, takes center stage. Our focus lies in contrasting the ecological implications of alternate pathways for synthesizing hydrogen through renewable origins, with a distinct spotlight on their potential integration within India's landscape [1–6]. The methodologies subjected to scrutiny encompass a spectrum of innovative endeavors—ranging from high-temperature electrolysis to harnessing wind and solar energy for electrolysis, biomass gasification, and the intricate thermochemical water cleavage through Cu-Cl and S-I cycles.

This scholarly endeavor endeavors to dissect, juxtapose, and meticulously assess the societal, environmental, and economic influences inherent in diverse hydrogen production avenues. An in-depth comparative exploration was undertaken, scrutinizing eight discrete technological stratagems. The analysis encompassed a multidimensional evaluation of their ecological ripples, quantifying acidification and global warming potentials (GWP and AP), delineating production expenditures, and gauging energy and exergy efficiencies. Furthermore, an empirical inquiry delved into the intricate interplay between plant capacity and the initial financial outlay entailed in the hydrogen-generating pursuits.

When examining emissions at the usage point, hydrogen demonstrates an elevated degree of purity. In the context of fuel cells, the amalgamation of hydrogen and oxygen culminates in the exclusive formation of water as its singular byproduct, devoid of any CO_2 emission. The merits linked to the hydrogen-centric economy can be succinctly encapsulated as follows: (i) bolstering energy security by curbing reliance on oil imports; (ii) championing sustainability through the harnessing of renewable energy reservoirs; (iii) alleviating pollution and augmenting urban air quality; and (iv) nurturing economic feasibility by potentially reshaping global energy markets in subsequent epochs [8, 9].

The present study undertakes a meticulous appraisal of the ecological ramifications intertwined with a plethora of techniques for hydrogen creation, spanning both renewable and non-renewable origins, all while accentuating their adaptability within the Indian context. The principal ambition is to proffer valuable, constructive counsel for research and development to pertinent governing bodies. The investigation centers its attention on eight unique methodologies, with the intention of furnishing a comprehensive panorama of the latent implications and advantages encapsulated within diverse hydrogen generation processes. The hydrogen-generating techniques examined in this work are shown in Figure 12.2.

Renewable energy			Primary energy source
Solar photovoltaic	Wind	Biomass	

Electrolysis	Gasification	Method

End use, hydrogen energy

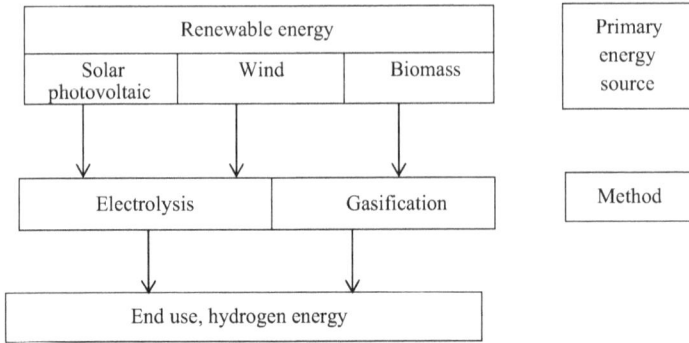

Figure 12.2 Selected hydrogen production methods. (Modified from Ref. [7].)

12.2 BACKGROUND: TECHNIQUES FOR PRODUCING HYDROGEN

12.2.1 Photovoltaic electrolysis

A photovoltaic (PV) water electrolysis system is constructed from several key components, including PV panels, a DC bus bar, an alternating current (AC) grid connection, an accumulator battery set, an electrolyzer, and storage canisters for the generated hydrogen. The fundamental concept behind this system is to address the inherent challenge of intermittent instability in solar power generation by harnessing water electrolysis. Through this approach, the system aims to ensure consistent and dependable power output within a predefined range. The schematic representation of the PV water electrolysis system is illustrated in Figure 12.3.

The system functions by capitalizing on the combined capabilities of its constituent parts. The PV panels absorb sunlight and generate direct current (DC) electricity, which is routed through a DC bus bar. This energy can be directed toward immediate consumption, charging an accumulator battery set for later use, or feeding into the AC grid for broader distribution.

The crucial role of the electrolyzer comes into play here. By utilizing the surplus or stored DC energy, the electrolyzer initiates the water electrolysis process. This process breaks down water molecules into hydrogen and oxygen gases, with hydrogen being collected and stored in specialized canisters.

The resulting hydrogen serves as a form of energy storage, which can be tapped into when solar power generation experiences intermittent dips or when additional power is required. This setup ensures a level of reliability that mitigates the impact of solar power fluctuations, contributing to a stable and consistent energy supply.

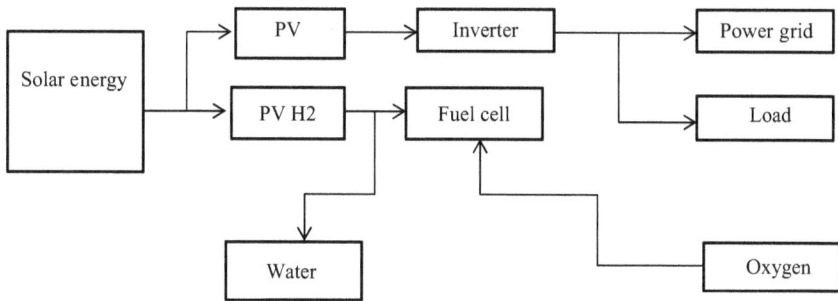

Figure 12.3 Structural diagram of the PV water electrolysis system.

The system's structural diagram, depicted in Figure 12.2, visually represents the interconnected components and their collaborative operation. This innovative integration of PV panels, energy storage, and electrolysis not only addresses the challenges posed by solar power intermittency but also offers a viable solution for sustaining dependable power generation over time.

12.2.2 Photovoltaic solar energy

This is one of the most expensive ways to make hydrogen. Currently, it costs around 25 times more to produce hydrogen using photovoltaic electrolysis than it does with fossil fuels.

12.2.3 Biomass gasification

Hydrogen has the capacity to be generated from a diverse array of biomass sources, encompassing crops, agricultural and forestry residues, animal and human waste, and byproducts derived from wood processing. However, currently, the large-scale and cost-effective production of hydrogen via biomass gasification remains unfeasible [10–14]. Nevertheless, this method presents a viable avenue for environmentally responsible energy recovery from household and agricultural waste. In contrast to the utilization of fossil fuels, this approach holds the potential to yield a reduction in net CO_2 emissions. This potential is rooted in the fact that the CO_2 generated during biomass oxidation can be effectively extracted from the environment and subsequently replenished through photosynthesis during the biomass growth phase [15–20]. It is crucial to acknowledge that concerns related to potential strain on natural resources and land utilization, stemming from expanding populations, stand as significant considerations associated with this strategic approach.

12.2.4 Wind-driven hydrogen production using water

Utilizing the untapped potential of an abandoned wind farm to harness hydrogen production presents an opportunity to enhance overall resource efficiency within the wind farm, consequently amplifying the contribution of grid-connected wind energy. This novel approach involves transforming the dormant electrical energy of the wind farm into direct current by means of rectification, facilitated by advanced power electronic control systems. Subsequently, this direct current is employed within electrolytic cells to initiate the hydrogen production process.

The generated hydrogen is subject to a series of refining steps, including separation and purification, followed by compression and storage within specialized modules. A comprehensive feasibility analysis, conducted at a site in Ghardaia, validates the viability of hydrogen production through wind power. The findings of this study unveiled the potential for system enhancement through the optimization of wind turbine tower height. Notably, this impact was evident in two scenarios involving wind farm clusters of 30 and 100 MW in Northeast China. Comparative assessments underscore the economic prowess of a collaborative infrastructure integrating wind farm clusters and hydrogen production systems. This symbiotic amalgamation promises substantial cost-efficiency gains, positioning it as a superior strategy when compared to standalone counterparts.

In conclusion, the strategic utilization of a dormant wind farm's latent capacity for hydrogen generation signifies an innovative means to bolster resource utilization and amplify grid-connected wind energy. The fusion of cutting-edge power electronic systems with hydrogen production processes holds promise, validated by a rigorous feasibility study. This paradigm shift, exemplified by the potential height-dependent enhancement of wind farm output, spotlights the economic benefits of coalescing wind farm clusters and hydrogen production systems.

12.2.5 Photo electrolysis and photocatalytic decomposition

Photo electrolysis represents a pioneering process that leverages the attributes of heterogeneous photocatalysts within a photovoltaic (PV) electrolytic cell. Positioned at one electrode, this photocatalyst is exposed to solar radiation to initiate the phenomenon. Specifically, the photo anode, situated within the cell, engages sunlight absorption, thus stimulating the semiconductor material to generate electrons at the anode. This pool of electrons subsequently undertakes external transmission to reach the cathode, facilitating the catalytic generation of hydrogen (H_2) at this electrode.

Nonetheless, despite the use of high-performance semiconductor materials like double-interface Ga and As electrodes, the achievable yield remains limited, with only approximately 13% success rate. In parallel, the principles governing the solar-driven photocatalytic decomposition of water to foster hydrogen production mirror those of solar photoelectrolysis. However, a pivotal distinction lies in the fact that the anode and cathode coexist within the same particle. This distinct setup induces the decomposition of water into hydrogen (H_2) and oxygen (O_2) gases. Yet, this simultaneous decomposition presents an intrinsic challenge: the electron holes produced within the same particle are susceptible to rapid recombination. An illuminating investigation conducted at the University of Science and Technology of China has engineered a distinctive "sandwich" structural material system grounded in quantum theory. This innovative approach effectively curbs the undesirable reverse reaction of O_2 and H_2—generated through photo electrolysis—from reverting back into water.

In essence, the strategic integration of heterogeneous photocatalysts with solar radiation through a PV electrolytic cell embodies the concept of photoelectrolysis. While remarkable strides have been taken, such as the "sandwich" material system, to optimize this process, its challenges and potentialities persist in the realm of sustainable hydrogen generation.

12.2.6 Hydrogen production via gasification of biomass

Biomass gasification stands as a transformative procedure aimed at converting hydrocarbons derived from biomass into gaseous fuel, accomplished through the utilization of a gasification medium. This intricate process hinges on the deployment of catalysts, instrumental in expediting gasification kinetics and lowering operational temperatures. At its core, biomass gasification represents a multi-step progression involving three key phases centered on hydrogen production.

The initial stage encompasses biomass gasification itself, a pivotal process wherein the biomass feedstock undergoes thermal degradation in the presence of a gasification medium. This thermal degradation facilitates the conversion of biomass constituents into a gaseous mixture containing hydrogen, carbon monoxide, and other gases [21–30].

Following this, the synthesis gas catalytic reforming phase comes into play. This crucial step entails the utilization of catalysts to further refine the gaseous mixture generated during gasification. Through catalytic reforming, the composition of the gas mixture is adjusted, enhancing the yield of desirable components, including hydrogen.

The concluding phase is centered on hydrogen separation and purification. Within this stage, the synthesized gas undergoes meticulous purification processes to isolate and extract hydrogen from the gas mixture. This involves the removal of impurities, unreacted components, and undesired byproducts, ultimately leading to the attainment of high-purity hydrogen gas.

In summation, biomass gasification orchestrates the intricate conversion of biomass hydrocarbons into gaseous fuel, facilitated by a gasification medium and catalysts. The hydrogen production aspect is intricately woven into three successive stages: biomass gasification, synthesis gas catalytic reforming, and hydrogen separation and purification. This innovative process bears substantial promise in contributing to sustainable energy solutions by harnessing the potential of biomass resources for hydrogen-rich fuel production.

During the mid-1970s, the concept of supercritical water gasification (SCWG) emerged as a pioneering method for hydrogen production, as initially proposed [31–37]. This innovative technique revolves around subjecting biomass to supercritical water, which triggers a sequence of intricate thermochemical conversions. These transformations encompass processes like pyrolysis, hydrolysis, condensation, and dehydrogenation. The outcome of these reactions is the generation of a diverse array of gases including hydrogen, carbon monoxide, carbon dioxide, methane, and others.

A distinctive advantage of the SCWG process is its unique ability to obviate the necessity for dry pretreatment of the biomass feedstock. Traditional biomass gasification often involves preparatory steps to reduce moisture content, which can be energy-intensive and time-consuming [38]. In contrast, SCWG stands out by enabling the use of moist or wet biomass directly, thereby eliminating the need for this preliminary drying phase.

Furthermore, the SCWG technique showcases the potential to curtail overall energy consumption. By circumventing the energy-intensive drying step and exploiting the unique properties of supercritical water, this approach exhibits promising efficiency gains compared to conventional gasification methods.

In conclusion, the groundbreaking concept of supercritical water gasification (SCWG) surfaced in the mid-1970s as an innovative means of hydrogen production. This technique leverages the intricate interactions between biomass and supercritical water, leading to a sequence of thermochemical transformations and the generation of various gases, including hydrogen. Notably, the SCWG process's advantage lies in its capacity to accommodate wet or moist biomass without prior drying, ultimately contributing to reduced energy consumption and enhanced process efficiency.

12.3 SUMMARY

The major advantages and major drawbacks of the chosen hydrogen-generating methods are summarized in Table 12.1. The security of the supply must be taken into account when providing energy services. Developing systems with high levels of efficiency, low emissions, and impurity levels,

Table 12.1 Key advantages and significant difficulties of particular hydrogen
generation techniques [39]

Biomass gasification	Thermochemical	Water electrolysis with renewable energy sources
Key benefits		
Economical synthetic fuel with abundant and cost-effective hydrogen	Clean and sustainable	No pollution with renewable sources
Critical challenges		
Feedstock impurities carbon capture and storage	Effective and durable materials of construction	Low efficiency Capital cost Integration with renewable energy resources

as well as increasing the use of renewable energy sources, are the key difficulties facing the hydrogen economy. One of them is lowering the cost of capital, operations, and maintenance.

12.3.1 Comparative evaluation of hydrogen generation techniques

12.3.1.1 Environmental impact comparison

Conducting a Life Cycle Assessment (LCA) for a given system entails a comprehensive evaluation of both its environmental ramifications and energy consumption advantages [7]. In this context, our focus centers on a meticulous analysis of the environmental implications linked to several well-established hydrogen production technologies. These technologies encompass diverse approaches, each with distinct attributes, including wind-driven hydrogen production, hydrogen generation through photovoltaic (PV) electrolysis, nuclear-assisted hydrogen production via thermochemical cycles and water thermolysis, and hydrogen production derived from the gasification of biomass.

Inclusive in our investigation is an exploration of hydrogen production via biomass electrolysis. This particular avenue holds significant promise as a viable contender for hydrogen generation. Its favorable scalability characteristics set it apart from alternative hydrogen production methodologies, rendering it a particularly attractive option.

As we delve into this study, the primary objective is to unravel the broader environmental implications associated with these hydrogen production technologies. Through rigorous analysis and comparison, we aim to shed light on their relative impacts, considering factors ranging from resource consumption to emissions.

In summation, the crux of this work lies in the comprehensive assessment of various established hydrogen production methods, drawing upon the lens of LCA. Through this endeavor, we strive to discern the environmental and energy-related nuances of these technologies, offering valuable insights into their overall sustainability and potential implications for future energy strategies.

12.3.1.2 Hydrogen production via PV electrolysis

In the research conducted by Cetinkaya et al. [40], the LCA methodology was employed to comprehensively investigate the advantages presented by photovoltaic (PV) power generation systems in the context of hydrogen production. The outcomes of their evaluation revealed a noteworthy finding: the equivalent greenhouse gas (GHG) emissions for these PV-based systems amounted to 2412 g of CO_2 per kilogram of hydrogen (gCO_2/kgH_2).

Dufour et al. 41], on the other hand, engaged in an in-depth analysis of the life cycle impacts associated with diverse hydrogen production methodologies. Their research culminated in a significant conclusion: within the framework of hydrogen production, PV power generation exhibited a life cycle energy consumption of 77.864 MJ/kg of hydrogen (MJ/kgH_2), concurrently manifesting a GHG emission rate of 6674 gCO_2/kgH_2.

These studies underscore the pivotal role that LCA plays in assessing the environmental and energy-related dimensions of hydrogen production systems. Both sets of research contribute valuable insights into the greenhouse gas emissions and energy consumption implications tied to PV-based hydrogen production, reinforcing the need for sustainable and efficient energy solutions.

12.3.1.3 Hydrogen production via gasification of biomass

In the work led by Susmozas et al. [41], an innovative approach was undertaken to harness hydrogen production through biomass gasification, employing the energy crop poplar as the primary material. Through meticulous evaluation, their findings unveiled that the net life-cycle greenhouse gas (GHG) emissions associated with this particular biomass gasification system amounted to 405 g of CO_2 per kilogram of hydrogen (gCO_2/kgH_2).

Another insightful study delved into the utilization of a downdraft gasifier (DG) for hydrogen production. Drawing upon data sourced from literature, the emissions of the DG system were methodically considered and analyzed, ultimately resulting in an emission estimation of 896.61 gCO_2/kgH_2 [42].

Furthermore, Hajjaji et al. [43] embarked on a journey employing the LCA methodology to probe hydrogen production through biomass gasification. Their efforts culminated in a significant revelation: the life cycle GHG emissions equivalent for the bio-waste gasification hydrogen production process stood at 5590 gCO_2/kgH_2.

These investigations collectively contribute to the comprehensive understanding of hydrogen production via biomass gasification, highlighting the significance of LCA in evaluating the environmental implications of such methodologies. By scrutinizing the GHG emissions throughout the life cycle, these studies provide valuable insights into the sustainability of hydrogen production systems, ultimately contributing to the discourse on environmentally responsible energy solutions.

12.3.1.4 Wind-driven hydrogen production

In the study conducted by Cetinkaya et al. [40], an extensive LCA was performed, encompassing a comprehensive evaluation of five distinct methodologies for hydrogen production. These methods encompassed steam reforming of natural gas, coal gasification, water electrolysis via wind and solar electrolysis, and thermochemical water splitting utilizing a Cu-Cl cycle. Their comprehensive analysis yielded significant revelations: notably, the quantification of carbon dioxide equivalent emissions linked to hydrogen production through wind power resulted in 970 g of CO_2 per kilogram of hydrogen (gCO_2/kgH_2).

Expanding upon this existing knowledge base, Ghandehariun et al. [44] embarked on an LCA with a specific focus on hydrogen production facilitated by wind energy in the Western Canada region. Through meticulous assessment, they approximated the cumulative greenhouse gas (GHG) emissions attributed to a wind-based hydrogen production facility at 680 gCO_2/kgH_2.

Furthermore, the work of Reiter [45] delved into the intricate interplay between the life cycle of renewable energy generation and its consequential impact on the efficiency of hydrogen production. By utilizing the GABI5 software for rigorous data analysis, their investigation yielded a significant outcome: an estimation of GHG emissions equivalent to 600 gCO_2/kgH_2.

In unison, these research endeavors underscore the pivotal role played by LCA in unraveling the intricate environmental footprint associated with diverse methods of hydrogen production. Through meticulous examination and quantification of greenhouse gas emissions throughout the entirety of the production process, these studies furnish invaluable insights into the viability and sustainability of hydrogen generation, particularly in the context of utilizing renewable energy sources such as wind power [46–49].

12.3.1.5 Hydrogen production via gasification of biomass

In the research conducted by Susmozas et al. [50], an innovative approach was undertaken to exploit hydrogen production through the biomass gasification method, using the energy crop poplar as the primary feedstock.

Their rigorous evaluation yielded compelling outcomes: specifically, the net life-cycle greenhouse gas (GHG) emissions associated with this particular biomass gasification system amounted to 405 g of CO_2 equivalent per kilogram of hydrogen (gCO_2/kgH_2).

Another notable study delved into the utilization of a DG for hydrogen production. Drawing upon data gleaned from the literature, the emissions of the DG system were meticulously considered and analyzed. As a result of their comprehensive assessment, these researchers calculated the emissions associated with the DG system to be 896.61 gCO_2/kgH_2 [51].

Furthermore, Hajjaji et al. [39] adopted the LCA methodology to scrutinize hydrogen production via biomass gasification. Their efforts yielded significant insights, revealing a life cycle GHG emission equivalent of 5590 gCO_2/kgH_2 for the process of hydrogen production through bio-waste gasification.

These investigations collectively underscore the intricate environmental considerations tied to hydrogen production via biomass gasification. Through the lens of LCA, these studies provide valuable insights into the net emissions associated with these methodologies, emphasizing the need for sustainable and environmentally conscious hydrogen production approaches.

12.3.1.6 Energy and exergy efficiency comparisons

The literature review undertaken by Dincer and Zamfirescu [52] lays the essential groundwork for the discourse surrounding the Energy and Exergy efficacy inherent in diverse methodologies for hydrogen synthesis. Their scrutiny of efficiency involves the measurement of the productive yield relative to the input expended. Essential factors such as the mass flow rate of the generated hydrogen, the energy input rate throughout the procedure, and the lower heating value of hydrogen (LHV) set at 121 MJ/kg are adopted to assess the energy efficiency of a given hydrogen production facility. The computation of energy efficiency is derived by employing the subsequent equation (Table 12.2):

$$\eta = (m^*\text{LHV})/E \qquad (12.1)$$

Table 12.2 Summary of GHG emission range of the hydrogen production approaches

Type of hydrogen production methods	GHG emission range (gCO_2/kgH_2)
Wind-driven electrolysis	600–970
Water thermolysis	412–860
Gasification of biomass	405–896.61
PV electrolysis	2412–6674

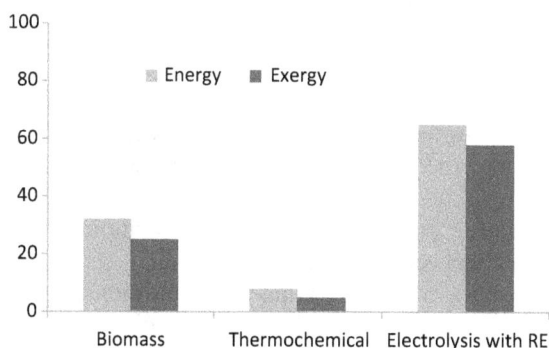

Figure 12.4 Energy and exergy efficiencies of selected hydrogen production methods. (Data from Ref. [53].)

Additionally, the chemical exergy of hydrogen (exch) and the rate of exergy input into the process ($\_ex_{in}$) are employed to formulate the exergy efficiency as depicted by the equation:

$$\varphi = (m^*ex)/\_ex_{in} \tag{12.2}$$

A graphical representation of energy and energy efficiency data for diverse hydrogen production methods is presented in Figure 12.4. The data underscores that among the considered methodologies, biomass gasification outperforms the alternatives in terms of efficiency. In contrast, solar-based electrolysis emerges as the least efficient performer among the selected manufacturing approaches.

12.4 CONCLUSIONS

This study orchestrates an all-encompassing scrutiny of conceivable ecological and technical implications tied to a multitude of hydrogen production methodologies, harmonized with the availability of requisite raw materials within the Indian context. The investigation harnesses the outcomes of comparison and evaluation to contrive pragmatic hydrogen generation strategies meticulously adapted to the India milieu. In the capacity of a case in point, the research delves into the evaluation of biomass gasification, electrolysis, and renewable energy sources [37,54]. This comprehensive assessment takes into careful consideration a spectrum of variables encompassing capital expenses, GWP, and acidification potential (AP) vis-à-vis hydrogen production.

Drawing insights from an array of studies within the literature, a consistent pattern emerges: renewable energy-driven methodologies for hydrogen production stand as more ecologically responsible when compared to fossil-fuel-driven counterparts. Nevertheless, the viability of large-scale implementation hinges on the ongoing reduction of hydrogen production costs through renewable means.

Amid the spectrum of renewable energy-based hydrogen production approaches, those rooted in biomass hold distinct advantages, both in economic feasibility and environmental impact. Within this domain, biomass-based hydrogen production technology emerges as a compelling contender, demonstrating promise in fostering a harmonious convergence of economic viability and ecological sustainability.

Notably, when assessed under comparable conditions as other established studies, hydrogen production via biomass electrolysis manifests a reduced environmental footprint compared to alternative methodologies. This distinctive attribute positions it as a noteworthy candidate for an environmentally conscious hydrogen energy production technique characterized by efficient resource utilization and reduced energy consumption.

In summary, renewable energy-based hydrogen production methodologies emerge as the frontrunners in the pursuit of environmentally-friendly alternatives to fossil-fuel-based approaches. While challenges persist, particularly in terms of production costs, the prospect of biomass-driven hydrogen production, and specifically biomass electrolysis, presents an attractive avenue that embodies the dual objectives of sustainable energy production and environmental preservation. The culminating findings of this research can be succinctly encapsulated as follows: In the context of GWP, the thermochemical-based method for hydrogen production exhibits the least impact, while biomass gasification emerges with the highest impact. Similarly, in terms of AP, the thermochemical-based method for hydrogen production registers the least potential for acidification, with biomass gasification demonstrating the highest potential.

In summation, this study delivers a nuanced understanding of the intricate environmental and technical implications inherent in diverse hydrogen production techniques, tailored to the specific circumstances of India. The comprehensive evaluation underscores varying impacts in terms of GWP and AP, elucidating key takeaways for sustainable hydrogen generation strategies.

REFERENCES

1. Yadav, A.K., Vinay, S.B., Optimization of Biodiesel Production from Annona Squamosa Seeds Oil Using Response Surface Methodology and Its Characterization, *Energy Sources A*, 2018; 40(9): 1051–1059.
2. Ahmed, A., Yadav, A.K., Singh, A., Enhancement of Biogas Yield from Dual Organic Waste Using Hybrid Statistical Approach and Its Effects on Ternary Fuel Blend (Biodiesel/n-butanol /Diesel) Powered Diesel Engine, *Environ. Progr. Sust. Energy*. doi:10.1002/ep.14163.

3. Dewangan, A., Yadav, A.K., Mallick, A., Current Scenario of Biodiesel Development in India: Prospects and Challenges, *Energy Sources A*, 2018; 40(20): 2494–2501.

4. Yadav, A.K., Khan, M.E., Pal, A., Biodiesel Production from Oleander (*Thevetia peruviana*) Oil and Its Performance Testing on a Diesel Engine, *Korean J. Chem. Eng.*, 2017; 34(2): 340–345.

5. Yadav, A.K., Khan, M.E., Pal, A., Performance, Emission and Combustion Characteristics of an Indica Diesel Engine Operated with Yellow Oleander (*Thevetia peruviana*) Oil Biodiesel Produced through Hydrodynamic Cavitation Method, *Int. J. Ambient Energy*, 2018; 39(4): 365–371.

6. Ahmed, A., Yadav, A.K., Singh, A., Process Optimization of Spirulina Microalgae Biodiesel Synthesis Using RSM Coupled GA Technique: A Performance Study of a Biogas-Powered Dual-Fuel Engine, *Int. J. Environ. Sci. Technol.*, 2023. doi:10.1007/s13762-023-04948-z 20.

7. Megía, P. J., Vizcaíno, A. J., Calles, J. A., Carrero, A., Hydrogen Production Technologies: From Fossil Fuels toward Renewable Sources. A Mini Review. Energy Fuels, 2021; 35(20): 16403–16415.

8. Ahmad, A., Yadav, A.K., Singh, A., et al., Multi-response Optimization of a Microalgae-spirulina-fueled VCR Diesel Engine: A Comprehensive RSM-GA Approach. *Environ Dev Sustain.*, 2023. https://doi.org/10.1007/s10668-023-04016-z

9. Yadav, A.K., Dewangan, A., & Mallick, A., Effect of n-Butanol and Diethyl Ether Additives on Performance and Emission Characteristics of a Diesel Engine Fuelled with Diesel-Pongamia Biodiesel Blends, *J. Energy Eng.*, 2018; 144(6): 04018062.

10. Ahmed, A., Yadav, A.K., Singh, A., Agbulut, U., A Hybrid RSM-GA-PSO Approach on Optimization of Process Intensification of Linseed Biodiesel Synthesis Using an Ultrasonication Reactor: Enhancing Fuel Properties and Engine Characteristics with Ternary Fuel Blends, *Energy*, 2023; 288: 129077. doi:10.1016/j.energy.2023.129077.

11. Dewangan, A., Mallick, A., Yadav, A.K., Islam, S., Agbulut, U., Production of Oxy-Hydrogen Gas and the Impact of Its Usability on CI Engine Combustion, Performance, and Emission Behaviors, *Energy*, 2023; 278: 127937.

12. Khan, T.A., Khan, T.A., Yadav, A.K., A Hydrodynamic Cavitation-Assisted System for Optimization of Biodiesel Production from Green Microalgae Oil Using a Genetic Algorithm and Response Surface Methodology Approach, *Environ. Sci. Pollut. Res.*, 2022; 29: 49465–49477. doi:10.1007/s11356-022-20474-w2022.

13. Singh, D., Singh, S., Yadav, A.K., Khan, O., Dewangan, A., From Theory to Practice: A Sustainable Solution to Water Scarcity by Using a Hybrid Solar Distiller with a Heat Exchanger and Aluminum Oxide Nanoparticles, *ACS Omega*, 2023; 37(8): 33543–33553.

14. Dewangan, A, Mallick, A, Yadav, A.K, Ahmad, A, Alqahtani, D., Combined Effect of Operating Parameters and Nano Particles on Performance of a Diesel Engine: Response Surface Methodology Coupled Genetic Algorithm Approach, *ACS Omega*, 2023; 8(27): 24586–24600.

15. Yadav, A.K., Pal, A., Experimental Studies on Utilization of *Prunus armeniaca* L. (Wild Apricot) Biodiesel as an Alternative Fuel for CI Engine, *Waste Biomass Valor.*, 2018; 9(10): 1961–1969.

16. Ahmed, A., Yadav, A.K., Singh, A., Application of Machine Learning and Genetic Algorithm for Prediction and Optimization of Biodiesel Yield from Waste Cooking Oil, *Korean J. Chem. Eng.*, 2023; 40: 2941–2956. doi:10.1007/s11814-023-1489-9.

17. Ahmed, A., Yadav, A.K., Singh, A., Enhancing Waste Cooking Oil Biodiesel Yield and Characteristics through Machine Learning, Response Surface Methodology, and Genetic Algorithm for Optimal Utilization in CI Engines, *Int. J. Green Energy*. doi:10.1080/15435075.2023.2253870.

18. Yadav, A.K., Khan, M.E., Pal, A., Ultrasonic Assisted Production of Biodiesel from Karabi Oil Using Heterogeneous Catalyst, *Biofuels*, 2018; 9: 101–112.

19. Yadav, A.K., Khan, T.A., Khan, T.A., Kumar, S., Methyl Ester of *Gmelina arborea* Oil as a Substitute for Petroleum Diesel: An Experimental Study on Its Performance and Emissions in a Diesel Engine, *Energy Sources A*, 2021: 43(11): 1307–1314.

20. Khan, O., Yadav, A.K., Khan, M.E., Sharma, D., The Ultrasonic Assisted Optimization of Biodiesel Production from Eucalyptus Oil, *Energy Sources A*, 2017; 39: 1323–1331.

21. Dewangan, A., Yadav, A.K., Wax Deposition during Production of Waxy Crude Oil and Its Remediation, *Petroleum Sci. Technol.*, 2017; 35(18): 1831–1838.

22. Khan, I.A., Singh, S.K., Yadav, A.K., Efficient Production of Biodiesel from *Cannabis sativa* Oil Using Intensified Transesterication (Hydrodynamic Cavitation) Method, *Energy Sources A*, 2020; 42(20): 2461–2470.

23. Dewangan, A., Mallick, A., Yadav, A.K., Comparative Study of *Manilkara zapota* and Karanja Based Biodiesel Properties and Its Effect on Multi-Cylinder Diesel Engine Characteristics. *Energy Sources A*, 2022; 44(2): 5143–5153.

24. Khan, O., Yadav, A.K., Khan, M.E., Characterization of Bio-Ethanol Obtained from *Eichhornia crassipes* Plant; Its Emission and Performance Analysis on CI Engine, *Energy Sources A*, 2021; 43(14): 1793–1803.

25. Ahmed, A., Yadav, A.K., Singh, A., Optimization of Cavitation-Assisted Biodiesel Production and Fuel Properties from *Neochloris oleoabundans* Microalgae Oil Using Genetic Algorithm and Response Surface Methodology, *J. Process Mech. Eng. E*. doi:10.1177/09544089231159832.

26. Yadav, M., Karimi, M.N., Yadav, A.K., Effect of Graphene Oxide Nanoparticles Dispersed Biodiesel on Combustion, Performance, and Emission Characteristics of a Diesel Engine, *J. Process Mech. Eng. E.*, doi:10.1177/09544089231209139.

27. Singh, S., Singh, B., Kumar, S., Yadav, A.K., Temperature-Dependent Dynamic Hysteresis Scaling of Ferroelectric Hysteresis Parameters of Lead Free [(Ba 0.825 + xCa0.175-x) (Ti1- x Snx)O3]Ceramics, *Ferroelectrics*, 2019; 551: 133–142.

28. Dewangan, A., Yadav, A.K., Mallick, A., Optimization of Biodiesel Production and Engine Performance from Underutilized Simarouba Oil in Compression Ignition Engine, *Int. J. Oil Gas Coal Technol.*, 2020; 25(3): 357.

29. Ahmed, A., Yadav, A.K., Singh, A., An Environmental Impact Assessment and Optimization Study of Biodiesel Production from Microalgae. *Int. J. Global* Warming, accepted 2023; 31: 294–313.

30. Yadav, A.K., Khan, M.E., Pal, A., A Comparative Study on Ultrasonic Cavitation and Mechanical Stirring Method towards Efficient Production of Biodiesel from Non-Edible Oils and Performance Testing on a C.I. Engine, *Int. J. Environ. Waste Manage.*, 2016; 18: 349–367.

31. Vinay, S.B., Yadav, A.K., Optimization of Performance and Emission Characteristics of CI Engine Fuelled with Mahua Oil Methyl Ester-Diesel Blend using Response Surface Methodology, *Int. J. Ambient Energy*, 2020; 44(6): 674–685.

32. Yadav, A. K., Pal, A., Ghosh, U., Gupta, S.K., Comparative Study of Biodiesel Production Methods from Yellow Oleander Oil and Its Performance Analysis on an Agricultural Diesel Engine, *Int. J. Ambient Energy*, 2019; 40(2): 152–157.

33. Khan, I. A., Singh, S. K., Yadav, A. K., Sharma, D., Enhancement in the Performance of a Diesel Engine Fueled with Pongamia Methyl Ester and n-Butanol as Oxygenated Additive, *Int. J. Ambient Energy*, 2019; 40(8): 842–846.

34. Ahmed, A., Yadav, A.K., Singh, A., Biodiesel Yield Optimization from a Third-Generation Feedstock (Microalgae Spirulina) Using a Hybrid Statistical Approach, *Int. J. Ambient Energy*, 2023; 44(1): 1202–1213.

35. Ahmed, A., Yadav, A.K., Biodiesel Production from Mahua Oil: Characterization, Optimization, and Modeling with a Hybrid Statistical Approach, *Int. J. Ambient Energy*, 2023; 44: 2618–2627.

36. Khan, T.A., Khan, M.E., Yadav, A.K., Experimental Studies on Utilization of *Neochloris oleoabundans* Microalgae Biodiesel as an Alternative Fuel for Diesel Engine, *Int. J. Ambient Energy*, 2023; 44(1): 115–123.

37. Singh, D., Yadav, A.K., Kumar, A., Samsher, Energy Matrices and Life Cycle Conversion Analysis of N-Identical Hybrid Double Slope Solar Distiller Unit Using Al_2O_3 Nanoparticle, *J. Water Environ. Nanotechnol.*, 2023; 8(3): 267–284.

38. Dewangan, A., Mallick, A., Yadav, A.K., Effect of Metal Oxide Nanoparticles and Engine Parameters on the Performance of a Diesel Engine: A Review, *Mater. Today: Proc.*, 2020; 21: 1722–1727.

39. Hajjaji, N., Martinez, S., Trably, E., et al., Life Cycle Assessment of Hydrogen Production from Biogas Reforming, *Int. J. Hydrogen Energy,* 2016; 41: 6064–6075, S0360319915310715.

40. Cetinkaya, E., Dincer, I., Naterer, G.F., Life cycle Assessment of Various Hydrogen Production Methods, *Int. J. Hydrogen Energy*, 2012; 37(3): 2071–2080.

41. Dufour, J., Serrano, D.P., Gálvez, J.L., González, A., Soria, E., Fierro, J.L.G., Life Cycle Assessment of Alternatives for Hydrogen Production from Renewable and Fossil Sources, *Int. J. Hydrogen Energy*, 2012; 37: 1173–1183.

42. International Energy Agency Technical Report, *2012 Key World Energy Statistics*; 2012. Website: https://www.iea.org/publications/freepublications/publication/kwes.pdf [accessed 10.01.13].

43. Ahmed, A., Yadav, A. K., Singh A., Singh D.K., A Comprehensive Machine Learning-Coupled Response Surface Methodology Approach for Predictive Modeling and Optimization of Biogas Potential in Anaerobic Co-Digestion of Organic Waste, *Biomass and Bioenergy,* 2024; 180: 106995.

44. Ghandehariun, S., Kumar, A., Life Cycle Assessment of Windbased Hydrogen Production in Western Canada, *Int. J. Hydrogen Energy*, 2016; 41: 9696–9704, S0360319916312654.

45. Reiter, G., Lindorfer, J., Global Warming Potential of Hydrogen and Methane Production from Renewable Electricity via Power-to-Gas Technology, *Int. J. Life Cycle Assess.*, 2015; 20(4): 477–489.

46. Yadav, A.K., Khan, M.E., Pal, A., Kaner Biodiesel Production through Hybrid Reactor and Its Performance Testing on a CI Engine at Different Compression Ratios, *Egypt. J. Petroleum*, 2017; 26: 525–532.

47. Ahmed, A., Yadav, A.K., Singh, A., Optimization of Biogas Yield from Anaerobic Co-Digestion of Dual Waste for Environmental Sustainability: ANN, RSM, and GA, Approach, *Int. J. Oil Gas Coal Tech.*, 2023; 33(1): 75.

48. Yadav, A.K., Khan, M.E., Pal, A., Experimental Investigations of Performance and Emissions Characteristics of Kusum (*Schleichera oleosa*) Biodiesel in a Multi-Cylinder Transportation Diesel Engine, *Waste Biomass Valor.*, 2017; 8(4): 1331–1341.

49. Yadav, A.K., Khan, M.E., Pal, A., Performance and Emission Characteristics of a Stationary Diesel Engine Fuelled by *Schleichera oleosa* Oil Methyl Ester (SOME) Produced through Hydrodynamic Cavitation Process, *Egypt. J. Petroleum*, 2018; 27: 89–93.

50. Susmozas, A., Iribarren, D., Dufour, J., Life-Cycle Performance of Indirect Biomass Gasification as a Green Alternative to Steam Methane Reforming for Hydrogen Production. *Int. J. Hydrogen Energy*, 2013; 38(24): 9961–9972.

51. Ahmed, A., Yadav, A.K., Singh, A., Modelling and Optimisation of VCR Diesel Engine Parameters Fuelled with Microalgae Methyl-Ester: A RSM-Coupled Taguchi Approach, *Int. J. Ambient Energy*, 2023, 44(1): 2498–2506.

52. Dincer, I., Zamfirescu, C., Sustainable Hydrogen Production Options and the Role of IAHE, *Int. J. Hydrogen Energy*, 2012; 37: 16266–16286.

53. Abanades, A., The Challenge of Hydrogen Production for the Transition to a CO_2-Free Economy, *Agron. Res. Biosystem Eng. Spec.*, 2012; 1: 11–6.

54. Yadav, A.K., Dewangan, A., Mallick, A., Synthesis and Stability Study of Biodiesel from Kachnar Seed Oil, *J. Energy Eng.*, 2018; 144(5): 04018053.

Chapter 13

Hydrogen production and storage
Technological advancements and applications

Kiumars Khani Aminjan
Malek Ashtar University of Technology Tehran

Milad Heidari, Sivasakthivel Thangavel, Pooyan Rahmanivahid, and Morteza Khashehchi
Global College of Engineering and Technology (GCET) Muscat

13.1 HYDROGEN

Hydrogen, denoted by the elemental symbol H with an atomic number of 1, holds the distinction of being the lightest of all elements. In typical conditions, hydrogen exists as a gas, comprised of diatomic molecules bearing the formula H_2. It is notable for its non-toxic, colorless, odorless, and remarkably combustible nature. Across the universe, hydrogen reigns as the most abundant chemical element, constituting approximately 75% of all ordinary matter [1]. The majority of stars, including our Sun, predominantly consist of plasma hydrogen. On our planet, hydrogen often takes the form of molecules within substances like water and organic compounds. The most common isotope of hydrogen symbolized as H, features a lone proton, one electron, and no neutrons within its atomic structure. Notably, the birth of hydrogen nuclei, the protons, commenced in the early universe merely a second after the Big Bang. It wasn't until around 370,000 years later, during the recombination epoch, that the cosmos had sufficiently cooled, allowing electrons to bind with protons and heralding the formation of neutral hydrogen atoms throughout the universe [2]. Hydrogen proves particularly captivating in the realm of energy. Despite possessing the lowest volumetric energy density among all molecules, it boasts the highest gravimetric energy density. However, to directly harness hydrogen as a fuel source, one must first enrich conventional fossil fuels. Hydrogen's utility extends to fuel cells and its role as an energy carrier in various energy storage devices. This versatility brings forth a plethora of benefits along with a corresponding set of drawbacks [3].

DOI: 10.1201/9781003472629-13

13.1.1 Hydrogen production

At present, hydrogen (H_2) finds its primary use in the chemical industry, particularly in the synthesis of methanol and ammonia. Nevertheless, it is foreseen that hydrogen will rapidly emerge as a significant fuel source, substantially enhancing environmental quality. Hydrogen (H), the most abundant chemical element on Earth, can be derived as molecular dihydrogen (H_2) from a diverse array of both renewable and non-renewable sources, utilizing a range of methodologies. Historically, the bulk of hydrogen production globally has relied on fossil fuels, with steam reforming of hydrocarbons, like natural gas, being the predominant contemporary technique. Additionally, the energy-intensive process of electrolyzing water can yield high-purity hydrogen. This article examines current techniques for hydrogen production, encompassing reforming methods (steam, partial oxidation, autothermal, plasma, and aqueous phase), as well as pyrolysis methods, applied to both fossil and renewable biomass resources. Furthermore, purification techniques, such as desulfurization and water-gas shift reactions, are explored alongside other hydrogen production approaches like electrolysis of water. Hydrogen, the simplest and most abundant element on Earth, is inherently present in various compounds, including water, hydrocarbons, and alcohols, due to its rapid reactivity with other elements. It is commonly regarded as an energy carrier rather than a primary energy source.

The production of hydrogen can draw upon a spectrum of domestic resources, ranging from nuclear power, natural gas, and coal to biomass and other renewable sources, encompassing geothermal, hydroelectric, solar, and wind power. The versatility of hydrogen as an energy carrier is instrumental in bolstering energy security, thanks to this diversity of domestic energy sources. Consequently, the ideal approach is to produce hydrogen through a gamut of materials and technological methods. This includes various process technologies, such as thermal methods (e.g., natural gas reforming, processing renewable liquids and bio-oil, biomass, and coal gasification), electrolytic methods (leveraging diverse energy sources to split water), and photolytic methods (employing biological and electrochemical materials to split water under sunlight exposure) [4].

The annual hydrogen production is estimated to surpass 55 million tons, with an annual demand growth rate of roughly 6%. Diverse raw materials allow hydrogen to be generated through several means. Presently, steam reforming of natural gas prevails as the principal method for hydrogen production, albeit it results in substantial greenhouse gas emissions [5,6]. The majority of the world's hydrogen needs are currently met through approximately 30% of oil/naphtha reforming from refinery/chemical industry off-gases, 18% from coal gasification, 3.9% from water electrolysis, and 0.1% from other sources [7]. While electrolytic and plasma-based hydrogen synthesis methods exhibit high efficiency, they are acknowledged to be energy-intensive processes [8].

The core issue at hand is the development of alternative hydrogen production methods, particularly those less reliant on fossil fuels, with a focus on their application as a transportation fuel. Addressing this challenge entails harnessing alternative renewable resources and associated production techniques, such as electrolysis, photolysis, thermal water cracking, and biomass gasification or pyrolysis. Nevertheless, a balanced approach is necessary, one that takes into account both economic viability and environmental considerations, as factors like photolytic water cracking, while ecologically beneficial, might exhibit limited industrial efficiency. Consequently, the chosen methodologies must strike a balance between economic and environmental priorities.

13.1.2 Hydrogen from fossil fuels

Hydrogen-containing materials obtained from fossil fuels, such as gasoline, hydrocarbons, methanol, or ethanol, are transformed into a hydrogen-rich gas stream using fossil fuel processing processes. Methane (natural gas) fuel processing is now the most widely used commercial method of producing hydrogen. The elimination of sulfur, which is a critical challenge in the development of a hydrogen-based economy, is present in the majority of fossil fuels. The desulfurization procedure will thus also be covered. Additionally, a new development in plasma reforming technology will be discussed. Green hydrogen, produced through renewable energy-powered electrolysis, is considered a cleaner and more sustainable alternative to hydrogen production from fossil fuels. It doesn't produce carbon emissions during the production process, making it a key component of efforts to decarbonize industries and energy systems.

13.1.2.1 Steam reforming

Steam methane reforming (SMR) is a widely used industrial process for producing hydrogen from natural gas. It is a key method for hydrogen production, particularly in applications like ammonia production, oil refining, and the production of hydrogen for fuel cells. SMR is an efficient method for producing hydrogen, but it has the drawback of generating carbon dioxide as a byproduct. However, efforts are being made to reduce the carbon footprint of this process by implementing carbon capture and storage (CCS) technologies, as well as exploring the use of cleaner sources of methane or renewable energy for the required heat.

The primary feedstock for SMR is methane (CH_4), which is typically derived from natural gas. Natural gas is a readily available and cost-effective source of methane. In the reformer, methane is combined with high-temperature steam (H_2O) in the presence of a catalyst, usually a nickel-based catalyst. This reaction is highly endothermic, meaning it requires a significant input of heat. The heat can be provided by burning a portion of the methane or by other external heat sources.

There are some chemical reactions. The mixture of carbon monoxide (CO) and hydrogen (H_2), known as synthesis gas or syngas, is generated. Syngas can be used as a feedstock for various industrial processes. The syngas may contain impurities such as carbon dioxide (CO_2), nitrogen (N_2), and sulfur compounds. These impurities are typically removed through various purification processes, depending on the intended use of the hydrogen. To reduce carbon emissions, carbon capture technology can be incorporated to capture and store the carbon dioxide (CO_2) byproduct. This is especially important in the context of reducing greenhouse gas emissions.

13.1.2.2 Partial oxidation

To produce hydrogen for use in fuel cells for cars and other commercial applications, partial oxidation (POX) and catalytic partial oxidation (CPOX) of hydrocarbons have been suggested [9,10]. The gasified source material is often heavy oil fractions (such as vacuum leftovers, and heating oil), which are difficult to further process and use [11]. It's important to note that while partial oxidation is a versatile process, it does produce carbon monoxide (CO), which can be toxic, and it generates heat. Controlling the process conditions is essential to ensure safety and optimize the desired syngas composition for specific applications. In addition, efforts are made to minimize the environmental impact, such as reducing carbon emissions or capturing and repurposing CO_2 emissions, depending on the application.

13.1.2.3 Autothermal reforming

Autothermal reforming (ATR) is a process used to produce hydrogen gas (H_2) from hydrocarbon fuels, such as natural gas or liquid hydrocarbons, while simultaneously generating heat for various industrial applications. ATR combines elements of both SMR and partial oxidation, making it a versatile and efficient method for hydrogen and syngas production. Steam is introduced during the CPOX process in ATR. ATR is a synthesis of partial oxidation (exothermic) and steam reforming (endothermic) processes. ATR is better than SR of methane since it doesn't need external heat and is easier and less costly. ATR can be further enhanced with the addition of advanced catalysts to improve reaction efficiency and control. Moreover, like other hydrogen production methods from fossil fuels, efforts can be made to minimize the environmental impact by implementing CCS technologies to reduce carbon dioxide (CO_2) emissions.

13.1.2.4 Water-gas shift, preferential oxidation, and methanation

Water-gas shift (WGS), preferential oxidation (PROX), and methanation are three important chemical processes used in the production and purification of hydrogen and the treatment of syngas, which is a mixture of carbon monoxide (CO) and hydrogen (H_2). These processes have different roles in syngas and hydrogen production and are often used in combination to tailor the composition of the final product. The WGS reaction is used to convert carbon monoxide (CO) and water vapor (H_2O) into carbon dioxide (CO_2) and hydrogen gas (H_2). This process is essential for purifying syngas because hydrogen gas is often the desired product while CO is an impurity.

PROX is a process used to selectively remove carbon monoxide (CO) from hydrogen-rich gas streams. It is particularly useful when the CO concentration needs to be reduced to very low levels, as in fuel cell applications. In PROX, a catalyst is used to preferentially oxidize CO while leaving the hydrogen gas unaltered. Oxygen is introduced into the gas stream to facilitate the oxidation.

Methanation is a process that converts carbon oxides (usually CO and CO_2) and hydrogen into methane (CH_4). It is used to remove residual CO and CO_2 from syngas or hydrogen streams, improving their purity. In practical applications, these processes are often integrated into a syngas or hydrogen production system to achieve the desired product composition and purity. For example, in the production of high-purity hydrogen gas, a typical sequence of processes might involve SMR to generate syngas, followed by the WGS reaction to remove CO, PROX to reduce CO to trace levels, and methanation to convert any residual CO or CO_2 to methane. The specific combination and order of these processes depend on the application and required gas quality.

13.1.2.5 Desulfurization

Desulfurization is the process of removing sulfur from various materials, including fuels, gases, and industrial processes. It is primarily used to reduce sulfur dioxide (SO_2) emissions, which are harmful air pollutants and contribute to environmental issues such as acid rain and air pollution. Desulfurization techniques are applied to a range of materials, and the choice of method depends on the specific application. Desulfurization is a critical aspect of environmental protection and safety in various industries, as it helps reduce the emission of harmful sulfur compounds and their impact on air and water quality. The specific desulfurization method chosen depends on the type and concentration of sulfur compounds and the desired level of removal for a given application.

13.1.2.6 Plasma reforming

Plasma reforming, also known as plasma gasification, is an advanced technology used for converting organic and inorganic materials into syngas (synthesis gas) through the use of a high-temperature, ionized gas called plasma. This process is particularly useful for waste disposal, energy generation, and the production of hydrogen and other valuable products. Plasma reforming offers several advantages:

Efficient Waste Disposal: It can convert a wide range of waste materials into valuable products while minimizing waste volume.

High Energy Efficiency: The high temperatures achieved in plasma reforming can efficiently break down complex materials, maximizing energy recovery from the feedstock.

Environmental Benefits: It can reduce emissions of pollutants, such as dioxins and furans, that may be produced by traditional waste incineration processes.

Syngas Production: The syngas produced can be used for various applications, including electricity generation, heat production, and hydrogen generation.

Resource Recovery: Valuable materials, such as metals, can be recovered from the residues, enhancing resource efficiency.

However, plasma reforming also faces challenges, including high capital and operational costs and the need for careful management of residues. Despite these challenges, it is a promising technology for managing waste, producing energy, and contributing to sustainable resource management.

13.1.3 Hydrogen from renewable sources

In addition to the reforming of fossil fuels, other processes can also be used to create hydrogen including hydrogen from water (e.g., electrolysis, photoelectrolysis, and thermochemical water splitting) and biomass-based methods (e.g., gasification, pyrolysis, and aqueous phase reforming). Hydrogen produced from renewable sources is often referred to as "green hydrogen." Green hydrogen is considered one of the cleanest and most sustainable forms of hydrogen production because it is produced using renewable energy sources, such as wind, solar, or hydropower, which do not produce greenhouse gas emissions. The primary method for producing green hydrogen is through a process called electrolysis, in which water (H_2O) is split into hydrogen (H_2) and oxygen (O_2) using electricity generated from renewable sources.

13.1.3.1 Electrolysis

Electrolysis is the most common and widely accepted method for producing green hydrogen. It involves using electricity to drive a chemical reaction that separates water into hydrogen and oxygen. Two main types of electrolysis are used: proton exchange membrane (PEM) electrolysis and alkaline electrolysis. In PEM electrolysis, an electric current is passed through a proton exchange membrane, which selectively allows protons (H^+) to migrate toward the cathode to form hydrogen gas. In alkaline electrolysis, the process takes place in an alkaline electrolyte solution, and hydroxide ions (OH^-) migrate toward the anode to form oxygen gas and hydrogen gas at the cathode.

13.1.3.2 Renewable energy sources

The key to producing green hydrogen is to use renewable energy sources to supply the electricity needed for electrolysis. These sources can include wind turbines, solar panels, and hydropower plants. The electricity generated from these renewable sources is clean and does not emit carbon dioxide or other greenhouse gases during the hydrogen production process.

13.1.3.3 Hydrogen production

The electricity is used to power the electrolysis process, which splits water molecules into hydrogen and oxygen. The hydrogen gas is collected and can be stored, transported, or used as a clean energy source for various applications.

Green hydrogen has a wide range of applications, including fuel cell vehicles, energy storage, industrial processes, and power generation. It can be used in sectors that are difficult to electrify directly. Green hydrogen offers several environmental and sustainability advantages:

Zero Carbon Emissions: Because it is produced using renewable energy, green hydrogen is considered a zero-emission fuel, as it does not contribute to carbon dioxide emissions.

Energy Storage: It can serve as a valuable energy storage medium, helping to balance energy supply and demand, especially in regions with intermittent renewable energy sources.

Versatility: Green hydrogen can be used in a variety of sectors and applications, contributing to the decarbonization of the economy.

Reduced Dependence on Fossil Fuels: It helps reduce the dependence on fossil fuels and promotes a transition to a more sustainable and clean energy system.

However, the adoption of green hydrogen faces challenges related to the high initial costs, the need for efficient renewable energy sources, and the development of infrastructure for storage, transportation, and utilization. As renewable energy sources become more widespread and cost-effective, the production of green hydrogen is expected to play a crucial role in reducing carbon emissions and achieving a more sustainable energy future.

13.1.3.3.1 Biomass gasification

Biomass gasification is a thermochemical process that converts biomass materials, such as wood, agricultural residues, energy crops, and other organic matter, into a gas mixture called "syngas" (synthesis gas). Syngas is a versatile fuel that contains hydrogen (H_2), carbon monoxide (CO), and traces of other gases. Biomass gasification has gained attention as a renewable energy technology because it can convert organic waste materials into energy, heat, and valuable chemicals while reducing greenhouse gas emissions. Gasification technology is widely utilized in various processes and is quite developed. It is frequently used with biomass and coal as fuel feedstocks. It is based on the partial oxidation of the feedstock material into a combination of hydrogen, methane, higher hydrocarbons, carbon monoxide, carbon dioxide, and nitrogen, known as "producer gas" [12]. This method is a version of pyrolysis. However, challenges in biomass gasification include the need for sophisticated gas cleaning and conditioning systems, potential issues with feedstock availability and quality, and the high initial capital costs of gasification plants. Despite these challenges, biomass gasification is a promising technology for sustainable energy and waste management.

13.1.4 Economic aspects of hydrogen production

Currently, the dominant and cost-effective method for hydrogen production involves the steam reforming of methane, primarily sourced from natural gas. The cost of hydrogen stands at approximately $7/GJ, with this approach constituting nearly half of the world's hydrogen production. Another comparable method for hydrogen generation is the partial oxidation of hydrocarbons. However, it's important to note that thermochemical processes like these produce greenhouse gases, necessitating capture and storage, which, in turn, introduces an additional cost of 25%–30% for hydrogen production [13]. Thermochemical processes, such as the gasification and pyrolysis of biomass, are still employed today. These methods, while environmentally beneficial, come at a cost roughly three times higher than the steam reforming process. Consequently, it is rare for these approaches to be considered more cost-effective than steam reforming. The cost of hydrogen derived from biomass gasification falls within the range of 10–14 USD/GJ, while that from pyrolysis varies from 8.9 to 15.5 USD/GJ, depending on factors such as equipment availability and feedstock costs. On the other hand, water electrolysis stands out as one of the simplest and waste-free means of hydrogen production. Electrolysis-based processes are characterized by high efficiency. However, the cost of the electricity input significantly impacts the final cost of hydrogen. Looking ahead to 2030, rapid biomass gasification and natural gas steam reforming are poised to be the two primary methods for hydrogen production, with coal gasification and electrolysis playing smaller roles. The application of solar energy, while a subject of debate, may find feasibility in specific scenarios. It is anticipated that solar energy's influence will likely grow by 2050 [5].

13.2 GREEN HYDROGEN PRODUCTION

The term "green hydrogen production" denotes the creation of hydrogen gas (H_2) utilizing sustainable energy sources and techniques that do not yield greenhouse gas emissions. It is widely recognized as one of the most environmentally sound methods for hydrogen generation, playing an integral role in endeavors to reduce carbon emissions across various economic sectors. Nonetheless, the adoption of green hydrogen faces obstacles, including the significant upfront expenses associated with electrolysis equipment, the imperative need for highly efficient renewable energy sources, and the establishment of suitable infrastructure for the storage, conveyance, and utilization of hydrogen. Diligent efforts are actively underway to surmount these challenges and advocate for green hydrogen as a pivotal component in the transition toward a sustainable, low-carbon energy system.

13.2.1 Potential for synergy with renewable energy power generation

Electrolysis-based hydrogen production offers opportunities for synergy with certain renewable energy sources marked by dynamism and intermittent power supply. For instance, while wind energy has become more cost-effective, its inherent unpredictability poses challenges for effective utilization. One promising solution involves integrating a wind farm's hydrogen fuel and electricity production, allowing for flexible adjustment of output to align with resource availability, system operational needs, and market dynamics. Moreover, the surplus power generated by wind farms can be harnessed for hydrogen production through electrolysis instead of curtailment, which is a common practice in such scenarios. It's worth noting that...

- As the majority of today's grid power is derived from energy-intensive and greenhouse-gas-emitting technologies, it may not be the optimal electricity source for electrolysis. To overcome these constraints in electrolysis-based hydrogen production, renewable or nuclear energy sources can be deployed independently from the grid or as an expanding component of the grid's energy mix.
- Ongoing efforts by organizations like the U.S. Department of Energy aim to reduce the cost of power generation from renewable sources and enhance the efficiency of fossil fuel-based electricity generation

while incorporating carbon capture, utilization, and storage. Notably, wind-based power production is experiencing rapid growth both domestically and internationally.

13.3 NEW METHODS AND DESIGNS

Cutting-edge technology and innovative approaches, exemplified by Siemens Energy's Silyzer product line, are crucial for efficiently generating green hydrogen from water and renewable energy sources. The Silyzer excels at harnessing the fluctuating energy output of solar and wind sources through PEM electrolysis. PEM electrolyzers deliver top-tier gas products, thanks to their blend of high efficiency and power density. They are user-friendly and demand minimal maintenance. By adopting a modular design that segments the electrolysis system into skids, we can reduce installation expenses, optimize pricing, and enhance the portability of the system.

13.3.1 PEM water electrolysis work

To combat the global increase in CO_2 emissions and mitigate climate change, it is imperative to develop strategies for producing environmentally friendly and sustainable fuels. Instead of relying on the carbon-intensive process of generating hydrogen from natural gas, which contributes significantly to CO_2 emissions, we can opt for green hydrogen production through electrolysis using renewable energy sources. Siemens has a specific focus on PEM technology for our hydrogen electrolyzer system. The distinctive feature of this membrane is its ability to allow protons to pass through while blocking gases like oxygen or hydrogen. Consequently, this membrane serves as a vital separator in the electrolysis process, preventing the mixing of product gases, among other functions.

13.4 GREEN HYDROGEN APPLICATIONS

Green hydrogen, produced using renewable energy sources without greenhouse gas emissions, has a wide range of applications across various sectors, contributing to a cleaner and more sustainable energy landscape. The shift to a sustainable future in the energy sector is increasingly dependent on green hydrogen. A push to reduce greenhouse gas emissions has combined with a drop in the cost of manufacturing green hydrogen to offer clean hydrogen an unheard-of boost. This component will be crucial to the decarbonization of of several industries, including transportation, industry, and energy storage. We can expect amazing things from green hydrogen, without a doubt. We previously described green hydrogen in a previous essay on sustainability, so we'll only outline some of its advantages and applications now. Green hydrogen's potential lies in its ability to serve as a versatile and clean energy carrier that can reduce greenhouse gas emissions across various sectors. It has the potential to contribute significantly to achieving a more sustainable and environmentally responsible energy landscape. However, the widespread adoption of green hydrogen depends on factors such as cost competitiveness, infrastructure development, and policy support.

Green hydrogen, produced using renewable energy sources and without greenhouse gas emissions, can serve as a clean and sustainable fuel in various applications. Green hydrogen can be used as fuel for fuel cell vehicles (FCVs). In an FCV, hydrogen is supplied to a fuel cell, where it reacts with oxygen from the air to produce electricity. This electricity powers the vehicle, and the only byproduct is water vapor, making FCVs a zero-emission transportation option. Beyond FCVs, green hydrogen can be used in various transportation modes, including buses, trucks, and trains, to reduce emissions and support sustainable mobility. Hydrogen can be used as a fuel for ships, particularly in applications where electrification is challenging, such as long-haul maritime routes. Hydrogen fuel cells or hydrogen-powered engines can provide an environmentally friendly alternative to traditional marine propulsion systems. Hydrogen can replace fossil fuels in industrial processes that require high-temperature heat or combustion, such as steel manufacturing, glass production, and petrochemical refining. Hydrogen can be used to store excess renewable energy for later use. Electrolysis systems can convert surplus electricity into hydrogen, which can then be used in fuel cells to generate electricity when needed. This helps balance energy grids with intermittent renewable sources.

Fuel cell electric vehicles (FCEVs) signify a transition toward adopting carbon-neutral energy sources in both the energy and transportation sectors. Currently, they account for just 0.5% of new low-emission vehicle sales. Nevertheless, the market for FCEVs is gaining momentum, as reported by the International Energy Agency. Paris is in the process of developing a fleet of hydrogen-powered taxis, while various European towns have incorporated this technology into their garbage collection vans. Hyundai has set an ambitious goal to manufacture 500,000 hydrogen cars by 2030. Furthermore, green hydrogen will play a crucial role in diverse modes of transportation, including trains, planes, trucks, buses, and maritime vessels.

At present, hydrogen finds primary applications in two key industries, aside from its role as an energy storage medium. One application is within the chemical sector, where it is used for producing fertilizers and ammonia. The petrochemical industry employs hydrogen to manufacture various petroleum products, representing its second major utilization.

Additionally, it is emerging as a viable option in the steel industry, which has faced criticism in Europe due to its environmental impact. Hydrogen enables the modification of certain industry processes to reduce environmental harm.

In the realm of residential applications, there are ongoing initiatives focused on transitioning from the current natural gas network to a clean hydrogen network, providing homes with clean and emissions-free power and heating. Green hydrogen undoubtedly plays a pivotal role in the broader effort to decarbonize the economy. The post-COVID-19 world is undoubtedly moving toward renewable energy sources. However, there remain hurdles on the path to widespread adoption, including the reduction of manufacturing costs, improvements in storage solutions, and the establishment of essential infrastructure. Overcoming these challenges will bring green hydrogen into the present as a prevailing energy source rather than a distant prospect.

13.5 BENEFITS OF GREEN HYDROGEN

In the ongoing battle between fossil fuels and renewable energy, the arguments against renewable sources are diminishing. There are numerous misconceptions surrounding renewable energy, but one of their major challenges is their heavy reliance on environmental conditions. Historically, wind power was absent when there was no wind, and solar power remained dormant during nighttime. The quest for a viable solution that allows surplus energy generated during favorable weather conditions to be stored for use during less favorable times has been a primary focus. After all, we cannot control the timing of the Sun's rays or summon the wind to turn windmill blades as needed. The frontrunner in this quest is green hydrogen. Through electrolysis, green hydrogen is produced from renewable resources, and this pure hydrogen can be stored in specialized tanks. When energy is needed, this hydrogen can then be directed into a fuel cell, where it is combined with oxygen from the air to generate electricity, with the only byproduct being water. Green hydrogen, which is produced using renewable energy sources without greenhouse gas emissions, offers numerous environmental, economic, and energy security benefits. The production of green hydrogen emits no carbon dioxide (CO_2) or other greenhouse gases. This makes it a clean and sustainable energy carrier that can significantly reduce carbon emissions in various sectors. Green hydrogen can be used in a wide range of applications, from fueling vehicles and generating electricity to providing heat for industrial processes and serving as a chemical feedstock. Its versatility makes it a valuable energy carrier for decarbonizing multiple sectors. Green hydrogen production can promote the efficient use of water resources by enabling the recycling of water in closed-loop systems, making it a sustainable solution in regions with water scarcity. Green hydrogen can store excess renewable energy for later use. This energy storage capability helps balance energy grids with intermittent renewable sources, improving grid stability and reliability. While green hydrogen offers numerous benefits, its widespread adoption depends on overcoming challenges such as cost competitiveness, infrastructure development, and policy support. However, as efforts to address these challenges continue, green hydrogen is poised to play a crucial role in reducing carbon emissions and achieving a more sustainable and environmentally responsible energy system.

REFERENCES

1. P. Boyd. (19 July 2014). "What is the chemical composition of stars?" NASA. Archived from the original on 15 January 2015. Retrieved 5 February 2008. https://africa-eu-energy-partnership.org/wp-content/uploads/2021/04/AEEP_Green-Hydrogen_Bridging-the-Energy-Transition-in-Africa-and-Europe_Final_For-Publication_2.pdf
2. Tanabashi et al. (2018). Chapter 21.4.1: "Big-Bang Cosmology." Archived 29 June 2021 at the Wayback Machine. Revised September 2017 by K.A. Olive and J.A. Peacock, p. 358. https://pdg.lbl.gov/2018/reviews/rpp2018-rev-bbang-cosmology.pdf
3. M. H. McCay and S. Shafiee. (2020). "Hydrogen: An energy carrier." In *Future Energy, third edition,* Improved, Sustainable and Clean Options for our Planet, edited by Trevor M. Letcher (pp. 475–493). https://doi.org/10.1016/B978-0-08-102886-5.00022-0
4. C. M. Kalamaras and A. M. Efstathiou, "Hydrogen production technologies: Current state and future developments", *Conference Papers in Science*, vol. 2013, Article ID 690627, 9 pages, 2013. https://doi.org/10.1155/2013/690627.
5. M. Balat and M. Balat, "Political, economic and environmental impacts of biomass-based hydrogen," *International Journal of Hydrogen Energy*, vol. 34, no. 9, pp. 3589–3603, 2009.
6. A. Konieczny, K. Mondal, T. Wiltowski, and P. Dydo, "Catalyst development for thermocatalytic decomposition of methane to hydrogen," *International Journal of Hydrogen Energy*, vol. 33, no. 1, pp. 264–272, 2008.
7. N. Z. Muradov and T. N. Veziroğlu, "From hydrocarbon to hydrogen-carbon to hydrogen economy," *International Journal of Hydrogen Energy*, vol. 30, no. 3, pp. 225–237, 2005.
8. J. D. Holladay, J. Hu, D. L. King, and Y. Wang, "An overview of hydrogen production technologies," *Catalysis Today*, vol. 139, no. 4, pp. 244–260, 2009.
9. K. L. Hohn and L. D. Schmidt, "Partial oxidation of methane to syngas at high space velocities over Rh-coated spheres," *Applied Catalysis A*, vol. 211, no. 1, pp. 53–68, 2001.
10. J. J. Krummenacher, K. N. West, and L. D. Schmidt, "Catalytic partial oxidation of higher hydrocarbons at millisecond contact times: Decane, hexadecane, and diesel fuel," *Journal of Catalysis*, vol. 215, no. 2, pp. 332–343, 2003.

11. A. Holmen, "Direct conversion of methane to fuels and chemicals," *Catalysis Today*, vol. 142, no. 1–2, pp. 2–8, 2009.
12. M. F. Demirbas, "Hydrogen from various biomass species via pyrolysis and steam gasification processes," *Energy Sources A*, vol. 28, no. 3, pp. 245–252, 2006.
13. H. Balat and E. Kirtay, "Hydrogen from biomass-present scenario and future prospects," *International Journal of Hydrogen Energy*, vol. 35, no. 14, pp. 7416–7426, 2010.

Chapter 14

Harnessing solar and hydrogen energy

Conserving for the generations to come

Puneet kumar and Apurva Goyal
IIMT University Meerut

14.1 INTRODUCTION

The risingdemand for energy, coupled along with the urgency to mitigate climate change, has prompted a significant shift towards renewable energy sources that are sustainable. Among the various renewable options, solar and hydrogen energy have emerged as prominent contenders, holding immense potential of meeting our future energy needs while plummeting greenhouse gas emissions. This book chapter aims to explore the pivotal role of solar and hydrogen energy in conserving resources for the benefit of future generations as India is working towards achieving 500 GW of installed electric capacity by 2030, generated from non-fossil sources.

Solar energy, consequentof the sun's radiant light and heat, has captivated scientists and researchers worldwide. It offers an abundant, clean, and virtually inexhaustible source of power. The development of innovative solar technologies, such as photovoltaic cells and concentrated solar power systems, has vastly improved energy conversion efficiency, making solar energy increasingly cost-effective and accessible.

Simultaneously, hydrogen energy has emerged as a promising avenue for achieving a sustainable energy future. Hydrogen, the most plentiful element in the universe, can be shaped from various renewable sources, including water and biomass. It serves as an energy carrier, capable of powering anextensive range of applications, including transportation, industry, and electricity generation. When utilized in fuel cells, hydrogen produces electricity without any harmful emissions, offering a clean and efficient alternative to conservative energy sources.

The integration of solar and hydrogen technologies presents a compelling solution to address the challenges of energy security, climate change, and environmental sustainability. By harnessing the sun's energy through photovoltaic systems, excess electricity can be used to fabricate hydrogen through electrolysis, thereby storing energy for later use or enabling its conversion back into electricity. This symbiotic relationship between solar and hydrogen energy holds immense promise for creating a reliable, decentralized and carbon-free energy infrastructure.

DOI: 10.1201/9781003472629-14

In this Scopus indexed book chapter, we delve into the multifaceted aspects of solar and hydrogen energy utilization. We explore the fundamental principles and technologies underpinning solar power generation and hydrogen production, highlighting their environmental advantages and potential applications. Furthermore, we examine the integration of solar and hydrogen systems, investigating their synergistic effects and exploring innovative approaches to maximize their combined benefits.

By conserving resources and reducing dependence on fossil fuels, solar and hydrogen energy offer a pathway towards a sustainable future for generations to come. Through comprehensive research and exploration of cutting-edge advancements, this book chapter aims to contribute to the collective knowledge and understanding of harnessing solar and hydrogen energy, ultimately fostering a greener and more resilient world for the benefit of future generations.

14.2 SOLAR ENERGY

Solar energy, derivative of the beaming light and heat of the Sun, is a formidable force in the realm of renewable energy. It holds immense potential as a clean, abundant, and virtually inexhaustible source of power. The utilization of solar energy has witnessed a remarkable evolution over the years, driven by advancements in technology, increasing efficiency, and declining costs. Today, solar energy has emerged as a key participant in the global energy landscape, transforming the way we generate electricity, heat our homes, and power our vehicles.

At the heart of solar energy conversion are photovoltaic (PV) systems, which harness the sun's energy and convert it into usable electricity. PV cells, typically composed of semiconductor materials, soak up sunlight and produce a flow of electrons, creating a direct current (DC). This DC electricity can be used straight away or stored in batteries for later use. The effectiveness of PV cells has steadily improved, with modern systems achieving conversion efficiencies exceeding 20%, enabling more energy to be harnessed from sunlight.

India's ambition to enhance capacity, advance energy safekeeping, unravel environmental issues, and dominate the considerable market for renewable energy sources can all benefit from the use of energy derived from the source sun. Concentrating solar power (CSP), commonly referred to as solar thermal electricity (STE), is a developing renewable energy technology that could be used in India to generate electricity in the future [1]. Although there is significant discussion regarding the future of solar technology and its markets, this power is currently one of the most popular energy investment sectors. In India, there is a continuous increase in the amount of electricity produced by CSP per year and huge potential for solar energy due to high solar radiation of 1700–1900 kWh per kilowatt peak for than 300 days with clear skies each year. 10,000 MW till 2017 and 100,000

MW until 2022 are the additional solar power generation goals established by the Indian government [2,3]. The government recently increased the goal for installing solar power capacity including the development of Ultra Mega Solar Power Projects by 2024 and solar parks by 2024. Therefore we can say that India has the potential of completing energy crisis through the most favourable utilization of solar energy if the policies are rendered effectively.

14.3 HYDROGEN ENERGY

Hydrogen energy is created by using hydrogen and compounds containing hydrogen to create energy with high energy efficiency, massively favourable environmental and social consequences, and competitive cost-effective rewards [4]. To meet goals for energy security, economic growth, and ecological protection on a national and worldwide scale, infrastructure development for mobile and fixed applications is becoming more and more crucial. Policies and technology related to hydrogen energy are still developing [5].

Hydrogen technologies that are viable and sustainable may be able to meet the world's energy demands. All energy-related industries, including those in transportation, construction, utilities, and business, stand to gain. It can provide storage options for intermittent (PV and wind), seasonal (hydroelectric), and baseload (geothermal) renewable energy sources, and when used in conjunction with state-of-the-art decarbonization technologies, it can decrease the climate impacts of prolonged fossil fuel use [6]. The International Energy Agency (IEA) established in 1974, also made significant contributions towards economic growth, energy security, environmental preservation, and the welfare of the group's members and the globe at large, andlaunched the Hydrogen Agreement in 1977 to promote technology for producing, storing, and using hydrogen as well as to hasten its adoption and wider use [6].

Also, India has set goals for achieving energy independence along with Net Zero by 2047 and 2070 respectively. Its Energy Transition is focused on maximizing the usage of renewable energy across all economic sectors in order to meet the above-mentioned goal. A potential substitute for facilitating this shift is Green Hydrogen. In addition to replacing fossil fuels in industry and providing clean transportation, hydrogen can also be used for decentralized power generation, aviation, and maritime transportation [7]. The government even launched a mission named as 'The National Green Hydrogen Mission' on January 4, 2022, with an objective of making India the leading producer as well as supplier of Green Hydrogen in the World under the mission "The National Hydrogen Mission" officially started by our Prime Minister Narendra Modi, on August 15, 2021, the 75th anniversary of India's independence. The Mission aspires towards making India a Green Hydrogen centre while assisting the government in achieving its climate goals. By 2030, it will be possible to produce 5 million tonnes of Green

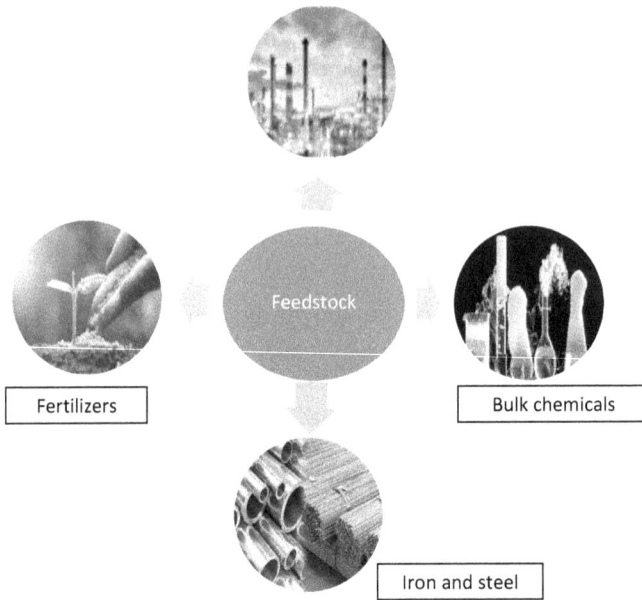

Figure 14.1 Feedstock can replace all above. (Source: Ministry of Information and Broadcasting Government of India.)

Hydrogen, whichin turn will improve the capacity for renewable energy [7]. It will fund pilot initiatives in additional difficult-to-abate industries, such as steel, long-range heavy-duty transportation, shipping, energy storage, etc., for substituting Green Hydrogen and its derivatives for fossil fuels and fossil fuel-based feedstocks [8] (Figure 14.1).

There are three types of hydrogen divided as per their extraction methods: Grey, Blue and Green Hydrogen. Few measures are being taken to increase the usage of hydrogen in the energy mix by the Ministry of Petroleum & Natural Gas (MoPNG). The first pilot uses Grey Hydrogen, also known as the Hydrogen CNG (H-CNG) programme, which blends hydrogen to a maximum of 18% with compressed natural gas (CNG) and other is Green Hydrogen which is produced and used as transportation fuel as well as an industrial yield for refineries [9].

14.4 INTEGRATION OF SOLAR AND HYDROGEN ENERGY

The integration of solar and hydrogen technologies represents a powerful symbiotic relationship that can revolutionize our energy landscape. Solar energy, harnessed through photovoltaic systems, provides a direct and abundant source of electricity generation, while hydrogen serves as a versatile energy carrier, capable of storing and converting energy for various applications. This integration holds immense potential in addressing the intermittent nature of solar power and overcoming the limitations of hydrogen storage, leading to a reliable and sustainable energy system.

One key aspect of solar-hydrogen integration lies in using excess electricity generated by solar photovoltaic systems to produce hydrogen through electrolysis. During times of peak solar power generation, when demand is low, the surplus electricity can be diverted to electrolyzers, splitting water into hydrogen and oxygen. This process not only allows for the efficient utilization of excess renewable energy but also facilitates energy storage in the form of hydrogen. The stored hydrogen can be used later during periods of low solar generation or high electricity demand, allowing for a continuous and reliable energy supply (Figure 14.2).

Moreover, fuel cells can be used to turn the hydrogen that has been stored back into energy or hydrogen combustion engines. This conversion

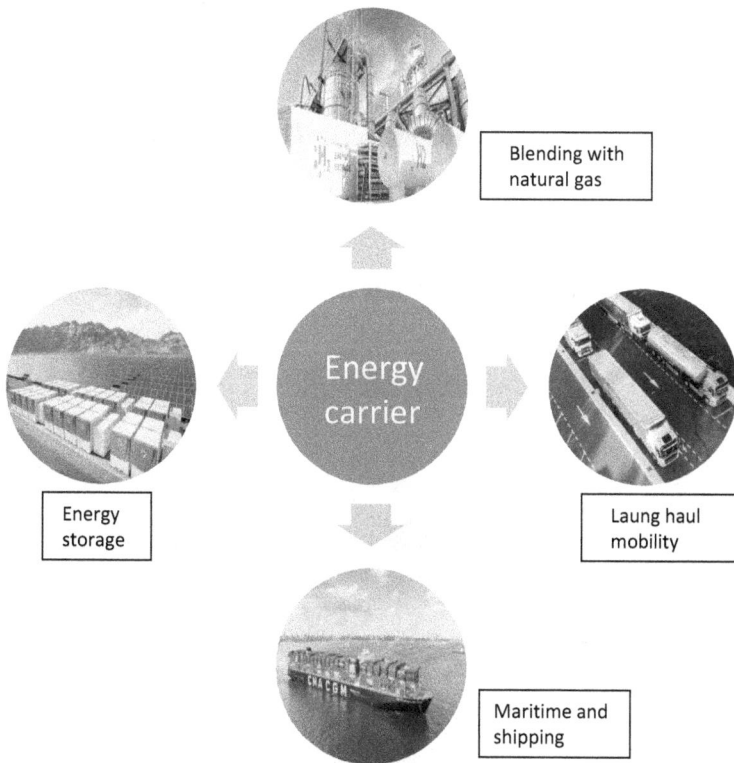

Figure 14.2 Green hydrogen can replace all above. (Source: Ministry of Information and Broadcasting Government of India.)

process enables the flexible utilization of hydrogen as an energy source for a wide range of applications, including transportation, heating, and power generation. Fuel cell electric vehicles (FCEVs), for instance, can utilize stored hydrogen to produce electricity, providing zero-emission transportation with long driving ranges and quick refuelling times. Similarly, fuel cells can power stationary applications, such as buildings or remote power systems, ensuring a continuous energy supply without relying solely on solar availability.

Research and development efforts are actively underway to optimize the integration of solar and hydrogen technologies. Advancements in electrolysis technologies, such as proton exchange membrane (PEM) electrolyzers, are improving the efficiency and cost-effectiveness of hydrogen production. In addition, the development of advanced materials, such as catalysts and membranes, aims to enhance the performance and durability of fuel cells, making them more viable for widespread adoption. Integration strategies are being explored to optimize system design, control, and operation, aiming for maximum efficiency and cost optimization.

Challenges remain in the integration of solar and hydrogen technologies, including the scalability of electrolysis and hydrogen production, the cost-effectiveness of storage and conversion systems, and the development of a hydrogen infrastructure. These challenges necessitate interdisciplinary collaborations among researchers, engineers, policymakers, and industry stakeholders to overcome technological, economic, and regulatory barriers (Figure 14.3).

Indian solar market

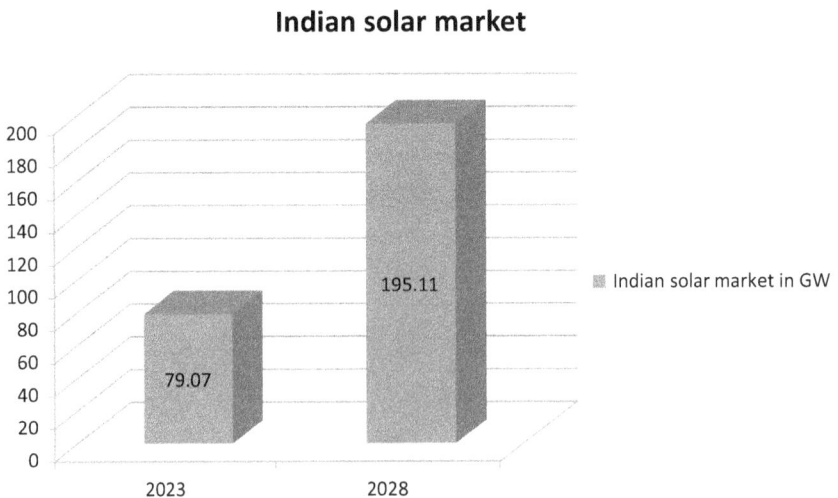

Figure 14.3 Estimated solar energy market size of India. (Source: Mordor Intelligence.)

14.5 RENEWABLE ENERGY ANALYSIS IN INDIA

By the end of this year, the Indian solar market is anticipated to be at 79.07 GW, and it is also expected to grow at a CAGR of 19.8% over the following 5 years to reach 195.11 GW. The Indian solar energy industry is anticipated to grow over the medium term as a result of declining solar power technology costs, more flexible solar systems, and solar energy's position as a cleaner source of electricity. Favourable government policies, particularly those from the Ministry of New and Renewable Energy (MNRE) that make it easier to generate power from renewable sources, can influence the market. The market for solar energy is, however, constrained by issues including transmission and distribution losses and erratic power supply continuity [10] (Figures 14.4 and 14.5).

India added 13,956 MW of solar power and 1847 MW of wind power capacity in 2022. In the states of Rajasthan, Gujarat, and Tamil Nadu, the majority of solar capacity was installed. According to the MNRE, the nation's total installed renewable energy capacity as of December 31st was 120.85 GW. In terms of the whole mix of renewable energy, solar energy makes up around 52%, followed by wind at 35%, biopower at 9%, and small hydro at 4% [11] (Figures 14.6 and 14.7).

To separate water from oxygen, electrolyzers driven by renewable energy are used to produce green or clean hydrogen. India authorized its National Green Hydrogen Mission earlier this year, which comprises an initial investment of 197.44 billion rupees ($2.4 billion) for research and development and experimental projects [12].

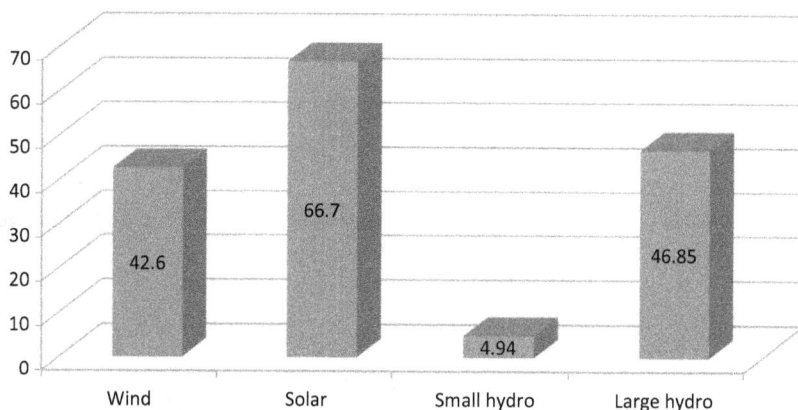

Figure 14.4 Indian solar energy market as of February 2023. (Source: Ministry of New and Renewable Energy.)

Solar capacity

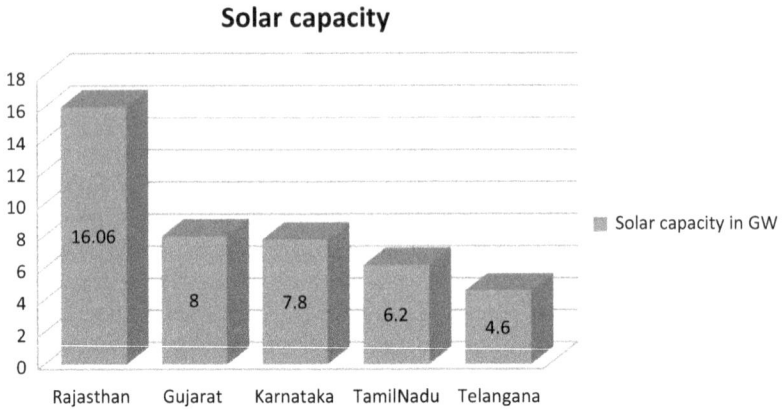

Figure 14.5 India's top cities with the most solar power capacity. (Source: Solar Energy Statistics. https://www.solarfeeds.com/.)

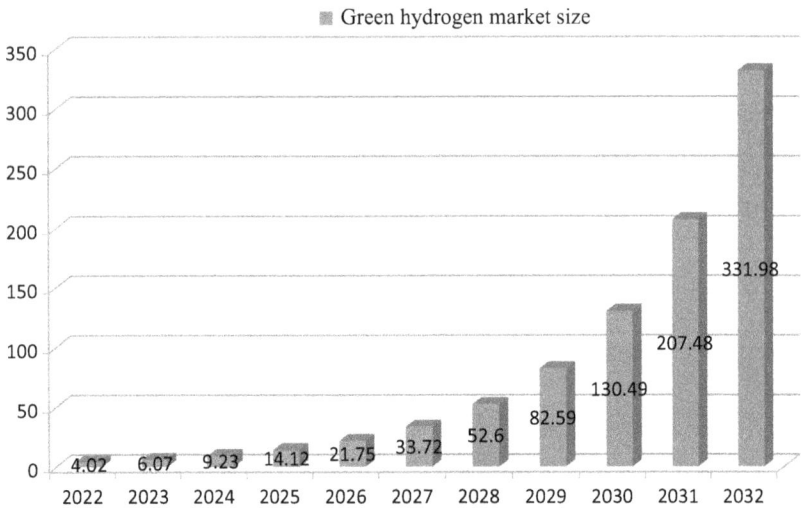

Figure 14.6 Estimated Green Hydrogen market size in $bn from 2023 to 2032. (Source: www.precedenceresearch.com.)

The increasing proportion of hydrogen in total emission reductions illustrates the significance of hydrogen in the Net Zero Emissions picture [13]. Strong hydrogen demand growth and the adoption of cleaner technologies for its production enable hydrogen and fuels derived from it to play a significant role in the decarbonization of sectors where emissions are difficult to reduce, such as heavy industry and long haul transportation, in the Net Zero Emissions Scenario [14] (Figure 14.8).

Cumulative mitigation measure emissions reduction in the Net Zero scenario, 2021–2050

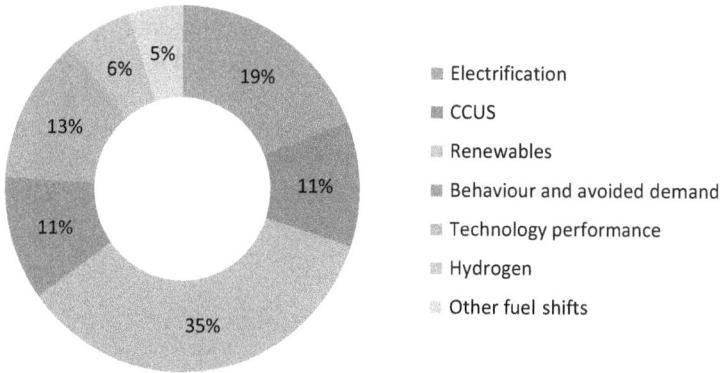

- Electrification
- CCUS
- Renewables
- Behaviour and avoided demand
- Technology performance
- Hydrogen
- Other fuel shifts

Figure 14.7 Cumulative emissions reduction by mitigation measure in the Net Zero Scenario, 2021–2050. (Source: IEA. Licence: CC BY 4.0.)

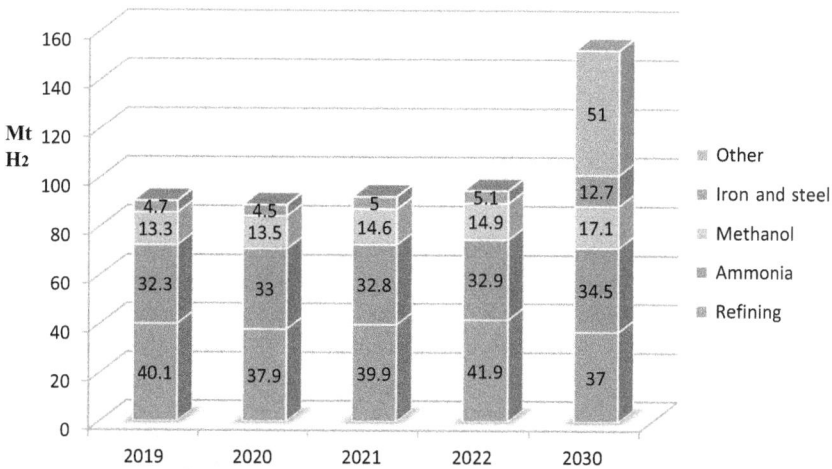

Figure 14.8 Global hydrogen demand by sector in the Net Zero Scenario, 2020–2030. (Source: IEA. Licence: CC BY 4.0.)

14.6 GLOBAL HYDROGEN DEMAND BY SECTOR IN THE NET ZERO SCENARIO, 2020–2030

In 2022, the universal demand for hydrogen increased by around 3%, but it still remained mostly focused on existing applications, with slow adoption of new uses. In 2022, the world's demand for hydrogen increased by approximately 3% to 95 Mt. With very little penetration in new applications, the demand for hydrogen is still primarily driven by established uses in the industrial and refining sectors (including chemicals and Direct Iron Reduction [DRI] using natural gas)[13].Less than 0.1% of worldwide demand is accounted for by demand in novel applications such transportation, high-temperature heat in industry, hydrogen-based DRI, power, and buildings. Although additional applications are beginning to gain traction, most of this demand is focused on road transportation. By 2030, there will be a more than 1.5-fold rise in hydrogen consumption, reaching more than 150 Mt, with new uses accounting for over 30% of total demand [14].

14.7 FUTURE PERSPECTIVES AND CHALLENGES

The integration of solar and hydrogen technologies presents exciting future perspectives for achieving a sustainable and decarbonized energy system. However, several challenges must be addressed to fully apprehend the potential of this integration. Understanding the future perspectives and overcoming these challenges is crucial for driving the adoption and scalability of solar-hydrogen systems.

- One of the key future perspectives lies in the continued advancement of solar energy technologies. Continuous development efforts and ongoing research aim towards enhancingthe efficiency, durability, and cost-effectiveness of photovoltaic systems. The use of innovative materials, such as perovskites and tandem solar cells, shows great promise in pushing the efficiency limits of solar cells and reducing manufacturing costs. Additionally, the integration of solar energy into building materials, such as solar windows and solar roof tiles, can enable seamless and widespread deployment of solar power generation.
- Another future perspective lies in the development of advanced electrolysis technologies for efficient and cost-effective hydrogen production. PEM electrolyzers are currently the most widely used, but advancements in alkaline electrolysis and solid oxide electrolysis offer potential breakthroughs in efficiency, durability, and scalability. Furthermore, the use of renewable energy sources, such as solar and wind, for electrolysis can ensure the carbon neutrality of hydrogen production, enhancing its environmental credentials [15].
- Grid integration and system optimization are crucial future perspectives for the integration of solar and hydrogen technologies. The development of smart grid technologies, energy management systems, and advanced control algorithms can enable seamless integration of solar power, hydrogen production, and electricity consumption.

- By leveraging digitalization and data analytics, intelligent energy management systems can optimize the utilization of solar energy, hydrogen storage, and grid balancing, enhancing overall system efficiency and stability.
- The scalability and cost-effectiveness of hydrogen storage technologies are essential future perspectives [16]. Research efforts focus on developing high-density and compact storage solutions, such as hydrogen compression, liquefaction, and solid-state storage materials.
- The development of storage technologies can enable the widespread use of hydrogen as an energy carrier and make it easier for it to be used in a variety of fields, such as manufacturing, transportation, and power generation [17].

14.8 CHALLENGES

Despite these future perspectives, several challenges need to be addressed, which are given as follows:

- Cost remains a significant barrier for widespread adoption of solar and hydrogen technologies. Continued research and development efforts are required to drive down the costs of photovoltaic systems, electrolyzers, fuel cells, and hydrogen storage technologies.
- A comprehensive hydrogen infrastructure, including facilities for production, storage, transportation, and refuelling, must be developed with cooperation from all relevant parties..
- Policy and regulatory frameworks play a vital role in shaping the future of solar-hydrogen integration [18]. Supportive policies, incentives, and regulations are needed to promote the deployment of solar and hydrogen technologies, encourage investments, and facilitate the integration of these systems into the existing energy infrastructure. Long-term policy commitments and international collaborations are necessary to provide a stable and predictable market environment for stakeholders in the solar and hydrogen sectors.
- Public awareness and acceptance are critical challenges that must be addressed. Educating the public about the benefits and potential of solar and hydrogen energy, dispelling myths, and addressing safety concerns can enhance public acceptance and support for the adoption of these technologies.
- Collaboration between academia, industry, and government agencies is essential in driving public awareness campaigns and fostering a supportive societal environment for solar-hydrogen integration [19].

14.9 CONCLUSION

India's power industry is undergoing a significant transformation, driven by electrification, rising demand, and a significant rise in the fabrication of renewable energy. The problems and opportunities of demand flexibility and VRE integration are made possible by high demand growth and renewable generation. The supply-side flexibility of thermal power plants, demand response, and battery energy storage are a few of the technical options that are getting more and more attention.

When the utilization factor decreases, the price of power-H2-power per unit rises more gradually than it does for other technologies like Li-ion batteries. Power-H2-power ranks among the most viable choices for long-term storage (together with other potential long-term chemical storage vectors, such as ammonia) because of this link between per-unit costs and utilization factor [20].

In addition to offering flexibility, hydrogen and solar energy are viewed as a viable fuel source for peaking power generation using current gas turbines, as well as for power generating and backup applications for distributed assets like mobile phone towers and the replacement of diesel gensets.

One important finding is that throughout the transition to a very high VRE system, only the final 10%–20% of dispatchable fossil generation will need to be eliminated using hydrogen and other seasonal storage options. According to the IEA's generation projection, the share of VRE in its sustainable development scenario will be at most 69% by 2040. Therefore, the promise for power-H2-power in India will only become apparent towards the end of our scenario period, and even then, it will have a minimal influence.

In conclusion, the future perspectives for the integration of solar and hydrogen technologies are promising, offering solutions for clean and sustainable energy systems. However, challenges related to cost, infrastructure, policy, and public acceptance need to be overcome. Continued research, technological advancements, policy support, and stakeholder collaboration are crucial in unlocking the full potential of solar-hydrogen integration and realizing a greener and more sustainable future.

REFERENCES

1. N. Sharma, P. Tiwari, Y.R. Sood (2012) Solar energy in India: Strategies, policies, perspectives and future potential, *Renewable and Sustainable Energy Reviews*, Volume 16, Issue 1, pp. 933–941. https://doi.org/10.1016/j.rser.2011.09.014.
2. A. Kumar, O. Prakash, A. Dube (2017) A review on progress of concentrated solar power in India, *Renewable and Sustainable Energy Reviews*, Volume 79, pp. 304–307. https://doi.org/10.1016/j.rser.2017.05.086.

3. A. Kumar, K. Kumar, N. Kaushik, S. Sharma, S. Mishra (2010) Renewable energy in India: Current status and future potentials, *Renewable and Sustainable Energy Reviews*, Volume 14, Issue 8, pp. 2434–2442. https://doi.org/10.1016/j.rser.2010.04.003.
4. P. E. V. de Miranda (2019) Chapter 1 - Hydrogen energy: Sustainable and perennial, Editor(s): P. E. V. de Miranda, In *Science and Engineering of Hydrogen-Based Energy Technologies*, Cambridge, MA: Academic Press, pp. 1–38. https://doi.org/10.1016/B978-0-12-814251-6.00001-0.
5. R. K. Dixon, J. Li, M. Q. Wang (2016) Chapter 13 - Progress in hydrogen energy infrastructure development-addressing technical and institutional barriers, Editor(s): R. B. Gupta, A. Basile, T. Nejat Veziroğlu, In *Woodhead Publishing Series in Energy, Compendium of Hydrogen Energy*. Sawston: Woodhead Publishing, pp. 323–343. https://doi.org/10.1016/B978-1-78242-362-1.00013-4.
6. C. C. Elam, C. E.Gregoire Padró, G. Sandrock, A. Luzzi, P. Lindblad, E. F. Hagen (2003) Realizing the hydrogen future: The International Energy Agency's efforts to advance hydrogen energy technologies, *International Journal of Hydrogen Energy*, Volume 28, Issue 6, pp. 601–607. https://doi.org/10.1016/S0360-3199(02)00147-7.
7. National Green Hydrogen Mission of Government of India. https://www.india.gov.in/spotlight/national-green-hydrogen-mission.
8. RU-44-01-0067-210322/EXPLAINER. https://static.pib.gov.in/WriteReadData/specificdocs/documents/2023/jan/doc2023110150801.pdf.
9. Hydrogen Initiatives of the Ministry of Petroleum & Natural Gas. https://mopng.gov.in/en/page/12#_ftn1.
10. Solar Industry in India Size & Share Analysis - Growth Trends & Forecasts (2023–2028). https://www.mordorintelligence.com/industry-reports/india-solar-energy-market.
11. SolarFeeds Marketplace (2022) https://www.solarfeeds.com/
12. What India Needs to Do to Become a Global Green Hydrogen Hub. https://www.thenationalnews.com/climate/road-to-net-zero/2023/04/10/what-india-needs-to-do-to-become-a-global-green-hydrogen-hub/.
13. IEA (2021) *Net Zero by 2050*, Paris: IEA. https://www.iea.org/reports/net-zero-by-2050, License: CC BY 4.0.
14. International Energy Agency (2023). *Global Hydrogen Review 2023*, Paris: IEA. https://www.iea.org/reports/global-hydrogen-review-2023, License: CC BY 4.0.
15. Energy Technology Perspectives (ETP) (2020) IEA. Track global progress towards the goal of tripling renewable energy capacity by 2030. https://www.iea.org/reports/energy-technologyperspectives-2020.
16. Hydrogen for Net-Zero - A Critical Cost-Competitive Energy Vector (2021) Hydrogen Council, McKinsey & Company. https://hydrogencouncil.com/en/wp-content/%20uploads/2021/11/Hydrogen-for-Net-Zero_FullReport.pdf/.
17. RENA (2020), Green Hydrogen Cost Reduction: Scaling up Electrolysers to Meet the 1.5 °C Climate Goal, International Renewable Energy Agency, Abu Dhabi. (December 2020). https://www.irena.org/-/media/Files/IRENA/Agency/Publication/2020/Dec/IRENA_Green_hydrogen_cost_2020.pdf.

18. IEA (2020), *Energy Technology Perspectives 2020*, Paris: IEA. https://www.iea.org/reports/energy-technology-perspectives-2020, License: CC BY 4.0.
19. Hydrogen on the Horizon: Ready, Almost Set, Go? World Energy Council. World Energy Council July 2021. www.worldenergy.org
20. W. Hall, T. Spencer, G. Renjith, S. Dayal (2020) *The Potential Role of Hydrogen in India - A Pathway for Scaling-up Low Carbon Hydrogen across the Economy*, New Delhi: The Energy and Resource Institute (TERI). https://www.teriin.org/sites/default/files/2021-07/Report_on_The_Potential_Role_of_%20Hydrogen_in_India.pdf.

Chapter 15

Solar power potential of India

Thermal, economical and technological advancements

Priyanka Gajare

G H Raisoni College of Engineering and
Business Management Jalgaon

Bipasa B. Patra

Padmashri Dr. V. B. Kolte College of Engineering Malkapur

15.1 INTRODUCTION

Emissions free and no energy consumption buildings are attracting attention rapidly as energy efficiency and renewable, non-polluting energy continue to be of great interest globally. To be energy-efficient or emission free, an entity must obtain energy from the environment, with solar energy being one of the most obvious sources. Building photovoltaic integrated systems (BIPV), which use solar radiation to generate electricity while integrating solar cells into building climate envelopes, may be a strong and adaptable tool for achieving these objectives from an aesthetic, financial, and technical perspective.

BIPV systems are viewed as either an architecturally integrated aspect of the building design or as a functional component of the building construction. Building-integrated photovoltaics (BIPV) is the installation of solar modules into the outermost layer of a building, such as the roof or the outer walls. BIPV systems may minimise the use of fossil fuels and the production of ozone-depleting gases while also saving money on materials and electricity costs. They could also improve the structure's artistic value. As a result, the BIPV system works as both a material for the building envelope and a source of energy [1–3].

The system architecture is mostly determined by PV efficiency variables, with aesthetics frequently coming in second. However, rising consumer interest in distributed PV technologies and industry competitiveness in reducing installation costs contribute to the development of multipurpose PV products that work together with building components. The President's 2011 State of the Union Address included a rooftop solar module product, reflecting the growing interest in this emerging solar production sector known as building-integrated PV (BIPV) (White House 2011).

DOI: 10.1201/9781003472629-15

239

The majority of BIPV systems, which are interfaced with the power grid, can also be used in standalone, off-grid systems. Grid-tied BIPV systems benefit from practically free storage because of collaborative service strategy. It is 100% efficient, and it has an infinite capacity. Grid-tied BIPV has advantages for both the utility and the building owner. The output of solar electricity tends to be at its highest when a building and the utility are using it the most. The building owner saves money on energy expenses because of the solar impact, and the imported solar electricity assists the grid system when it is greatest in demand [4–7].

15.1.1 Economical aspects of BIPV

The global market for BIPV is expected to grow in 2016 according to consultancy firm NanoMarkets, New York. Furthermore, as reported by NanoMarkets, the volume of BIPV based on polysilicon will drop from 75% of the overall market to 33% by 2016, and the amount of BIPV based on copper indium gallium selenide (CIGS) solar cells will account for 17% of the market in 2016. Room-saving technologies like BIPV may be the greatest solution for the linked financial difficulty because PV panels require a lot of room to install. Products including materials for roofing, windows awnings, and glass walls offer the possibility for cost reductions by changing traditional building supplies with PV materials at marginal prices.

In contrast to glass, steel, or other more typical covering materials, installing BIPV only adds a small amount (2%–5%) to the overall building expenses of a commercial project. Although it is still somewhat pricey, BIPV technology is a developing field of study. One may also observe that the construction sector is extremely sensitive to price. Additionally, ambiguity about BIPV and its implementation is a critical factor to take into account because the technology is still in its infancy. Installation, electrical aspects, safety concerns, problems with integration, building physical features, protection from exposure to the environment, durability, maintenance, deconstruction, evaluation of the life cycle, ability to sell excess electricity to the grid or not, architectural aspects, and other variables are some of the various variables that contribute to this uncertainty. Of course, all these considerations could result in higher costs [8,9].

A PV technique's cost will eventually drop with the development of technology, resulting in a reduced cost per installed kW. This is a crucial milestone in the effort to make the installation and building integration of PV products economical without the need for subsidies, laying the groundwork for the following stage, which involves exploring possibilities for research and improvements in order to create the BIPV of the future. When comparing various renewable energy systems, it is crucial to consider the energy payback time, which measures how long it takes a solar cell system to produce the same amount of energy that it consumed to build it. It is necessary to estimate the system's embodied energy in order to calculate the energy payback period.

15.2 SOLAR POTENTIAL OF THE BUILDING ENVELOPE

An increase in PV area is necessary to decarbonise the energy system. This development must be achieved with minimal new land usage across various nations. Fortunately, there are economically viable locations for installing PV on the roofs and walls of both new and current structures. The facade area becomes more important as structures rise in height. Since roof space is usually needed for the installation of ventilation, heating, and air conditioning systems, the surface area of the roof in high-rise buildings is frequently highly small and constrained. This suggests that the importance of facade panels grows with the height of the building.

Furthermore, even in mega towns where high-rise structures are frequently surrounded by related to high-rise buildings, higher buildings' facade panels are usually less often cast in shadow by nearby trees or structures as they go taller. German's envelopes for buildings were shown to offer a lot more economic potential than the minimum 200 GWp for PV system installation in environmentally friendly places.

In the present research, the solar energy of the roofs and facades of common urban areas was investigated during-depth using 3D architecture and city designs, and the findings were expanded using clustering techniques to include all of Germany. Due to the building models' neglect for window spaces, balconies, chimneys, and other geometric characteristics, roofs and facades have a greater economic value than 1400 GWp. A detailed investigation of all building envelope areas of the entire German building assets, including facades, shows that there are significant regions on structures, proving that there is a very high potential for the installation of PV modules on buildings, which exceeds 200 GWp.

It goes without saying that not all PV systems require to be attached to or built into building facades. In many countries, there isn't enough room for ground-mounted systems, thus photovoltaic panels can be installed in very big, attractive spaces on buildings. It is essential that PV in structures (such as BIPV) be correctly integrated both technically and aesthetically if it is to become more widely accepted by the building industry and the general public [10–13].

15.3 TECHNOLOGICAL DESIGN OF BIPV SYSTEM

A BIPV module's design must be both aesthetically pleasing and capable of producing the greatest amount of electricity yield at an affordable price. The visual design is typically not permanent, though, and if a higher power output or lower cost can be obtained, small changes may be implemented. Knowing and comprehending the many technical design possibilities for the various components that make up a module is therefore a must for good module design.

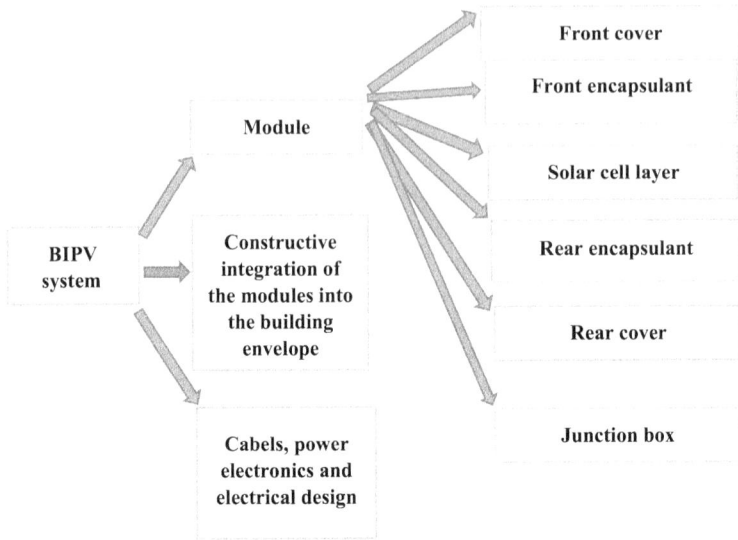

Figure 15.1 A review of the technical design for BIPV modules.

The main components of a photovoltaic panel are the solar cell layer, two encapsulated layers entirely encasing the cell layer, and a rear and front cover. The design of a PV module is shown in Figure 15.1. In the case of a basic module, a junction box and a structure are additionally connected to the composite to enable mechanical and electrical connection, respectively.

In such cases, the module is a vital component of the whole system and is used in PV that is built into the device (like solar street lighting). The junction box supplies cables or connections for electrically connecting the modules. It is possible to minimise the size of the junction box by including bypass diodes within the laminate. In this case, the junction box serves as a connection box.

The installation system delivers a mechanical link to the building and, when necessary, further mechanical support. The fixing system can also serve as the back cover, replacing the material completely or in part while supplying an attachment foundation.

While the laminate is normally composed of the encapsulant and front cover, as well as the rear cover and encapsulant, it is possible to integrate the cover and encapsulant into a single material. Examples of front covers with low-reflective module surfaces or certain device integration scenarios that use a very thick polymeric layer for high efficiency applications. Due to the embedding material's inherent high vulnerability to moisture, stresses from mechanical contact, ultraviolet (UV) radiation, and chemical interactions, utilising a separate impermeable top layer is typically more cost-effective [14–17].

15.4 COVER MATERIALS (FRONT AND RARE)

Cover of front side made of glass and a back cover made of stacked polymer sheets are two examples of radically distinct cover materials. They can, however, be divided into three groups based on their configuration: bulk substance, and the interior surface.

All surfaces of module may be patterned, coated, or even given extra layers of finishing. Coatings and structure can be used in combination. Coatings can take many different forms as additional layers on the bulk material of the cover layer, including sputtered coatings, enamel coatings, printed coatings, as well as various varnishes and lacquers (Figure 15.2).

Structures along the outside surface change how reflections appear and behave, which, for example, can prevent glare from the sun's specular reflection. Glare control of the surface cannot be achieved by conventional flat anti-reflective layers alone, such as interference layers or nanoporous structures (k/4 layers), due to the reason that the luminance of the sun can only be reduced by 99% (i.e., a factor of 10^{-2}), which essentially makes no difference to perceived glare due to the logarithmic sensitivity of the human eye. Diffusing surface textures can be used to provide very good antiglare properties [18–20].

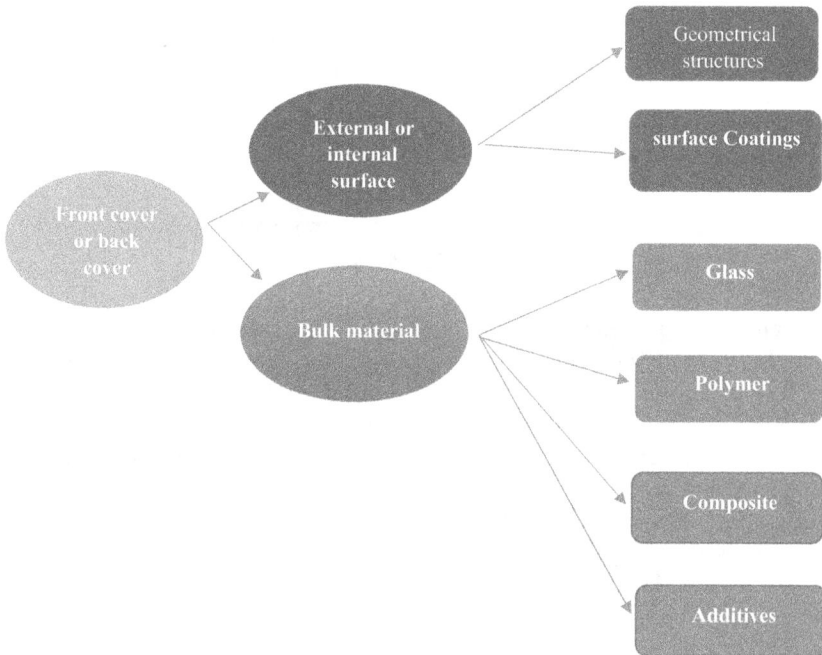

Figure 15.2 Overview on design parameters for the front and back cover.

Typically, either the entire rear cover material is changed or more glazing components are added. The module is specifically constructed inside a double glazing module. The rear cover in this case is a whole glazing assembly, as opposed to a single polymer film or glass pane.

External or interior surfaces can both receive functional coatings or structuring (such as adhesion strengthener on internal surfaces versus AR coating on external surfaces). Given that environmental factors have less of an impact on the internal surface, aesthetic coatings are typically applied there.

There are numerous types of bulk materials. The most popular material for BIPV is glass, which has a number of benefits like being very transparent as well as stable mechanically, being a material that is well-known in the construction industry, having a low thermal expansion coefficient, being incombustible, sustainable, and recyclable. Additionally, it is frequently possible to classify glass used as the main material for the front and rear covers as laminated glass or even laminated safety glass, which has benefits for structural integration, safety, sound insulation, and extended service life with warranties of up to 30 years. Because of this, many BIPV modules include glass front and back covers.

Polymers or polymer-based materials are widely used for the front and back covers or sheets, including ethylene tetrafluoroethylene copolymer (ETFE), polyamide (PA), polypropylene (PP), polyethylene terephthalate (PET), and polyvinyl fluoride (PVF). Even very thin metal layers, such foil made of aluminium, can be included into polymer layer compounds. For lightweight applications, such as photovoltaics installed on buildings or the roofs of commercial buildings, modules with at least one non-glass cover are frequently utilised [21–23].

15.5 DESIGN FOR ADDITIONAL INTERLAYERS AND EMBEDDING MATERIALS

The embedding material in a solar module provides the solar cells with airtight containment and stickiness to the outer material. The embedding substance is commonly a cross-link material like acetate from ethylene vinyl (EVA) or a thermoplast like polyvinyl butyral (PVB) that connects the cell coat to the covering layer.. Due to its low cost, EVA is most frequently utilised to produce standard modules. PVB is frequently employed for BIPV applications since it is frequently utilised for laminated glass in the construction sector (Figure 15.3).

The encapsulating layer may contain additional sheets or interlayers. Obtaining a needed colour, either reflection or dispersion of light is the main goal of embedding. For instance, coloured nets can be added to the during the installation, the module manufacturing process to include them [24–26].

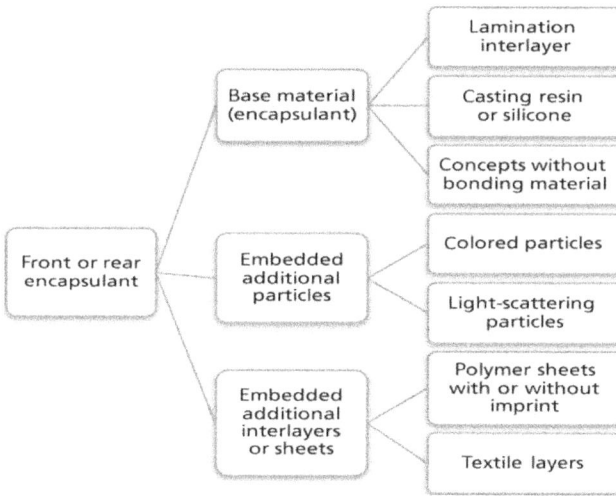

Figure 15.3 Design for front or rear embedding materials.

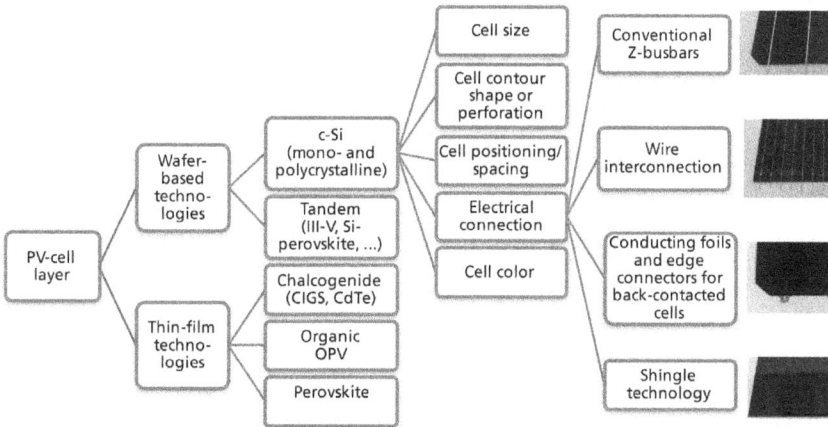

Figure 15.4 Structure for the solar cell layer.

15.6 DESIGN FOR THE LAYOUT OF THE ELECTRICAL MODULE WITH PV CELL LAYER

The functional structure possibilities for the BIPV systems' PV cell layer are summarised in Figure 15.4. Cell size, form (contour and perforation), location, spacing, and colour can all be changed for any wafer-based technologies. The integration of wafer-based technology considerably affects module aesthetics.

For back-contacted solar cells, the first is the metal wrap-through (EWT/MWT) or emitter and IBC technological innovation for solar cells. In each scenario, the interconnectors are only visible in the space between the solar cells and are located solar cell of back surface. But, the primary disadvantage of this method is the solar cells' much greater price when compared to the traditional busbar design. The shingled interconnection technology is a new method for achieving a homogenous module appearance. Similar to how roof tiles are stacked, the solar cells in this system are cut into small strips and connected in series by placing one cell on top of a small section of another cell. The usage of conductive adhesives.

The PV cell layer controls the module's power production, service life, and mostly in quality of optics. There are two types of cell technologies: wafer-based ones like crystalline and thin-film ones.

The layer of PV cells controls amount of electricity generated, how long it will last and most cases, how will be optical quality of the module. There are now primarily two cell technology groupings. Thin-film technologies come next, then wafer-based technologies like crystalline.

15.7 PV CELL TECHNOLOGY

15.7.1 Wafer-based technologies

15.7.1.1 Crystalline silicon (monocrystalline/multicrystalline)

The most popular type of solar cell used in PV systems is crystalline silicon (c-Si). The current efficiency record for cells designed specifically for terrestrial usage is 26.7%, and it dates back to the early 1980s. Approximately 95% of the overall manufacturing of PV technology based on Si wafers was in 2017. c-silicon glass cells are used., a 30-year warranty is feasible. Solar cells made on silicon chips may be used in low-curvature curved modules, keeping the individual cells almost flat. Lead is used in the majority of systems for unit the metallisation process and connectivity using solar cells made from Si wafers.

15.7.1.2 (Single/multi-junction) III and V cells

Multi and single junction of GaAs cells with efficiencies of 38.8% and 29.1%, respectively, for III–V cells set new records. III–V cells have an established, albeit modest, market. They were created for use with planetary and air concentrators. The long-term stability is on par with c-Si solar cells and much more suitable for space applications where high energy particles harm the materials and reduce performance. However, it is mostly found in bound form and in very little concentrations in III–V cells.

15.7.1.3 Perovskite Si tandem

On large size wafers, perovskite on Si tandem cells demonstrated a on large size wafers, perovskite on Si tandem cells demonstrated a historic performance of 28%. It's possible that the temperature and radiation dependence is comparable to that of solar cells containing c-silicon. Primary content of the module is a little bit more than for c-Si technology because of the lead-based perovskite layer now in use. It is found in an unbound, water-soluble form.

15.7.2 Tiny films based technology

15.7.2.1 Chalcogenide (CIGS, CdTe) cell

Chalcogenide (CIGS, CdTe) cell is the second most relevant technology after c-Si solar cells, solar cells for BIPV applications. They offer architects and building owners a more aesthetically pleasing appearance because of their uniform black appearance. Chalcogenide solar cells' record efficiency, which is currently between 22% and 23%, has also risen recently. While CIGS and c-silicon technology have similar temperature dependences, CdTe has a less irradiated impact. All currently applicable technologies also contain trace amounts of dangerous substances like lead and cadmium.

15.7.2.2 Organic/dye-sensitised (OPV) cells

The development of organic (OPV) cells, also known as dye-sensitised cells, is now accelerating. Production companies for OPV modules and material manufacturers have forged significant relationships. BIPV modules have been developed and put in place in test structures. Current records for organic solar cell efficiency stand at 17.4% and 12.3%, respectively. The available module efficiencies, however, are only about 8%. Even yet, this cell technology is more severely damaged by UV light than any other cell technology. The modules can even be bent repeatedly without significantly harming the cells.

15.7.2.3 Perovskite cells

The new cell technology of the future is solar cells. Even greater efficiencies (now 25.2%) than those of chalcogenide solar cells have been observed. Promising is also the possibility for cost savings. To yet, no modules have been commercially available, though. The extreme susceptibility to humidity affects long-term stability as well, necessitating additional measures like glass soldering to encapsulate the modules. The dependence on temperature and radiation might be marginally worse. Perovskite solar cells, which are now promising, contain a high quantity of soluble lead, necessitating significant encapsulation precautions.

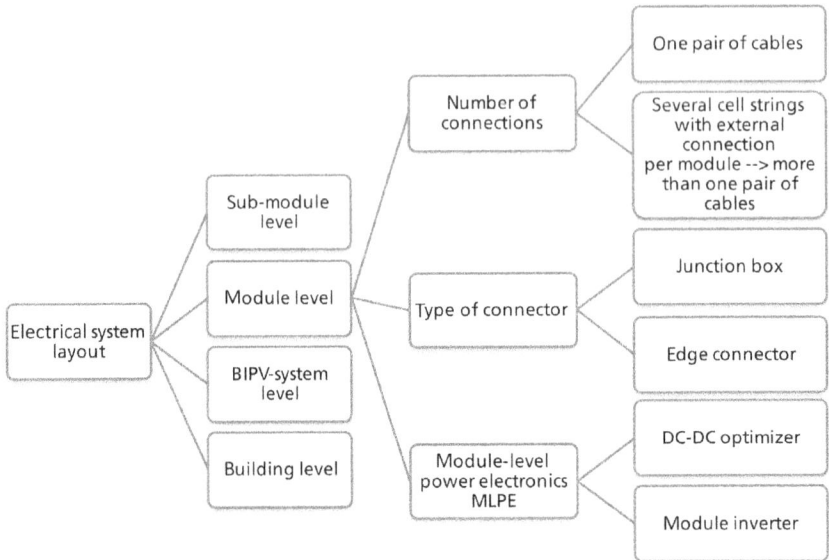

Figure 15.5 Structure the parameters for BIPV systems.

15.8 DESIGN FOR ELECTRICAL BIPV SYSTEM

The general goal of conventional PV systems that are mounted on rooftops or the ground is to provide the most electricity for the least amount of money. On the other hand, BIPV systems must be compatible with the geometry, layout, and construction of an existing or uniquely constructed building.

This section looks at various design options for BIPV systems' electrical architecture. The technical design parameters are divided into four levels, as shown in Figure 15.5: the substructure inside individual modules, which includes electronic components integrated into those modules (sub-module level); the individual module with its external connections and possibly additional external electronic components, like DC/DC converters or micro-inverters; the BIPV system with several modules and additional power electronic components, like inverters.

15.8.1 Sub-module level

PV modules are included at the sub-module level. All of the electrical characteristics of the PV system are significantly influenced by the PV cell technology. In addition to picking right away from solar cells, options include the amount of cells in a module's lateral and vertical directions, the space with respect to neighbouring cells and the edges of the perspective of the glass the cells (such as symmetrical, oblique, or straight), and the technology used for cell interconnection. It is significant to remember that several electrical subsystems may use the cells that produce photovoltaic energy concealed beneath a single glass cover.

15.8.2 Module level

Once the entire BIPV system has been developed and optimised, power electronics parts can be beneficial at the module level. Each module has a junction box with a bypass diode placed directly to it, and there are various types available (Edge connected junction boxes or standard ones mounted on the module's back cover, for reference), which are crucial for both mechanical installation and framing and can have an impact based on the visual appeal of the system.

Each system needs one of the stages of MPP tracking, as well as the bypassing, whether it be active or passive, turned subsections or modules. MPP monitoring is utilised by common PV generators. Decentralised MPP tracking can be helpful for BIPV as a result of typically enhanced geometric layout, various module directions, as well as some shading. DC/DC converters or micro-inverters at the module level are two examples of how to accomplish this.

15.8.3 BIPV system level

Several modules are linked in series and parallel at the BIPV system level. Usually, a single or multiple inverters transform DC electricity into AC power. The inverter, which regulates and runs the entire system, is the key element of the electrical system. The operational voltage of the system is often determined by tracking the maximum power point. Transformerless inverters, which are often smaller and more effective, can be used to operate the majority of modules on the market today.

15.8.4 Building level

BIPV systems are only utilised as feed-in and are linked to the public AC energy grid. In this instance, the highest level for a single building is the level of the BIPV system, while the grid level handles everything else. However, BIPV systems are becoming larger and more crucial component elements of the self-consumption and self-sufficiency of building energy systems. In this scenario, the demand side and potential storage capabilities must also be considered in the electrical layout of BIPV systems. A BIPV system with multiple orientations may make it easier in order to balance supply and demand.

Degrees of freedom within the building include the option to add a power source, include the system with automobiles powered by electricity, and add along with manage substantial heat pumps along with other electrical loads. The structure and the dimensions of its parts should be optimised comprehensively due to the BIPV, power sources, heating and cooling systems, and other electrical appliances are interdependent on one another.

15.9 IMPORTANT OPTIONS FOR THE CONSTRUCTIONAL INTEGRATION

15.9.1 Position of the modules

Each year's harvest and the timing energy from the sun production are determined by potential shadowing, the BIPV modules' azimuthal (compass) orientation and inclination. Throughout the span of the day, a more consistent power production profile can be attained whenever a BIPV system is designed to include a number of subsystems with varying orientations and various recommendations. This can reduce the demand for power, and increase the use of produced electricity for oneself.

The distribution of both direct and diffuse irradiance depends on the angle impinge within a module the cause of the reliance on positioning. This is brought on the Sun's position, with its daily and seasonal variations, and the corresponding positioning of a module's surface to yearly variations, as well as the influence of shadowing, such as that induced by nearby structures, vegetation, or protruding building components. The impact of shadowing extends beyond only a decrease in irradiation. Partially shadowing BIPV systems or individual modules can have a considerable negative impact on the electricity output. As a result, the amount of energy produced is much more uniformly distributed over time when a building is oriented so if the building's corner faces south and if the southeast and Southwest after that, BIPV modules are used to cover the façade completely system on a roof or façade that faces south. The efficiency of the module is also affected by its angular dependence, which is primarily a result of optical losses brought on by the front cover and the solar cells' angle-dependent reflectance. To maximise yield, BIPV modules must be able to tilt in relation to the exterior (roof or facade) of the building. Particularly systems that are built into the facade can be slanted in relation to the vertical wall.

15.9.2 Construction integration of the modules into the building envelope

Glass or plastic can be used to create BIPV modules' front panel. Materials having an exterior made of polymer are frequently employed either in roofing membranes or are strongly attached to other parts. This module is frequently created as an optical module and glass front covers are typically used in BIPV applications.

The issue is that BIPV systems must be prepared cross-disciplinarily for different trades, that as a result, accountability is frequently shared among many people, and that certain players are still unfamiliar with such components. Therefore, it is not necessarily quick, dependable, or economical to plan and implement a BIPV system. An additional factor that may cause problems with planning, production, and installation is criteria that conflict resulting from instructions and rules provided by several independent agencies [27–30].

BIPV modules can be mounted using popular techniques for supporting glass elements.

- **Analytical Considerations for BIPV**
  Higher system costs are probably the cause of the limited implementation of BIPV globally, although other variables could also have an impact on market potential. Less evident problems may have an impact on the BIPV sector's growth rates, particularly when PV and BIPV cost differences are small. The following are some factors related to BIPV:

- **The Importance of Aesthetics**
  The possibility of better addressing the aesthetic interests of numerous stakeholders is one of BIPV's main advantages. Although consumer receptivity to payment of a premium for BIPV has been indicated by market data, it is still too soon to tell how much emphasis consumers place on visual appeal and whether or not they will continue to tolerate premium prices as the solar market develops.

- **Standards and Codes**
  Compared to competing PV products, the rules and standards environment for BIPV is more complex. The standards for roofing goods and PV devices differ in terms of performance, safety, and durability, and the certification processes for PV and BIPV are sometimes disjointed. The goal of standards bodies is to expand BIPV-specific recommendations and harmonise codes. But BIPV may still find it difficult to navigate challenges with codes and standards.

- **Regulation and Policy Concerns**
  There is more and more potential for policy support of BIPV specialised incentive programmes, actions to encourage sustainable structures, and initiatives to discuss building regulations and other pertinent guidelines. Increased market potential may result from focused, cost-based incentive programmes, like premium-rate BIPV feed-in rates. A favourable environment for BIPV devices can also be created through solar access rules.

- Fragmentation of the Market

    The dynamics of the acceptance of BIPV may be different from those of rack-mounted solar energy because goods are frequently made for more specialised potential in the marketplace for new residential roofing or certain structures items (like ceramic tiles). Although solar panels can be used in utility, business, and housing applications, they are far greater transferable than BIPV devices, whose layouts might differ significantly even between applications. The price of a considerably more developed PV placed on a rack business must be met by BIPV opportunities, notwithstanding the fact that they are more fragmented.

15.10 SOLAR THERMAL ENERGY STORAGE SYSTEM

Solar thermal energy storage systems are innovations that can capture electricity, transform it into a different form of energy (chemical, thermal form, or mechanical), store it, and then release it as needed. These include batteries with lithium-ion. Technologies that can harness electricity, convert it into another type of energy (chemical, thermal, or mechanical), store it, and then release it when needed are referred to as solar thermal energy storage systems. One such technology is lithium-ion batteries. Nevertheless, there is never a 100% efficiency with energy storage. When converting and recovering energy, some is constantly lost. Energy can be used flexibly at periods other than when it was created thanks to storage. By balancing supply and demand, reliability, and efficiency, storage may improve power quality.

Energy capability, or the total amount of energy that can be conserved, and power capacity, or the total quantity of energy that can be released at a certain time, are two different concepts (typically measured in kilowatts or megawatts), are two different aspects of storage facilities. To fulfil various requirements, different energy and power storage capacities might be used. While more long-term storage, lasting only a few minutes, can help ensure a solar plant runs smoothly throughout output changes brought on by passing clouds, short-term storage, lasting only a few minutes, can help provide supply over days or weeks when solar energy production is low or during a significant weather event, for example [31,32].

The idea of storing solar thermal energy is not new. Early people developed numerous strategies for harnessing solar energy after realising its abundance. The Greek historian Xenophon recorded Socrates' advice on how to orient a building so that it will remain summertime is refreshing, while wintertime is warmer. To heat their showers and consume less firewood for their hypocaust, often known as their bathwater fire, Romans would add numerous windows to a bath house's south wall. Native Americans in Arizona's canyons utilising the southern canyon exposure to the cliff to warm their adobe dwellings, which were cunningly situated in caves such winter time's low sun position would drench them in sunshine while the summer angle would not.

When people first began to live in natural caves, thermal energy storage existed. Caves are warmer in comparison to their exterior environment the December and frigid in the summertime. The deep subterranean caves (structures located deep beneath) that have hardly any seasonal climate fluctuations were used by cave inhabitants. Collecting icy snow and ice for cold drinks, preservation of nourishment and space, chilling is the oldest known method of storing thermal energy. According to historical accounts, the Romans, Greeks, and Chinese experimented with using rounded reflections focus sunshine's rays, which could result in fires and explosions. A Greek myth claims that lenses were employed by Archimedes in 212 B.C. to direct sunlight onto the ships of a Roman navy that was attacking Syracuse, decimating the fleet.

Horace Benedict de Saussure, a French-Swiss scientist, in 1767 saw the invention of the first solar heat collector that was suitable for cooking. Adams added a De Saussure's solar heat is produced through a solar energy extractor catcher more than a century the year 1876, to create the first solar-powered thermal gadget built in bulk. According to reports, the octagonal solar oven of Adams, which had a total of eight mirrors solar concentrators, took 2 hours to prepare food for seven soldiers. According to Adams, "the troops declare the meal to be prepared considerably better than in the usual technique," and in January, Bombay's coldest month of the year, it takes just 2 hours to adequately boil the meat and vegetable rations for seven men. His solar oven was widely distributed and fairly well-liked in India.

The Adams' solar oven had gained popularity in the US as a camping item and a teaching tool for teenagers. The idea of the use of solar ovens for military food preparation was also investigated by Mouchot, a French mathematician. For French soldiers in Algeria, Mouchot created solar cookers in 1877, a year after Adams' discovery. These cookers had a bright steel conical fashioned from a 105.5° portion of a circle. He also created the initial publication on solar energy's industrial use as well as a separate burner for steaming vegetables. A speaker at the 1992 Stockton, California hosts the first global workshop on solar energy for cooking, Wang Xiping showed a Peking duck that had been roasted with solar energy in Chengdu, China, in 1894.

The market for thermal energy storage is currently expanding quickly. According to Ice Energy CEO Mike Hopkins, "thermal loads are the biggest loads," thus I anticipate that thermal energy will surpass battery energy. The troublesome loads and the loads are these for which using electrical storage is not appropriate. The majority of thermal energy storage technologies are currently either in the demonstration and development stages or have reached their full development and commercialisation potential.

The cost of high-density storage is now being reduced by research and development in the field of phase change materials (PCM) and thermochemistry in thermal energy archiving. Systems for thermal storage are used all around the globe. To provide commercial and industrial refrigeration capacity contexts, for instance, tanks for holding cold water have been constructed up all over the world. Thermal storage of energy in aquifers and boreholes is used to generate cooling and heating in Canada, Germany, the Netherlands, and Sweden. There has been an estimated 1 GW of storage of ice installed in the US to lower peak energy usage in regions having a lot of cooling-degree days.

In Canada, the Netherlands, and Norway, commercial-scale deployments of borehole and aquifer systems to provide heating capacity have been effective. The first significant district heating application storage of seasonal thermal energy in boreholes in North America is in Canada. In addition, this is the initial system of its kind operating in such a frigid climate and is designed to provide more than 90% of the space heating needs [33,34].

15.10.1 Storing thermal energy method

Thermal energy can be kept in three different ways, with the initial two being the most popular.
    TES systems:

1. Latent heat storage
2. Thermochemical
3. Sensible storing heat

15.10.1.1 Latent heat capacity

It includes warming a substance before it undergoes a stage shift, usually solidifying from liquid. Once the substance achieves that temperature is fusion, it takes in a significant wattage of heat, known as fusion's latent heat, to continue the shift in phase. Solid to liquid phase changes are most frequently used because the associated volumetric expansion is relatively small and a significant amount of latent heat is available. Solid to liquid phase changes also contain massive heat from phase changes, but the system is extremely complicated and impractical due to large density changes.

It can deliver the stored energy while maintaining a nearly constant temperature. Given that the latent heat of fusion is generally substantially more than specific heat in the majority of materials, its compactness is another benefit. PCM are used most often in programmes where weight and space are limited with a substantial energy density that is needed, or if there is a thermal load in operation that demands a heat-resistant energy source.

15.10.1.2 Thermochemical storage

Reversible endothermic chemical processes are used in the third technique. The energy needed to dissociate a chemical product is known as chemical heat; all of this energy (or almost all of it) can be later recovered when a synthesis process occurs. Although this type of storage has many alluring benefits, fully reversible endothermic chemical processes are still in their infancy.

15.10.1.3 Sensible heat storage

The quantity kept energy is determined by specific heat of the substance and temperature variation it underwent. It involves heating a liquid or solid without causing a change in phase. The materials utilised for this form of storage span a wide spectrum, from solids like rocks and refractories to liquids such as water, certain oils, and molten inorganic salts. The application's required temperature level has a significant impact on the material choice. When water is utilised when the temperature is below 100°C, and when it is close to 1000°C, refractory blocks are employed. Sensible heat storage systems are easier to design, but they require greater volumes and cannot be used to retrieve energy at constant temperature [26].

Numerous useful thermal energy storage systems have been tested and assessed since 1985. These comprise the two-tank direct system, the two-tank indirect system, and the single-tank thermocline system.

In a concentrating solar power (CSP) system, the sun's rays are reflected onto a receiver, providing heat that is utilised to produce energy that can be either used immediately or saved for later use. CSP facilities can therefore be adjustable or transportable sources of clean, green energy.

Two-Tank Direct System: Under this configuration, the fluid that was utilised to capture solar thermal energy is also used to store it. There are two tanks, one of which is heated to a high temperature while the other is chilled are used to hold the fluid. After being heated to a high temperature by solar energy in solar accumulator or receiver, fluid from the tank with low temperatures is transferred to be stored in the high-temperature tank. An exchanger of heat uses fluid from the tank for extreme temperatures to produce steam, which is then utilised to generate power. This liquid returns after that, to the low-temperature tank leaving the heat exchanger operating with low heat.

Two-Tank Indirect System: Dual tanks indirect systems work in the same way as systems with two tanks but with distinct heat-transfer and storage fluids. When using heat transfer fluid in facilities because it is too expensive or unsuitable for use as the storage fluid, this technique is employed.

The temperature is high, heat-transfer liquid heats the storage as it moves through liquid an additional heat transfer after leaving the low-temperature tank. Back in the a high degree storage tank, the fluid travels at a rapid rate. When the fluid it is heated to when it returns to the solar collector or receiver, a high degree once more. The liquid enters a low temperature in this heat exchanger. Similar to the dual tank direct system, steam is produced utilising the high-temperature tank's storage liquid. Another heat exchanger is needed for the indirect system, which raises the system's cost.

This system has been proposed for multiple parabolic plants in the United States as well as several Spanish parabolic power stations. The heat-transfer fluid for the factories will be natural oil, and the storage liquid will be molten salt.

Single-Tank Thermocline System: The thermal energy is stored in a solid medium using thermocline systems with a single tank, most frequently silica sand. A fragment of the media is always operating at a high temperature while another section is always operating at a low temperature. The hot and cold temperature zones are separated by a temperature gradient or thermocline. High-temperature heat-transfer fluid is poured into the thermocline's top, which exits the bottom at a low temperature. The thermocline is lowered as a result of this process, this also provides the system with thermal energy for storage. The thermocline advances and the thermal energy of the system increases when the flow is reversed and is drawn out, producing steam and power. The fluid inside the tank is thermally stratified due to buoyancy factors, which aids in maintaining and stabilising the thermocline.

In comparison to systems that require two tanks, this one is less expensive because it simply requires one tank and employs a solid storage medium. At the power plant Solar One, this system was demonstrated using mineral oil as the storage fluid and steam as the heat-transfer fluid. Early parabolic trough power plants featured two-tank direct storage. The medium for storing and transferring heat utilised by the trough plants was mineral oil; by contrast, Solar Two used molten salt.

15.11 MOLECULAR SOLAR THERMAL SYSTEM

The world population is projected to expand by 25% to 99.9 billion individuals during the 30 years from now. The consumption of energy will rise steadily along with the population growth. Over the past few decades, fossil fuels have served as the main means of supplying the world's energy needs. Because they have the potential to be environmentally friendly and sustainable alternatives, renewable energy sources such as solar, wind, and geothermal power are advancing. Despite the fact that combustible energy source emissions have negative consequences on the environment, such as air pollution, which is also associated with health issues, and global warming.

A plentiful source of energy on Earth is the Sun. Each year, the surface of the Earth receives about 3.4×10^6 sun energy in exajoules as a result incoming photons are absorbed and scattered. The world's annual primary energy usage is 7000–8000 times more than this. So, without emitting harmful pollutants, solar energy technologies may be able to meet the increasing global energy needs. Compared to alternative energy sources, however, solar energy usage is still quite low. However, in comparison to other energy sources, solar usage is still quite low. Where 3.6% or less of generation of energy globally in 2019 came from solar power. Solar energy is sporadic in nature production, and storage is one of its drawbacks. Therefore, innovative, scalable technology for storing solar energy is required for greater adoption of solar energy. New solar energy storage techniques are required in order to deliver adequate and dependable electrical energy.

A molecular solar thermal system, also known as a MOST, is another promising method of storing solar energy for use in the creation of heat and electricity. By using this method, a molecule is photoisomerised into a more energetic isomer. One isomer is converted into another during the process of photoisomerisation, which is powered by solar radiation. This isomer has the capacity to store energy from the Sun until it is released by a catalyst or heat source (at which point the isomer becoming its original isomer once more.). Norbornadiene (NBD) is a potential contender for such a MOST. The reason for this occurs as a result of the substantial energy differential between the NBD and quadricyclane (QC) photoisomer. Approximately 96 kJ/mol is the energy difference. It is also known that in such systems, the donor-acceptor replacements offer a useful way to red shift the absorption at the longest wavelength. Solar spectrum corresponds to is improved by doing this.

Having a high energy storage density that is suitable and, preferably greater than 300 kJ/kg is a significant problem for a functional MOST system. Light can be captured in the visible range, which presents another difficulty for a MOST system. To modify this absorption maxima, the NBD is functionalised with the donor and acceptor units. A larger molecular weight, however, makes up for this beneficial impact on solar absorption. It follows that the energy density is smaller. There is another drawback to this beneficial impact on solar absorption. Specifically, when the absorption is redshifted, the energy storage time is shortened. Connect a single chromophore to several photo switches as a potential way to get around the anti-correlation between energy density and red shifting. It is desirable to create so-called dimers or trimers in this situation. There is a shared donor and/or acceptor among the NBD. The quantum yield of photoconversion for this system could eventually increase to 94% per NBD unit. An indicator of photon emission efficiency is a quantum yield. With this technology, recorded energy densities exceeded the aim of 300 kJ/kg and amounted to

559 kJ/kg. Thus, molecular photo switches have great potential not just in order to store solar thermal energy but also for other uses.

Researchers reported generating electricity from the MOST in 2022 by merging it with a thermoelectric generator the size of a chip. According to reports, the device can have 18 years of solar energy storage, making it a potential choice for renewable energy storage.

REFERENCES

[1] Zondag, H. A., De Boer, R., Van Helden, W. G. J., Van Zolingen, R. J. C., & Van Steenhoven, A. A. (2003). The performance of a low-cost solar water heater with integrated storage. *Solar Energy*, 74(3), 253–269. https://doi.org/10.1016/S0038-092X(03)00041-3.

[2] Allouhi, A., Kousksou, T., Belarbi, R., Jamil, A., Mourad, Y., & Zeraouli, Y. (2017). Solar thermal energy storage in a two-tank system: A review. *Renewable and Sustainable Energy Reviews*, 75, 1367–1377. https://doi.org/10.1016/j.rser.2016.11.050.

[3] Zalba, B., et al. (2003). Review on thermal energy storage with phase change: Materials, heat transfer analysis and applications. *Applied Thermal Engineering*, 23(3), 251–283.

[4] Meena, C. S., Prajapati, A. N., Kumar, A., & Kumar, M. (2022). Utilization of solar energy for water heating application to improve building energy efficiency: An experimental study. *Buildings*, 12, 2166. https://doi.org/10.3390/buildings12122166.

[5] Meena, C. S., Kumar, A., Jain, S., Ur Rehman, A., & Mishra, S. (2022). Innovation in green building sector for sustainable future. *Energies*, 15, 6631. https://doi.org/10.3390/en15186631.

[6] Dutt, N., Hedau, A. J., Kumar, A., Awasthi, M. K., Singh, V. P., & Dwivedi, G. (2023). Thermo-hydraulic performance of solar air heater having discrete D-shaped ribs as artificial roughness. *Environmental Science and Pollution Research*. https://doi.org/10.1007/s11356-023-28247-9.

[7] Kushwaha, P. K., Sharma, N. K., Kumar, A., & Meena, C. S. (2023). Recent advancements in augmentation of solar water heaters using nanocomposites with PCM: Past, present, and future. *Buildings*, 13, 79. https://doi.org/10.3390/buildings13010079.

[8] Gómez-Villarejo, R., et al. (2019). Cryogenic energy storage for renewable sources integration: A review. *Renewable and Sustainable Energy Reviews*, 105, 295–312.

[9] Sharma, A., Tyagi, V. V., Chen, C. R., & Buddhi, D. (2009). Review on thermal energy storage with phase change materials and applications. *Renewable and Sustainable Energy Reviews*, 13(2), 318–345.

[10] Lund, J. W. (2015). *Geothermal Energy Systems: Exploration, Development, and Utilization*. Cambridge, MA: Academic Press, Elsevier.

[11] Acuna, J. A., Munoz, F. D., & Uihlein, A. (2016). Thermal energy storage in deep geothermal systems: A review. *Renewable and Sustainable Energy Reviews*, 61, 127–137.

[12] Singh, V. P., Jain, S., Karn, A., & Kumar, A. (2022). Experimental assessment of variation in open area ratio on thermohydraulic performance of parallel flow solar air heater. *Arabian Journal for Science and Engineering*, 48, 11695–11711. https://doi.org/10.1007/s13369-022-07525-7.

[13] Singh, V. P., Jain, S., Kumar, A., Mishra, S., & Sharma, N. K. (2022). Heat transfer and friction factor correlations development for double pass solar air heater artificially roughened with perforated multi-V ribs. *Case Studies in Thermal Engineering*, 39, 102461. https://doi.org/10.1016/j.csite.2022.102461.

[14] Singh, V. P., Jain, S., Karn, A., Kumar, A., Dwivedi, G., & Meena, C. S. (2022). Recent developments and advancements in solar air heaters: A detailed review. *Sustainability*, 14, 12149. https://doi.org/10.3390/su141912149.

[15] Singh, V. P., Jain, S,, Karn, A., Kumar, A., Dwivedi, G., Meena, C. S., & Cozzolino, R. (2022). Mathematical modeling of efficiency evaluation of double-pass parallel flow solar air heater. *Sustainability*, 14, 10535. https://doi.org/10.3390/su141710535.

[16] Moreno, E., Sánchez, L., Daza, L., Sánchez-Aguilar, R., & Romero, M. (2019). Thermal energy storage technologies for concentrated solar power plants. *Energies*, 12(11), 2201.

[17] Preuster, P., Papp, C., Wasserscheid, P., & Libuda, J. (2017). Liquid organic hydrogen carriers (LOHCs): Toward a hydrogen infrastructure based on sustainable fuels. *Accounts of Chemical Research*, 50(1), 74–85.

[18] Singh, V. P., Jain, S., & Kumar, A. (2022). Establishment of correlations for the thermo-hydraulic parameters due to perforation in a multi-V rib roughened single pass solar air heater. *Experimental Heat Transfer* 35(5), 1–20. https://doi.org/10.1080/08916152.2022.2064940.

[19] Saxena, A., Prajapati, A. N., Pant, G., Meena, C. S., Kumar, A., & Singh, V. P. (2023). Water consumption optimization of hybrid heat pump water heating system. In: Shukla, A. K., Sharma, B. P., Arabkoohsar, A., Kumar, P. (eds) *Recent Advances in Mechanical Engineering. FLAME 2022. Lecture Notes in Mechanical Engineering*. Springer, Singapore, pp. 721–732. https://doi.org/10.1007/978-981-99-1894-2_61.

[20] Jradi, M., Veje, C., & Jørgensen, B. N. (2017). Performance analysis of a soil-based thermal energy storage system using solar-driven air-source heat pump for Danish buildings sector. *Applied Thermal Engineering*, 114, 360–373.

[21] Patra, B., Nema, P., Khan, M. Z., & Khan, O. (2023). Optimization of solar energy using MPPT techniques and industry 4.0 modelling. *Sustainable Operations and Computers*, 4, 22–28.

[22] Verma, V., Meena, C. S., Thangavel, S., Kumar, A., Choudhary, T., & Dwivedi, G. (2023). Ground and solar assisted heat pump systems for space heating and cooling applications in the northern region of India - A study on energy and CO2 saving potential. *Sustainable Energy Technologies and Assessments*, 59, 103405. https://doi.org/10.1016/j.seta.2023.103405.

[23] Awasthi, M. K., Dutt, N., Kumar, A., & Kumar, S. (2023). Electrohydrodynamic capillary instability of Rivlin-Ericksen viscoelastic fluid film with mass and heat transfer. *Heat Transfer*, 53(1), 1–19. https://doi.org/10.1002/htj.22944.

[24] Meena, C. S., Kumar, A., Roy, S., Cannavale, A., & Ghosh, A. (2022). Review on boiling heat transfer enhancement techniques. *Energies*, 15, 5759. https://doi.org/10.3390/en15155759.

[25] Patra, B., & Nema, P. (2022). Comparative design analysis of modified solar based 15 level multi level inverter for power quality improvement. *Journal of Algebraic Statistics*, 13 (3), 1084–1095.

[26] Kundu, A., Kumar, A., Dutt, N., Meena, C. S., & Singh, V. P. (2023). Introduction to thermal energy resources and their smart applications. In: Kumar, A., Singh, V. P., Meena, C. S., Dutt, N. (eds) *Thermal Energy Systems: Design, Computational Techniques and Applications*. CRC Press, Boca Raton, FL, pp. 1–15, Chapter 1. https://doi.org/10.1201/9781003395768-1.

[27] Dewangan, A. K., Moinuddin, S. Q., Cheepu, M., Sajjan, S. K., & Kumar, A. (2023). Thermal energy storage: Opportunities, challenges and future scope. In: Kumar, A., Singh, V. P., Meena, C. S., Dutt, N. (eds) *Thermal Energy Systems: Design, Computational Techniques and Applications*. CRC Press, Boca Raton, FL, pp. 17–28; Chapter 2. https://doi.org/10.1201/9781003395768-2.

[28] Kundu, A., Kumar, A., Dutt, N., Singh, V. P., & Meena, C. S. (2023). Modelling and simulation of thermal energy system for design optimization. In: Kumar, A., Singh, V. P., Meena, C. S., Dutt, N. (eds) *Thermal Energy Systems: Design, Computational Techniques and Applications*. CRC Press, Boca Raton, FL, pp. 103–140; Chapter 7. https://doi.org/10.1201/9781003395768-7.

[29] Patra, B., & Nema, P. (2021). Analysis of solar integrated multilevel inverter for smart grid power filters. In: *International Conference on Advances in Electrical, Computing, Communication and Sustainable Technologies*, pp. 1–5.

[30] Pant, G., Meena, C. S., Saxena, A., Kumar, A., Singh, V. P., & Dutt, N. (2023). Study the temperature variation in alternate coils of insulated condenser cum storage tank: Experimental study. In: Sikarwar, B. S., Sharma, S. K., Jain, A., Singh, K. M. (eds) *Advances in Fluid and Thermal Engineering. FLAME 2022. Lecture Notes in Mechanical Engineering*. Springer, Singapore, pp. 627–638. https://doi.org/10.1007/978-981-99-2382-3_52.

[31] Kumar, A., Singh, V. P., Meena, C. S., & Dutt, N. (2023). *Thermal Energy Systems: Design, Computational Techniques and Applications*. Boca Raton, FL: Taylor & Francis (CRC Press). https://doi.org/10.1201/9781003395768.

[32] Singh, S., Kumar, A., Behura, S. K., & Verma, K. (2022). Challenges and opportunities in nanomanufacturing. In: Singh, S., Behura, S. K., Kumar, A., Verma, K. (eds) *Nanomanufacturing and Nanomaterials Design: Principles and Applications*. CRC Press, Boca Raton, FL, pp. 17–30; Chapter 02. https://doi.org/10.1201/9781003220602-2.

[33] Srivastava, S., Verma, D., Thusoo, S., Kumar, A., Singh, V. P., Kumar, R. (2022). Nanomanufacturing for energy conversion and storage devices. In: Singh, S., Behura, S. K., Kumar, A., Verma, K. (eds) *Nanomanufacturing and Nanomaterials Design: Principles and Applications*. CRC Press, Boca Raton, FL, pp. 165–173; Chapter 10. https://doi.org/10.1201/9781003220602-10.

[34] Singh, V. P., Dwivedi, A., Karn, A., Kumar, A., Singh, S., Srivastava, S., & Srivastava, K. (2022). Nanomanufacturing and design of high-performance piezoelectric nanogenerator for energy harvesting. In: Singh, S., Behura, S. K., Kumar, A., Verma, K. (eds) *Nanomanufacturing and Nanomaterials Design: Principles and Applications*. CRC Press, Boca Raton, FL, pp. 241–242; Chapter 15. https://doi.org/10.1201/9781003220602-15.

Chapter 16

Green hydrogen productions
Methods, designs and smart applications

Mira Chitt and Sivasakthivel Thangavel
Global College of Engineering and Technology (GCET) Muscat

Vikas Verma
Tezpur University Assam (Central
University, Government of India)

Ashwani Kumar
Technical Education Department Uttar Pradesh Kanpur
(Under Government of Uttar Pradesh)

16.1 INTRODUCTION

Hydrogen, the most abundant element in the universe, holds immense promise as a clean and versatile energy carrier (Abdalla et al., 2018). Electrolysis, a well-established process, plays a central role in producing hydrogen in a sustainable and environmentally friendly manner. Through the application of electricity to water, electrolysis efficiently splits water molecules into its constituent elements: hydrogen and oxygen. This method, known as water electrolysis, can be powered by various energy sources, including renewable energy, enabling the production of green hydrogen, free from harmful greenhouse gas emissions. In this comprehensive exploration, we delve into the intricacies of hydrogen production through electrolysis, its technologies, advancements, and potential applications (Chi and Yu, 2018). Green hydrogen is a cutting-edge and environmentally friendly energy solution that has garnered increasing attention in recent years as the world seeks to transition toward a more sustainable future. Unlike conventional hydrogen production methods that rely on fossil fuels and emit greenhouse gases, green hydrogen is generated through a clean and renewable process, making it an essential player in the fight against climate change. At its core, green hydrogen is produced by using renewable energy sources, such as wind, solar, or hydroelectric power, to electrolyze water, splitting it into hydrogen and oxygen. The process involves passing an electric current through water, which results in the separation of hydrogen gas from the oxygen in a device called an electrolyzer. The hydrogen produced through this method is entirely carbon-free and emits only water vapor when used

DOI: 10.1201/9781003472629-16

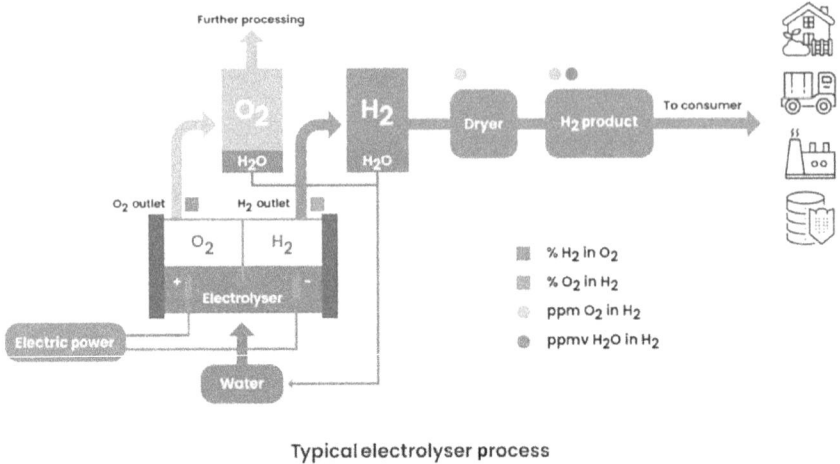

Typical electrolyser process

Figure 16.1 A typical electrolyzer process.

as a fuel or in various industrial applications. In this chapter, the methods, designs, and applications of hydrogen production through electrolysis are discussed in details.

Hydrogen production through electrolysis involves the splitting of water (H_2O) into its constituent elements, hydrogen (H_2) and oxygen (O_2), using an electric current. This is a fundamental process in which the water molecules are broken down into their elemental components at the anode and cathode of the electrolyzer, driven by the application of an external electric potential.

Several commercial techniques for water electrolysis are currently in use or under development to produce hydrogen on various scales. This chapter discusses the techniques that differ in their electrolyzer designs, operating conditions, and applications (Figure 16.1).

16.2 GREEN HYDROGEN PRODUCTION METHODS

16.2.1 Proton exchange membrane electrolysis

This method, also known as Solid Polymer Electrolyte (SPE), is widely used for small- to medium-scale hydrogen production. These electrolyzers operate at relatively low temperatures (typically below 100°C) and are known for their rapid response to load changes, making them suitable for applications like hydrogen refueling stations and energy storage systems. Companies like Nel Hydrogen, ITM Power, and Ballard Power Systems produce commercial PEM electrolyzers. The proton exchange membrane (PEM) electrolysis is a type of water electrolysis that utilizes a

SPE membrane to facilitate the splitting of water into hydrogen and oxygen (Shiva Kumar and Himabindu, 2019). The principle of operation of PEM electrolysis involves several key steps.

The heart of PEM electrolysis is the SPE membrane, which is often made of a perfluorinated polymer material like Nafion. This membrane is selectively permeable, allowing only protons (H^+) to pass through while blocking the passage of electrons and other ions. At the anode (positive electrode), water molecules (H_2O) undergo an electrochemical oxidation process. The anode is typically coated with a catalyst, often based on platinum or iridium oxide. The anode reaction involves the following steps:

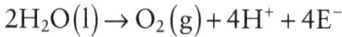

$$2H_2O(l) \rightarrow O_2(g) + 4H^+ + 4E^-$$

This reaction releases oxygen gas (O_2), protons (H^+), and electrons (e^-) at the anode.

At the cathode (negative electrode), the protons (H^+) generated at the anode migrate through the PEM to reach the cathode. The cathode is also coated with a catalyst, usually based on platinum. The cathode reaction involves the reduction of protons to form hydrogen gas (H_2):

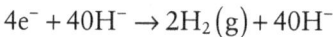

$$4e^- + 40H^- \rightarrow 2H_2(g) + 40H^-$$

This reaction produces hydrogen gas (H_2) at the cathode.

The SPE membrane acts as a barrier that prevents the mixing of gases and electrons between the anode and cathode compartments. This separation ensures that the produced hydrogen and oxygen gases remain separate and can be collected independently.

The overall reaction of PEM electrolysis can be represented as the combination of the anode and cathode reactions:

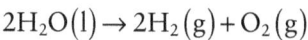

$$2H_2O(l) \rightarrow 2H_2(g) + O_2(g)$$

In general, the electrolysis process requires an external source of electrical energy to drive the electrochemical reactions. Applying a voltage across the electrodes provides the necessary energy to split water molecules. PEM electrolyzers are known for their high efficiency, rapid response to load changes, and the ability to operate at low temperatures (typically around 50°C–80°C). The system can be controlled by adjusting the applied voltage and current to regulate the rate of hydrogen production.

PEM technology offers several advantages for hydrogen production. It can achieve high efficiency levels, especially at partial loads, making it suitable for a wide range of applications. It can also respond quickly to changes in demand, making them suitable for applications with varying power requirements. PEM systems are typically compact and have a relatively small footprint, making them suitable for on-site hydrogen production. Since PEM electrolysis only produces hydrogen and oxygen, it does not produce greenhouse gases or other pollutants during operation. PEM electrolyzers can be easily scaled up or down based on the desired hydrogen production capacity.

PEM technology, although promising for hydrogen production through water electrolysis, confronts several notable challenges. One of the primary concerns lies in optimizing the efficiency of PEM electrolysis systems, which involves striking a delicate balance between effective proton exchange within the membrane, catalyst activity, and electrical conductivity. Material durability is another crucial hurdle, as PEM components must withstand the harsh chemical environment and high operating temperatures without degradation. The reliance on costly catalyst materials, such as platinum, poses economic challenges and necessitates the development of more affordable alternatives. Furthermore, ensuring consistent and uniform water distribution across the membrane surface to prevent local dry-out and facilitate even proton transport remains a technical obstacle (Wang et al., 2022). Finally, the integration of PEM electrolyzers within larger energy systems requires addressing issues related to grid stability, dynamic load management, and efficient renewable energy utilization. As researchers work diligently to address these challenges, PEM technology holds the potential to contribute significantly to efficient and sustainable hydrogen production, powering a range of applications and industries.

16.2.2 Alkaline electrolysis

The principle of operation of alkaline electrolysis involves the use of an alkaline electrolyte solution to facilitate the splitting of water into hydrogen and oxygen. This type of electrolysis operates at relatively higher temperatures (around 60°C–90°C) and is characterized by its use of a potassium hydroxide (KOH) electrolyte. In an alkaline electrolysis, a potassium hydroxide (KOH) solution is used as the electrolyte. The electrolyte facilitates the movement of ions and enables the electrochemical reactions to occur at the anode and cathode.

At the anode (positive electrode), water molecules (H_2O) undergo an electrochemical oxidation process. The anode is typically made of a conductive material coated with a catalyst, often based on nickel or nickel oxide. The anode reaction involves the following steps:

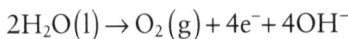

$$2H_2O(l) \rightarrow O_2(g) + 4e^- + 4OH^-$$

This reaction releases oxygen gas (O_2), electrons (e^-), and hydroxide ions (OH^-) at the anode (Figure 16.2).

At the cathode (negative electrode), the hydroxide ions (OH^-) generated at the anode migrate through the electrolyte solution to reach the cathode. The cathode is also made of a conductive material coated with a catalyst,

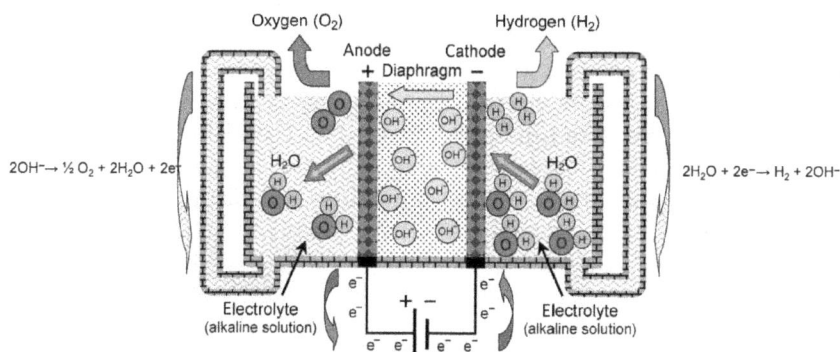

Figure 16.2 Alkaline electrolysis method.

usually based on nickel or nickel oxide. The cathode reaction involves the reduction of hydroxide ions to form hydrogen gas (H_2):

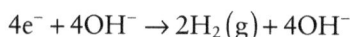

$$4e^- + 4OH^- \rightarrow 2H_2(g) + 4OH^-$$

This reaction produces hydrogen gas (H_2) at the cathode.

The alkaline electrolyte solution not only provides a medium for ion transport but also helps maintain the electrical neutrality within the electrolysis cell. It allows for the migration of hydroxide ions (OH^-) from the anode to the cathode, balancing the charges generated during the anode and cathode reactions. Like other electrolysis methods, alkaline electrolysis employs a physical barrier to separate the hydrogen and oxygen gases produced at the cathode and anode, respectively. The overall reaction of alkaline electrolysis can be represented as the combination of the anode and cathode reactions:

$$2H_2O(l) \rightarrow 2H_2(g) + O_2(g)$$

Similar to other electrolysis processes, alkaline electrolysis requires an external source of electrical energy to drive the electrochemical reactions. Applying a voltage across the electrodes provides the necessary energy to split water molecules.

Alkaline electrolysis technology, a promising avenue for hydrogen production, is not without its challenges. One key concern is the high energy consumption required to drive the electrolysis process efficiently. This results from the necessity of maintaining elevated temperatures and concentrations of the alkaline electrolyte solution. Corrosion of electrodes and other system components due to the caustic nature of the electrolyte presents a significant durability challenge, necessitating the use of resistant materials and periodic maintenance. Alkaline electrolysis systems often exhibit slower response times compared to other methods, which can hinder their ability to adapt to dynamic energy demand or intermittent renewable energy sources. Moreover, the bulky design and infrastructure requirements of alkaline electrolyzers can impede their integration into existing energy grids and industrial processes. To ensure the viability of alkaline electrolysis technology, ongoing efforts are focused on improving energy efficiency, extending system durability, enhancing response times, and streamlining system design for better compatibility with various applications and energy landscapes.

16.2.3 Solid oxide electrolyzer cell

It is an advanced electrolysis technology that operates at high temperatures, typically above 800°C. It utilizes a solid oxide electrolyte to facilitate the splitting of water into hydrogen and oxygen (Figure 16.3).

Unlike other electrolysis methods that operate at lower temperatures, SOEC operates at high temperatures, typically in the range of 800°C–1000°C. These elevated temperatures enable the solid oxide materials to efficiently conduct ions and promote the electrochemical reactions.

The key component of an SOEC is the solid oxide electrolyte, which is a ceramic material that can efficiently conduct oxygen ions (O_2^-) at high temperatures. Common electrolyte materials include yttria-stabilized zirconia (YSZ) and doped ceria. This solid electrolyte separates the anode and cathode compartments and allows only oxygen ions to migrate through it.

At the anode (positive electrode), oxygen ions (O_2^-) are generated as a result of the electrolyte's ionic conductivity. The anode is usually made of

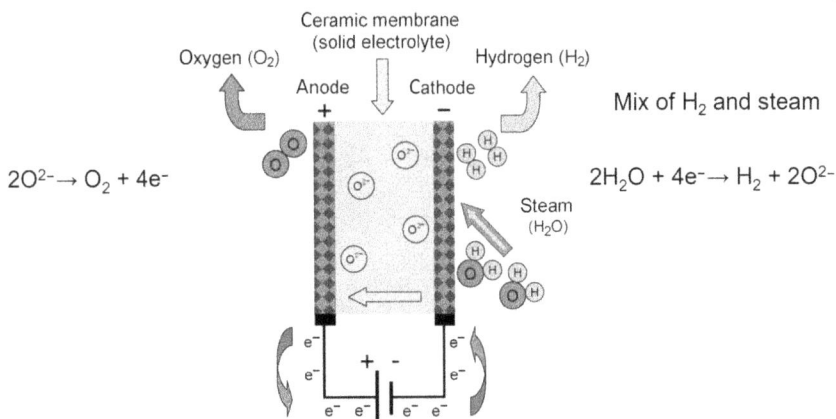

Figure 16.3 Solid oxide electrolyzer cell (SOEC) method.

a porous ceramic material coated with a catalyst, often based on nickel or nickel-cermet. The anode reaction involves the following steps:

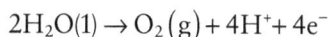

$$2H_2O(l) \rightarrow O_2(g) + 4H^+ + 4e^-$$

This reaction releases oxygen gas (O_2) and electrons (e^-) at the anode.

At the cathode (negative electrode), water vapor (H_2O) molecules are split into hydrogen ions (H^+) and electrons (e^-) by the supply of high-temperature thermal energy. The cathode is also made of a porous ceramic material coated with a catalyst, typically based on nickel or nickel-cermet. The cathode reaction involves the following steps:

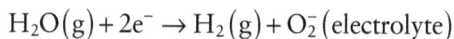

$$H_2O(g) + 2e^- \rightarrow H_2(g) + O_2^- (\text{electrolyte})$$

This reaction produces hydrogen gas (H_2) and oxygen ions (O_2^-) at the cathode.

The solid oxide electrolyte facilitates the migration of oxygen ions (O_2^-) from the anode to the cathode, while electrons (e^-) flow externally through an electrical circuit from the anode to the cathode. This separation of ionic and electronic conduction is a fundamental aspect of SOEC operation.

The overall reaction of SOEC involves the combination of the anode and cathode reactions:

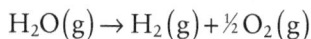

$$H_2O(g) \rightarrow H_2(g) + \frac{1}{2}O_2(g)$$

The solid oxide electrolyzer cell (SOEC) offers a range of significant advantages that make it a promising technology in the field of energy conversion and storage. One of its key benefits is its exceptional efficiency in electrolyzing water to produce hydrogen and oxygen. SOECs operate at high temperatures, typically around 500°C–900°C, which allows for faster electrochemical reactions and reduced energy losses compared to traditional low-temperature electrolysis methods. This inherent efficiency translates to lower electricity consumption and, consequently, reduced greenhouse gas emissions. Furthermore, SOECs are versatile and can play a crucial role in facilitating the integration of renewable energy sources into the grid. During periods of excess electricity generation, such as when renewable sources like solar and wind produce more power than is immediately needed, SOECs can efficiently convert this surplus electricity into hydrogen, which can be stored and used later for power generation, heating, industrial processes, or even as a clean fuel for transportation. SOECs also exhibit excellent durability and long operational lifetimes due to their robust solid oxide electrolyte materials and relatively simple design. This durability reduces maintenance and replacement costs over time, making

SOECs economically viable for various applications. In addition, SOECs have the potential to support carbon capture and utilization efforts by utilizing the produced hydrogen in various sectors, thus aiding in the reduction of carbon dioxide emissions.

In addition, SOEC systems currently have relatively slow startup and shutdown times due to the time required for reaching and cooling down from high operating temperatures. This limits their ability to respond quickly to fluctuations in electricity supply and demand, which is crucial for grid stability and supporting renewable energy integration. Another challenge lies in the cost of manufacturing SOECs at scale. The complex manufacturing processes and the need for high-quality materials contribute to higher upfront costs, making it essential to find innovative ways to reduce production expenses and improve cost-effectiveness. Moreover, while SOECs offer the potential for efficient hydrogen production, the overall hydrogen infrastructure and distribution network are still underdeveloped in many regions. Establishing a comprehensive hydrogen ecosystem, including storage, transportation, and utilization, presents a significant challenge that requires coordinated efforts across various sectors (Adjah-Tetteh et al., 2022). Finally, despite the environmental benefits of hydrogen produced through SOECs, the overall environmental impact also depends on the source of electricity used for electrolysis. If fossil fuels continue to dominate the energy mix, the potential carbon footprint associated with hydrogen production could undermine its sustainability goals.

16.2.4 Multi-step electrolysis

It is also known as multi-step water splitting or thermochemical water splitting. It involves a series of electrochemical and chemical reactions to break down water into hydrogen and oxygen using multiple intermediate compounds. This approach is aimed at improving the overall efficiency of the water-splitting process by taking advantage of different reaction thermodynamics at various temperature ranges. While multi-step electrolysis is more complex than traditional single-step electrolysis, it has the potential to achieve higher efficiencies and enable the use of low-grade or waste heat sources. Multi-step electrolysis consists of two or more sequential reaction steps, each occurring at different temperature ranges. These steps involve the conversion of water and intermediate compounds to eventually produce hydrogen and oxygen.

The intermediate compounds, often metal oxides or other chemical species, play a crucial role in the multi-step electrolysis process. These compounds undergo reversible chemical reactions that release or absorb oxygen during the process. The first step involves the high-temperature reduction of an intermediate compound using electrical energy or high-temperature heat sources. This reduction releases oxygen gas and generates an oxygen-deficient compound. The oxygen-deficient compound from the high-temperature reduction step is then subjected to a second step at a lower temperature.

In this step, the compound reacts with water vapor to regenerate the original intermediate compound, releasing oxygen gas.

In the final step, the regenerated intermediate compound is subjected to low-temperature electrolysis. This involves traditional electrochemical water splitting, where water is directly electrolyzed into hydrogen and oxygen using an electric current.

The overall multi-step electrolysis reaction is represented as follows:

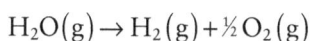

$$H_2O(g) \rightarrow H_2(g) + \tfrac{1}{2}O_2(g)$$

Compared to single-step electrolysis, multi-step electrolysis can achieve higher energy efficiencies, particularly when waste heat or low-grade heat sources are available for certain reaction steps. This approach allows the utilization of various heat sources, such as waste heat from industrial processes or concentrated solar energy, to drive the high-temperature reactions.

While this approach offers the advantage of generating higher-value products and improving overall energy efficiency compared to single-step electrolysis, it also comes with several notable challenges. One of the primary challenges is optimizing the sequence of electrochemical reactions and designing efficient catalysts for each step. Coordinating multiple reactions in a sequential manner while maintaining high selectivity and minimizing energy losses requires careful catalyst selection and engineering (Nasser et al., 2022). Developing catalysts that are stable, selective, and cost-effective for each reaction step is a complex task that demands interdisciplinary research efforts. Furthermore, multi-step electrolysis systems often involve intricate reactor designs and intricate separation and purification processes to isolate the desired products from intermediate compounds or byproducts. These aspects add complexity to the overall system and can impact both efficiency and scalability.

In addition, the integration of renewable energy sources into multi-step electrolysis setups introduces challenges related to intermittency and fluctuating electricity supply. Efficiently coupling the electrolysis process with variable renewable sources requires sophisticated control and energy management strategies to ensure stable and continuous operation. Another significant challenge is scaling up multi-step electrolysis technologies from the laboratory to industrial scales. Factors such as reactor scalability, mass transfer limitations, and safety considerations become more pronounced as the process is magnified, necessitating careful engineering solutions and technology validation.

16.2.5 Photoelectrochemical (PEC) hydrogen production

It is an innovative approach to directly convert solar energy into hydrogen gas through a process that involves semiconductors and electrochemical reactions. PEC technology aims to utilize sunlight to drive the water-splitting reaction, producing hydrogen and oxygen.

PEC hydrogen production begins with a semiconductor material that is capable of absorbing photons (particles of light) from the Sun. When photons are absorbed by the semiconductor, they generate electron-hole pairs, creating mobile charge carriers within the material. The excited electrons are then separated from the positively charged holes. The electrons can flow through the semiconductor material, while the holes are typically consumed by a separate oxidation reaction (Clarizia et al, 2023).

At the anode, a redox reaction takes place in which water molecules (H_2O) are oxidized. The positively charged holes from the semiconductor contribute to this oxidation reaction, releasing oxygen gas (O_2), protons (H^+), and electrons.

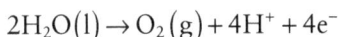

$$2H_2O(l) \rightarrow O_2(g) + 4H^+ + 4e^-$$

The electrons generated by the absorbed photons at the semiconductor's cathode side are used to reduce protons (H^+) to form hydrogen gas (H_2). This reduction reaction occurs at the cathode.

$$2H^+ + 2e^- \rightarrow H_2(g)$$

The protons (H^+) generated at the anode migrate through an electrolyte or a proton-conductive membrane to the cathode compartment, where they combine with electrons to form hydrogen gas.

The overall reaction of PEC hydrogen production involves the combination of the anode and cathode reactions:

$$2H_2O(l) \rightarrow 2H_2(g) + O_2(g)$$

PEC cells often require an external bias or electrical potential to drive the reactions and ensure efficient charge transfer at the electrode-electrolyte interfaces. In addition, catalysts are often applied to the electrode surfaces to enhance reaction rates and improve overall efficiency.

In addition to this, PEC hydrogen production effectively captures and stores solar energy in the form of chemical bonds in hydrogen gas. This stored hydrogen can later be used as a clean fuel for various applications, including energy generation and transportation.

PEC technology, while holding great promise for renewable hydrogen production, faces several formidable challenges. Efficiently harnessing sunlight to drive the electrolysis of water demands intricate material design and optimization to ensure optimal light absorption, electron excitation, and catalytic activity. The stability of photoelectrode materials under corrosive water-splitting conditions remains a hurdle, requiring the development of robust and long-lasting components. In addition, achieving the delicate balance of band gap tuning to match the solar spectrum is essential for maximizing energy conversion. Integrating various components into a cohesive and scalable system, complete with efficient catalysts, presents complex engineering and design considerations. Moreover, PEC technology must strive for cost competitiveness with conventional hydrogen production methods while ensuring minimal environmental impact throughout its lifecycle (Huang et al., 2023). As researchers and engineers work collaboratively to overcome these challenges, PEC technology holds the potential to revolutionize hydrogen production and contribute significantly to a sustainable energy future.

16.2.6 Hybrid electrolyzer systems

It combines different electrolysis technologies or integrates electrolysis with other processes to enhance overall efficiency, flexibility, and performance. These systems leverage the strengths of each component to optimize hydrogen production and address specific challenges.

Hybrid electrolyzer systems can integrate two or more types of electrolysis technologies, such as PEM electrolysis and alkaline electrolysis, within a single system (Brauns and Turek, 2020). Each electrolysis technology operates optimally under certain conditions, and by combining them, the system can capitalize on their respective advantages. For example, PEM electrolysis can provide a rapid response to load changes, while alkaline electrolysis may offer higher efficiency at certain production rates.

Hybrid electrolyzer systems often integrate with renewable energy sources like solar or wind power (Egeland-Eriksen et al., 2021). The variability of renewable energy generation can be managed by adjusting the electrolysis process. Excess energy generated during periods of high renewable energy output can be used for hydrogen production, effectively storing the energy in the form of hydrogen.

Power-to-Gas (P2G) Integration

Hybrid electrolyzer systems can be integrated with power-to-gas (P2G) systems, where excess renewable electricity is used to produce hydrogen via electrolysis. The generated hydrogen can then be injected into the natural gas grid, stored, or used for various applications. In this way, renewable energy is effectively converted into a storable and transportable form.

High-Temperature Process Integration

Some hybrid systems integrate electrolysis with high-temperature industrial processes, such as steelmaking or cement production (Marocco et al., 2023). The waste heat from these processes can be utilized to drive high-temperature electrolysis reactions, enhancing overall energy efficiency and reducing emissions.

Electrolyzer and Fuel Cell Coupling

Hybrid systems can also couple electrolyzers with fuel cells. Excess hydrogen produced by the electrolyzer can be fed into a fuel cell to generate electricity when demand is high or when renewable energy generation is low (Yue et al., 2021). This effectively converts the stored hydrogen back into electricity, providing a two-way energy conversion pathway.

Overall System Control

Hybrid electrolyzer systems require sophisticated control strategies to optimize the operation of the integrated components. These controls manage factors such as the flow of electricity, the rate of hydrogen production, and the interaction with external energy sources.

Hybrid electrolyzer systems represent a significant advancement in hydrogen production, offering a range of compelling advantages. By synergistically combining different electrolysis technologies, such as PEM and SOEC, these systems can harness the unique strengths of each component. This results in enhanced energy efficiency, as high-temperature electrolysis units, like SOECs, can utilize waste heat from industrial processes or renewable sources to drive the electrolysis reaction, reducing the overall energy input required. The coupling of diverse technologies also provides greater operational flexibility, allowing hybrid systems to respond effectively to varying energy inputs and demand fluctuations. Moreover, the integration of complementary electrolysis units can lead to improved system stability and durability, as well as optimized utilization of catalysts and materials. Hybrid electrolyzer systems offer a pathway to capitalize on renewable energy sources more effectively, address challenges associated with intermittency, and contribute to the development of a more resilient and sustainable hydrogen infrastructure.

Hybrid electrolyzer systems, which combine different electrolysis technologies, offer a promising avenue for efficient hydrogen production. However, they also face several challenges that need to be addressed for widespread adoption (Shiva Kumar and Lim, 2022). One of the primary hurdles is the integration of diverse components with varying operating conditions and materials compatibility. Achieving seamless interaction between high-temperature and low-temperature electrolysis units, such as SOEC and PEM electrolysis, demands precise engineering to optimize energy transfer and avoid thermal inefficiencies. Balancing the electrical and thermal requirements of different components within a hybrid system is essential to ensure overall efficiency. In addition, hybrid electrolyzer systems must overcome technical obstacles related to system control, dynamic response, and transient behavior during load changes or intermittent energy sources. Managing the interfaces and interactions between the different electrolysis units while maintaining stability and reliability poses a complex challenge. Furthermore, the cost-effectiveness of hybrid systems must be carefully evaluated, considering the investment in multiple technologies and the associated maintenance requirements. Despite these challenges, ongoing research and innovation in hybrid electrolyzer systems hold the potential to revolutionize hydrogen production, offering a flexible and sustainable solution that addresses the energy demands of the future.

16.3 GREEN HYDROGEN APPLICATIONS

Green hydrogen, a revolutionary clean energy solution, has ignited a global wave of enthusiasm due to its diverse range of applications across various sectors. Produced through the process of electrolysis, which splits water into hydrogen and oxygen using renewable energy sources like wind, solar, and hydroelectric power, green hydrogen offers a promising path to decarbonize economies and address pressing environmental concerns (Zhou et al., 2022). Its applications span an array of industries, each poised to benefit from its potential in unique and impactful ways.

In the realm of energy storage, green hydrogen emerges as a game-changer. With the ability to convert excess renewable energy generated during off-peak hours into hydrogen, this innovative solution holds the key to balancing energy supply and demand. The stored hydrogen can be reconverted into electricity through fuel cells or combustion as needed, effectively contributing to grid stability and minimizing the wastage of clean energy resources.

Transportation, a significant contributor to carbon emissions, stands to undergo a transformation through the utilization of green hydrogen. Hydrogen fuel cell vehicles (FCVs) and hydrogen-powered buses offer a cleaner alternative to traditional internal combustion engines (Mendez et al., 2023). The process of converting hydrogen and oxygen into electricity within fuel cells produces only water vapor as a byproduct, eliminating harmful emissions and reducing the carbon footprint of transportation networks. Industries dependent on hydrogen for their processes, such as steel, ammonia, and chemical production, can experience a paradigm shift toward sustainability by adopting green hydrogen. By replacing conventionally sourced hydrogen derived from fossil fuels, these sectors can drastically reduce their environmental impact, helping to mitigate climate change and drive a transition to greener industrial practices. In the pursuit of sustainable heating and cooling solutions, green hydrogen emerges as a versatile contender (Sharma et al., 2023). Its direct use or conversion into heat and power via fuel cells can revolutionize residential, commercial, and industrial heating and cooling systems, offering environmentally friendly alternatives that prioritize the reduction of greenhouse gas emissions. The power generation landscape also stands to benefit from the integration of green hydrogen. The combustion of hydrogen in gas turbines presents an efficient method of electricity generation, especially during peak demand periods. This not only contributes to grid stability but also provides a clean and renewable source of power, aligning with the global shift toward reducing reliance on fossil fuels. Beyond domestic applications, the potential of green hydrogen extends to the realm of international energy trade. Regions rich in renewable energy resources can produce surplus green hydrogen and export it to regions facing energy deficits. This energy carrier concept could revolutionize global energy dynamics, enabling countries to collaborate on a scale previously unattainable and promoting equitable access to sustainable energy. Challenging sectors like maritime and aviation, notorious for their high carbon emissions, also find a glimmer of hope in green hydrogen. With ongoing research and technological advancements, hydrogen has the potential to revolutionize propulsion systems, paving the way for cleaner and more sustainable modes of transport in these traditionally difficult-to-decarbonize sectors. In remote and off-grid locations, where traditional energy sources are often inaccessible, green hydrogen emerges as a beacon of reliable energy supply. By harnessing renewable sources and employing electrolysis, these communities can establish self-sustaining energy systems, reducing their dependence on environmentally harmful alternatives and improving their quality of life. The chemical and refining industries, significant contributors to global emissions, can undergo a transformative shift toward sustainability with the adoption of green hydrogen (Bauer et al., 2023). By substituting fossil-fuel-derived hydrogen with its green counterpart, these industries can substantially curtail their carbon footprints, contributing to the global fight against climate change. Perhaps one of green hydrogen's most remarkable attributes lies in its potential to facilitate energy system integration. By bridging sectors such as electricity, transportation, and heat, green hydrogen can create an interconnected energy ecosystem that maximizes the efficiency of clean energy use and accelerates the transition to a low-carbon future.

16.4 CONCLUSIONS AND FUTURE SCOPE

Green hydrogen production represents a pivotal step toward achieving a more sustainable and environmentally conscious energy landscape. As outlined in this exploration, the utilization of renewable energy sources, such as solar and wind power, to power the electrolysis process ensures that hydrogen is generated without emitting harmful greenhouse gases. This clean and versatile fuel holds immense potential across various sectors, including transportation, industry, and energy storage. The advancements in green hydrogen production methodologies have underscored the feasibility and viability of this technology. Electrolysis technologies have become increasingly efficient, driving down production costs and enabling larger-scale implementation. Integration with renewable energy systems not only enhances the environmental benefits but also contributes to energy resilience and grid stability. Looking ahead, the future of green hydrogen production is marked by both promising developments and critical challenges that warrant attention (Falcone et al., 2021). Continued research and innovation will play a crucial role in refining electrolysis processes, optimizing catalysts, and enhancing overall efficiency. Advancements in materials science and engineering will lead to more cost-effective and scalable electrolyzers, making green hydrogen production increasingly competitive. Green hydrogen production stands at the forefront of sustainable energy solutions, poised to play a transformative role in our transition to a low-carbon future. As research, innovation, and collaboration continue to drive this field forward, green hydrogen holds the potential to reshape global energy systems, mitigate climate change, and secure a cleaner, more resilient world for generations to come.

REFERENCES

A. M. Abdalla, S. Hossain, O. B. Nisfindy, A. T. Azad, M. Dawood, and A. K. Azad. (2018) Hydrogen production, storage, transportation and key challenges with applications: A review. *Energy Conversion and Management*, 165, 602–627.

C. Adjah-Tetteh, Y. Wang, Y. Sun, Z. Jia, X. Yu, and X.-D. Zhou. (2022) *A Solid Oxide Electrolysis Cell (SOEC) with High Current Density and Energy Efficiency for Hydrogen Production*. The Electrochemical Society. IOPscience.

F. Bauer, J. P. Tilsted, S. Pfister, C. Oberschelp, and V. Kulionis. (2023) Mapping GHG emissions and prospects for renewable energy in the chemical industry. *Current Opinion in Chemical Engineering*, 39, 10088.

J. Brauns and T. Turek. (2020) Alkaline water electrolysis powered by renewable energy: A review. *Processes*, 8 (2), 248.

J. Chi and H. Yu. (2018) Water electrolysis based on renewable energy for hydrogen production. *Chinese Journal of Catalysis*, 39 (3Mar), 390–394.

L. Clarizia, M. N. Nadagouda, and D. D. Dionysiou. (2023) Recent advances and challenges of photoelectrochemical cells for hydrogen production. *Current Opinion in Green and Sustainable Chemistry*, 41, 100825, 1–12

T. Egeland-Eriksen, A. Hajizadeh, and S. Sartori. (2021) Hydrogen-based systems for integration of renewable energy in power systems: Achievements and perspectives. *International Journal of Hydrogen Energy*, 46 (63), 31963–31983.

P. M. Falcone, M. Hiete, and A. Sapio. (2021) Hydrogen economy and sustainable development goals: Review and policy insights. *Current Opinion in Green and Sustainable Chemistry*, 31 (100506), 100506.

X. Huang, Y. Qu, Z. Zhu, and Q. Wu. (2023) Techno-economic analysis of photovoltaic hydrogen production considering technological progress uncertainty. *Sustainability*, 15 (4), 3580.

P. Marocco, M. Gandiglio, D. Audisio, and M. Santarelli. (2023) Assessment of the role of hydrogen to produce high-temperature heat in the steel industry. *Journal of Cleaner Production*, 388, 135969.

C. Mendez, M. Contestabile, and Y. Bicer. (2023) Hydrogen fuel cell vehicles as a sustainable transportation solution in Qatar and the Gulf cooperation council: A review. *International Journal of Hydrogen Energy*, 48(99), 38953–38975.

M. Nasser, T. F. Megahed, S. Ookawara, and H. Hassan. (2022) A review of water electrolysis-based systems for hydrogen production using hybrid/solar/wind energy systems. *Environmental Science and Pollution Research*, 29 (58), 86994–87018.

G. D. Sharma, M. Verma, B. Taheri, R. Chopra, and J. S. Parihar. (2023) Socio-economic aspects of hydrogen energy: An integrative review. *Technological Forecasting and Social Change*, 192, 122574.

S. Shiva Kumar and V. Himabindu. (2019) Hydrogen production by PEM water electrolysis – A review. *Materials Science for Energy Technologies*, 2 (3), 442–454.

S. Shiva Kumar and H. Lim. (2022) An overview of water electrolysis technologies for green hydrogen production. *Energy Reports*, 8, 13793–13813.

T. Wang, X. Cao, and L. Jiao. (2022) PEM water electrolysis for hydrogen production: Fundamentals, advances, and prospects. *Carb Neutrality*, 1, 21.

M. Yue, H. Lambert, E. Pahon, R. Roche, S. Jemei, and D. Hissel. (2021) Hydrogen energy systems: A critical review of technologies, applications, trends and challenges. *Renewable and Sustainable Energy Reviews*, 146, 111180.

Y. Zhou, R. Li, Z. Lv, J. Liu, H. Zhou, and C. Xu. (2022) Green hydrogen: A promising way to the carbon-free society. *Chinese Journal of Chemical Engineering*, 43, 2–13.

Chapter 17

Experimental investigation of thermohydraulic performance of solar air collector with double sided roughened absorber plate

Ravi Kant Ravi

Dr. B.R. Ambedkar National Institute of
Technology Jalandhar, Punjab

Rajesh Kumar

Dayalbagh Educational Institute, Dayalbagh, Agra

Manoj Kumar

National Institute of Technology, Srinagar, Jammu and Kashmir

Mukesh Kumar

IIEST Shibpur, Howrah, West Bengal, India

Chandraveer Singh

G.B. Pant Institute of Engineering and
Technology, Pauri Garhwal, Uttarakhand

17.1 INTRODUCTION

Solar air heaters (SAHs) are the specific type of heat exchangers that can be used for drying of agricultural crops, space heating, timber seasoning and other applications by utilizing solar radiation [1]. The main disadvantage of SAHs is their low heat transfer coefficient between the absorber plate and air flowing through the duct. Counter flow solar air heater (CFSAH) with a double-sided roughened absorber plate is the way to enhance the heat transfer rate with minimum pumping loss in a SAH. As reported in the literature [2–4], artificial roughness considerably enhances the energy extraction coefficient between air and heat collecting surface. In addition, the counterflow type has the advantage of minimizing heat losses and provides an increase in thermal performance with a minimum increase in the cost and size of the collector [5].

A two-pass SAH was first investigated by Satcunanathan and Deonarine [6]. They reported that the collectors' efficiency was 10%–15% more than

DOI: 10.1201/9781003472629-17

the single pass collector. This enhancement in the thermal efficiency of the collector was attained due to the reduction in thermal losses by operating the system in double flow mode. Wijeysundera et al. [7] also enhanced the performance of the conventional solar air collector by placing the two glass covers over the absorber plate of SAH. An experimental investigation on artificially roughened DPSAH duct was conducted by Sharma et al. [8] to see the effect of ribbed parameters on thermal performance. The statistical correlation for Nusselt number and friction factor has also been developed as a function of flow and rib parameters of double pass duct with V rib on both sides of the absorber. The thermohydraulic performance of DPSAH was investigated by one of the author's previous publications [9]. These publications concluded that the significant enhancement in thermal performance of SAH duct could be obtained by employing the rib geometry on both longitudinal faces of the absorber and by operating the duct in double-pass mode. Further in other investigations, the correlations for Nusselt number and friction factor as a function of roughness and operating parameters have been developed by Ravi and Saini [10].

17.2 THE NOVELTY AND OBJECTIVES OF THE WORK

In the literature, numerous investigators have studied different shapes and sizes of roughness components, PCMs and extended surfaces to boost the collector surface's heat transfer and thermal efficiency. Among all these studies, very few studies have been available to evaluate the thermohydraulic performance (heat and friction) of DPSAH with artificial roughness. Further, the use of simple v-ribs geometries is also accountable for the formation of wake zones in front and behind the rib element. These wake domains inhibit additional enhancements to the heat transfer in these zones of interest. To optimize the heat transmission across the absorber surface of the SAH, the wake zones must be eliminated. Therefore, these factors inspired the authors to conduct the current investigation and as a result, a thorough parametric analysis of rib roughness geometry and double pass solar air collector to improve thermohydraulic performance. It is hypothesized that the current roughness geometry will reduce the wake zone by allowing to pass the air through the gaps in multi v-ribs with increased velocity, thereby reducing the wake zones.

17.3 ROUGHNESS GEOMETRY AND PARAMETERS

The material used for making a roughness pattern is aluminium wire of 2 mm diameter. An epoxy solution is used as adhesive to attach the wire ribs over the collector surface. Figure 17.1 shows the photograph of discrete multi V-shaped and staggered rib roughness topology terminology and

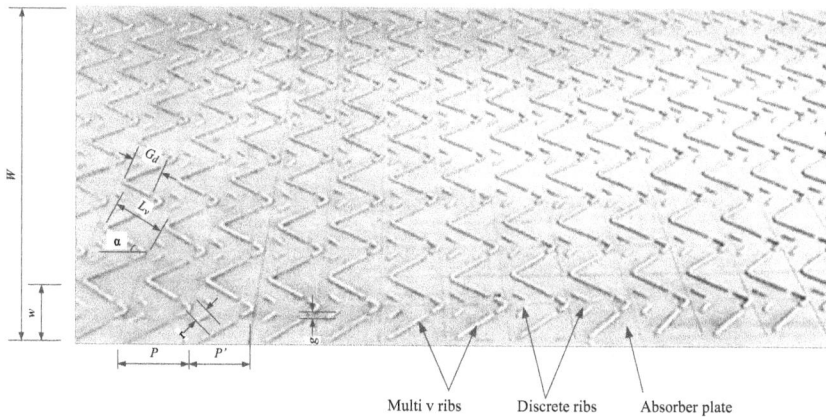

Figure 17.1 Photographs of absorber plate with roughness.

Table 17.1 Range/values of rib and duct parameters

S. no.	Parameters	Symbol	Value/range
1.	Relative gap distance	G_d/L_v	0.24–0.80 (5 steps)
2.	Relative roughness gap	g/e	0.5–1.5 (5 steps)
3.	Reynolds number	Re	2000–20,000 (10 steps)
4.	Relative roughness width	W/w	7
5.	Relative staggered rib size	r/e	1
6.	Width of each duct	W	300 mm
7.	Height of each duct	H	25 mm
8.	Length of absorber	L	1250 mm

configuration. The ribbed pattern is created on each side of the heated plate, i.e., made up of 0.8 mm Galvanized Iron (G.I.) sheet. Table 17.1 shows all configurations of designing of duct and roughness constraints that are utilized to carry out experimentation.

17.4 EXPERIMENTATION

The whole experimental test rig has been developed and constructed in the laboratory to investigate the heat transfer and fluid flow characteristics of the duct. The photograph of the experimental setup and different components is shown in Figure 17.2, which comprises the entry, test and exit section made up of wood. A 20 mm thick plywood covering with insulation sheets is being provided for the bottom and sides of the duct. The minimum length of the inlet and exit sections of duct is taken as per ASHRAE (1977) [11] standard, i.e., $5\sqrt{W \times D}$ and $2.5\sqrt{W \times D}$, respectively. The length of the

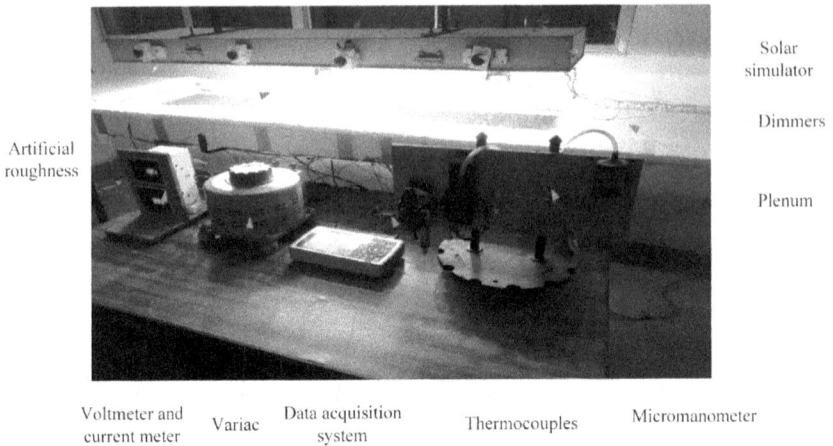

Figure 17.2 Photograph of the test rig.

test portion has been chosen to assume that a fully developed flow will be generated in its stated length, and the length of the exit section has been chosen so that the end effect of the heated airflow is nullified. To circulate the air inside the duct, a 3-phase electrical centrifugal blower is coupled with a flexible hose, and its flow is controlled by a valve attached to it. A U tube manometer filled with kerosense as manometric fluid is used for the mass flow rate measurement of flowing air through the pipe whereas a micromanometer is used to measure the pressure drop through the test section. The heat to absorber plate has been provided with the help of the solar simulator that contains 24 halogen tubes with each rated power of 150 W. The solar power metre has also been used to check the intensity of the radiation falling on the absorber plate. Temperature at various locations of the duct and absorber plate has been measured with the help of calibrated T-type thermocouples attached to the data logger.

17.5 PERFORMANCE PARAMETERS

In the present work, Nusselt number is calculated as the envoy of thermal performance and the friction factor as representative of pressure loss. The other intermediate parameters are calculated by using the different relations used by Ravi and Saini [2] and are given below as:

The bulk temperature of the heated plate (T_{pm}) is calculated by using the following equation as

$$T_{pm} = \frac{1}{19}\left(\sum_{k=1}^{19} T_k\right) \tag{17.1}$$

The bulk mean temperature of the air can be evaluated as

$$T_{am} = \frac{T_{in} + T_{out}}{2} \tag{17.2}$$

The heat transfer coefficient (h) for the SAH duct is estimated as

$$h = \frac{mc_p(T_{out} - T_{in})}{A_p(T_{pm} - T_{am})} \tag{17.3}$$

where m is air mass flow rate, A_p is the area of absorber plate, and C_p is specific heat of air,

The Nusselt number is calculated by using heat transfer coefficient (h), hydraulic diameter of duct (D) and thermal conductivity of air (K) in the relation as;

$$Nu = \frac{hD}{K} \tag{17.4}$$

The friction factor of the duct can be determined by using the equation below as;

$$f = \frac{2(\Delta p)_{duct} D}{4\rho_{air}LV^2} \tag{17.5}$$

where $(\Delta p)_{duct}$ is pressure loss in duct, ρ_{air} is density of air and V is the velocity in m/s.

17.6 VALIDATION TEST AND UNCERTAINTY ANALYSIS

The validation of the test rig has been carried out by comparing the experimental and theoretical values of Nusselt number and friction factor obtained for the smooth absorber plate and Dittus–Boelter equation (17.6) and modified Blasius equation (17.7) respectively.

$$Nu_s = \frac{0.024 \times Re^{0.8}}{Pr^{-0.4}} \tag{17.6}$$

where Pr is Prandtl number

$$f_s = \frac{0.085}{Re^{0.25}} \tag{17.7}$$

The results of the above analysis have been plotted in Figure 17.3 which depicts that the maximum deviation between theoretical and experimental data of Nu and f lies between ±3.26% and ±4.15% respectively. Consequently, a reasonable relationship between two sets of values verifies the accuracy of the information gathered through experimentation.

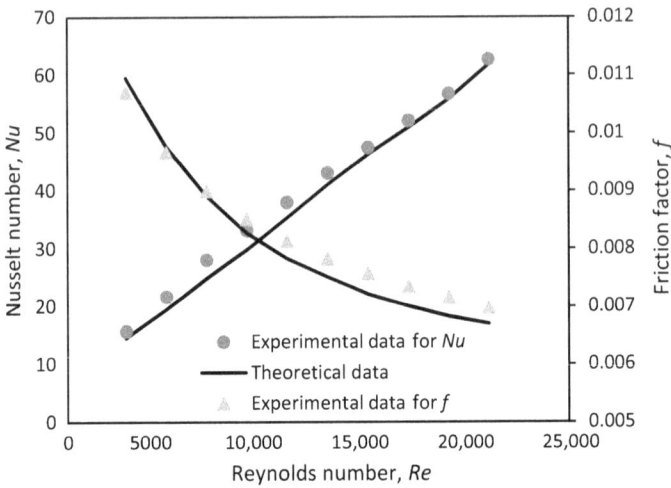

Figure 17.3 Comparison of experimental and theoretical data in lower flow channel.

The uncertainty analysis has been conducted by using Kline and McClintock [12] and Holman [13] equations involving the inaccuracy in experiments, dimensions, measurements and calculations, etc. The maximum error occurred for Q_u, Nu, h, Re and f was found to be as ±3.22%, ±3.51%, ±3.21, ±4.27%, and ±4.05%, respectively.

17.7 RESULTS AND DISCUSSION

In this section, the Nusselt number (Nu) and friction factor (f) characteristics have been presented and compared with smooth collector based on experimental data obtained for investigated rib parameters.

17.7.1 Effect of Reynolds number

Figures 17.4 and 17.5 represent the influence of Reynolds number (Re) on Nusselt number (Nu) and friction factor (f) for some selected values of roughness parameters. It is evident from Figure 17.4 that Nu increases sharply with increase in Re and found the maximum values at $Re=20,000$. It may be due to the fact, as Re increases, it gives rise to the turbulent intensity due to the increased value of turbulent kinetic energy and dissipation rate of air turbulence. The increased intensity of turbulence reduces the length of reattachment points of the flow streams between the two consecutive rib pitches. Due to this, the formation of low heat transfer zone in between two reattachment points gets reduced and leads to rise in Nu. On another side of experimental findings, Figure 17.5 reflects that f decreases with an increase in Re for all selected values of roughness parameters. It may be explained

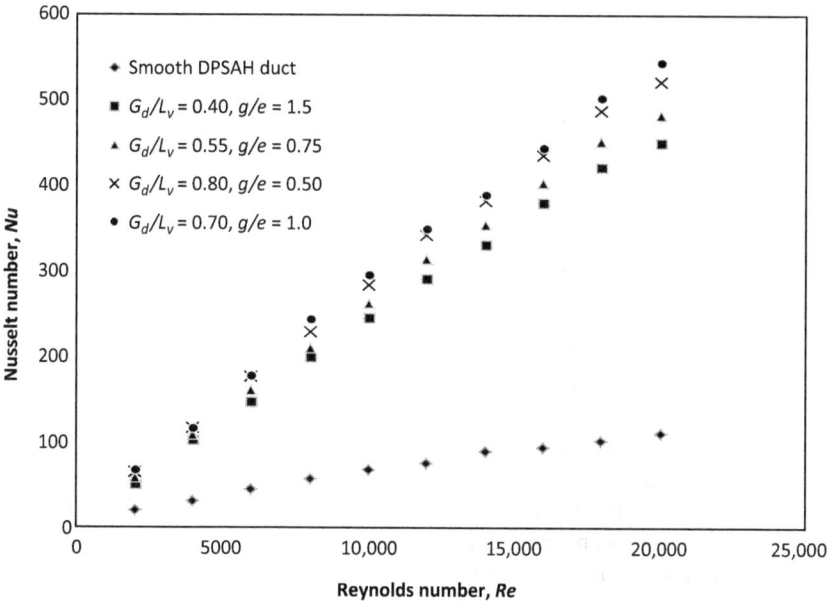

Figure 17.4 Effect of Re on Nu.

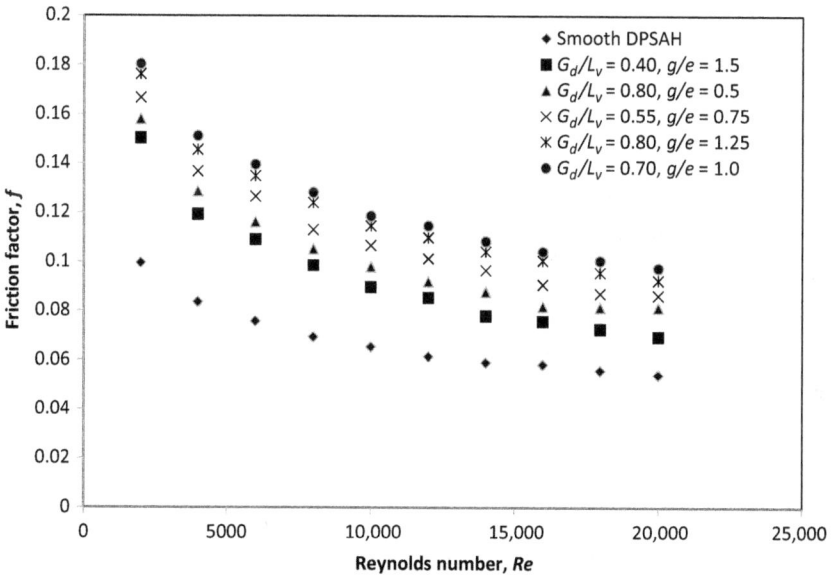

Figure 17.5 Effect of Re on f.

based on the concept of viscous sublayer. As the Re increases, the viscous sublayer gets thinner for all values of roughness parameters, resulting in reduced f. It is obvious that in comparison with the smooth plate, both the Nu and f are increased for the roughened plate of DPSAH. The highest Nu is reported to be as 588.06 at $G_d/L_v=0.70$ and $g/e=1.0$ for $Re=20,000$, while the peak value of f is found to be as 0.18032 at $G_d/L_v=0.70$ and $g/e=1.0$ correspond to $Re=2000$.

17.7.2 Effect of relative gap distance

Figure 17.6 shows the variation of relative gap distance (G_d/L_v) with Nusselt number (Nu) for some selected values of Reynolds number (Re). It can be observed that the Nu rises with rise in G_d/L_v from 0.55 to 0.70, attains a peak at G_d/L_v of 0.70 and then declines with a further rise in the G_d/L_v. This kind of trend may be attributed to the variation of gap distance, i.e., changes in auxiliary flow cells brought about by the existence of gaps at distinct positions may be used to explain gap position. It may also be observed that the too narrow gap and too much gap did not trigger the flow segregation and detachment points. However, the heat transfer is found to be maximum in the vicinity of reattachment points. Figure 17.7 shows the variation of friction factor (f) with that of G_d/L_v for some opted values of Re. It is clearly visible that the peak value of f is obtained at $G_d/L_v=0.70$ for $Re=4000$.

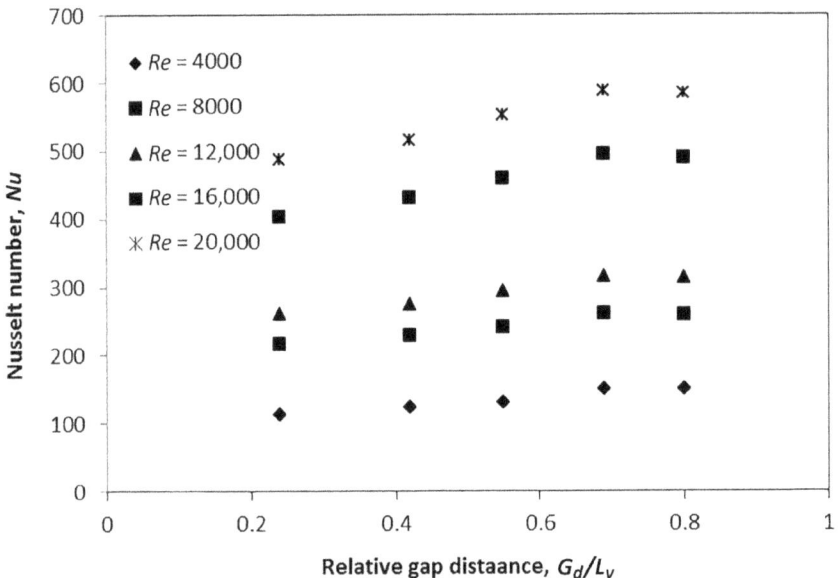

Figure 17.6 Effect of G_d/L_v on Nu.

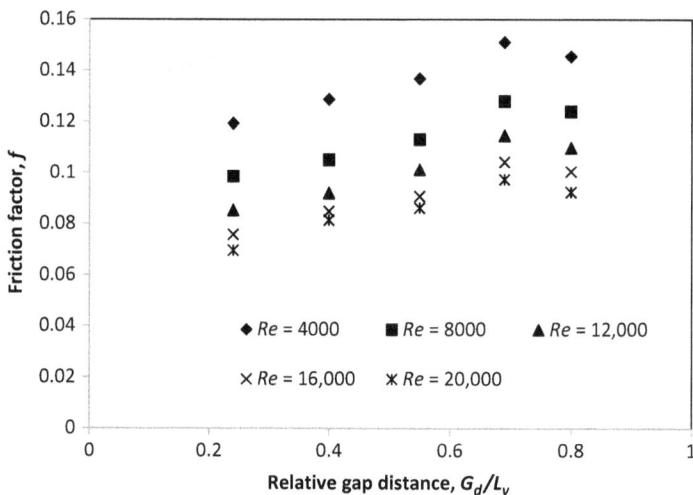

Figure 17.7 Effect of G_d/L_v on f.

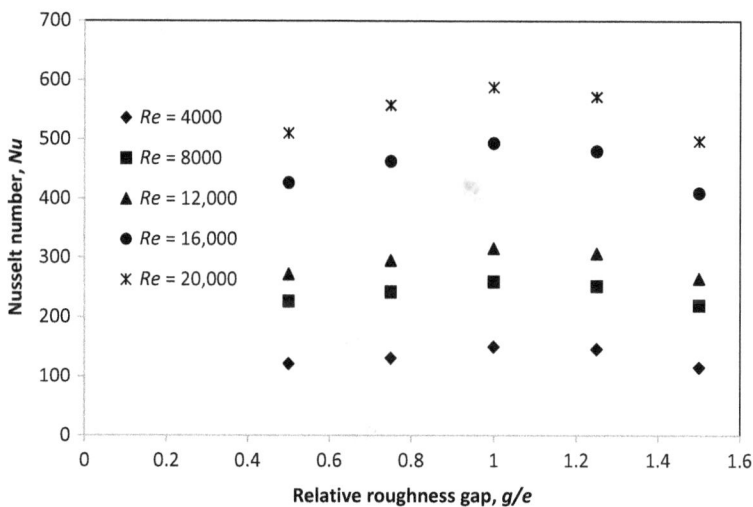

Figure 17.8 Effect of g/e on Nu.

17.7.3 Effect of relative roughness gap

The variation of Nu for different values of g/e and some selected values of Re is shown in Figure 17.8. It is found that the Nu increases as g/e increases from 0.5 to 1.0 and then decreases from 1.0 to 1.5. The peak value of Nu has been obtained as 588 at g/e=1.0 corresponding to Re=20,000. The

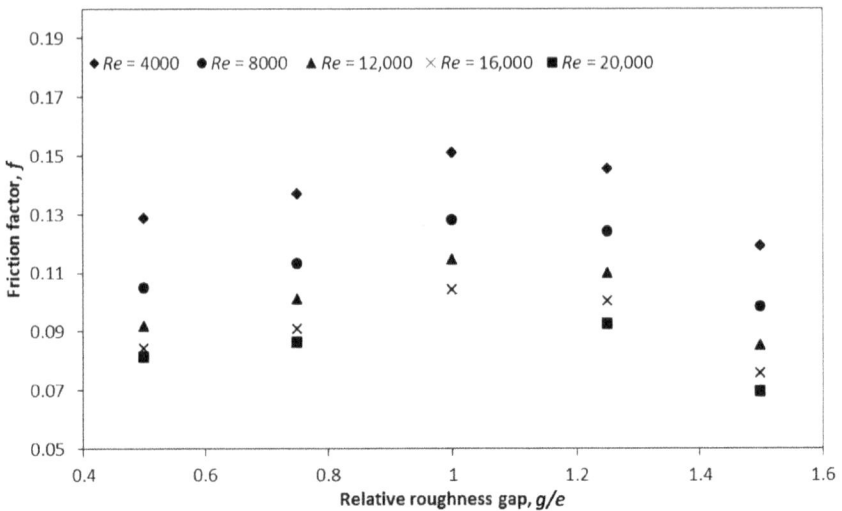

Figure 17.9 Variation of *f* with *g*/*e*.

reason is that this relative roughness gap is quite much supportive and favourable for the reattachment of flow cells that are taking place over the absorber plate. The free shear layer (sometimes also called laminar shear layer) allows the flow cells to reattach over the roughened plate. At this gap distance, most of free layers impose over there and most of the heat is transferred to the fluid cells. Hence, the temperature of roughened plate gets diminished and the rate of convective heat transfer rises.

Figure 17.9 depicts the influence of *f* with *g*/*e*, for some selected values of *Re* where peak value in plot can be recognized. It is seen that for all values of *g*/*e*, *f* declines with rise in *Re*. Addition of *g*/*e* in the ribs produces recirculation loops, which are accountable for larger disturbance and therefore larger friction factor. It can also be observed that the strength of auxiliary flow cells is exhausted for *g*/*e* of 1.5 as related to *g*/*e* from 0.5 to 1.25 therefore pressure drop is diminished as compared to other cases. The results are greatly agreed with earlier works available in the literature [14,15].

17.8 COMPARISON WITH THE PREVIOUS RESEARCH WORK

In this section, an effort has been made to compare the performance of the proposed roughness pattern and two-pass duct with the previous studies available in the literature. To compare the performance, a thermohydraulic performance parameter (TPP) given by Lewis [16] and presented in Eq. (17.8) is used. According to this, a TPP value greater than 1 will be beneficial to use the roughness geometry. The comparison results are shown

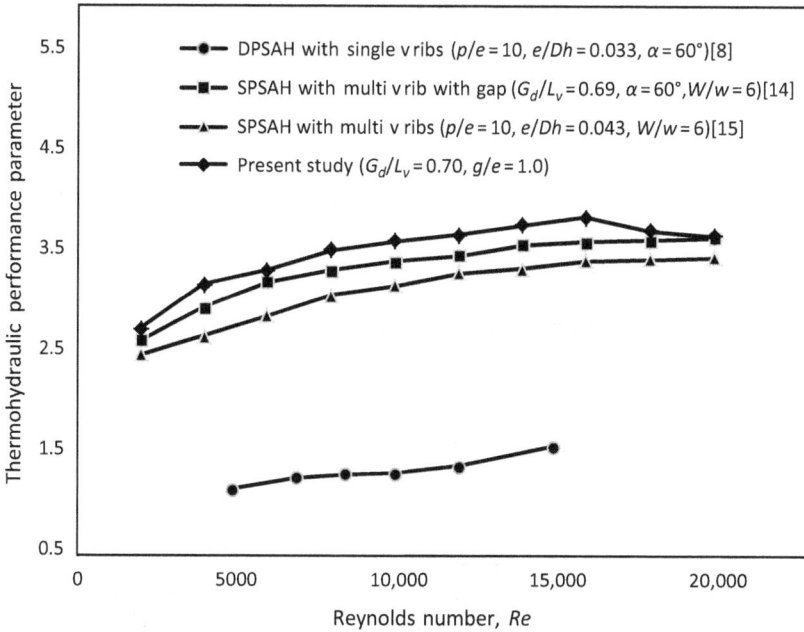

Figure 17.10 Comparison of thermohydraulic performance parameter with previous investigations.

in Figure 17.10, which depicts that the maximum value of TTP has been achieved at G_d/L_v=0.70 and g/e=1.0 at Re=16,000. Further, the values of TPP are found to be more as compared to other roughness geometries.

$$TPP = \frac{Nu/Nu_s}{(f/f_s)^{\frac{1}{3}}} \qquad (17.8)$$

17.9 CONCLUSIONS

In the present work, based on the experimental work, the following conclusions are drawn.

1. The highest increment in Nusselt number has been found to be as 4.52 times as compared to non-roughened duct for the selected range of parameters. However, this enhancement in Nu has been obtained with a simultaneous increase in friction factor as 3.13 times as compared to the smooth one.
2. Maximum increase in heat transfer rate and friction factor takes place for G_d/L_v=0.70 and g/e=1.0.

3. Thermohydralic performance parameter (TPP) is found to be greater than one for all values of roughness parameters ensuring the benefit of using roughness elements. The maximum value of TPP is found to be 3.88 for G_d/L_v=0.70 and g/e=1.0 at Re=16,000. Therefore, these values of rib parameters can be considered as optimum values.

4. The study broadly concluded that Nusselt number as envoy of thermal performance has increased with increase in Reynolds number (Re) and the use of artificial roughness. Also, the increment in friction factor (f) has also been observed, which represents the pumping power requirement.

5. The study concluded that the performance of air heating collector can be calculated by including the mechanical power required to circulate the air through the collector.

REFERENCES

[1] Prasad, K., & Mullick, S. C. (1983). Heat transfer characteristics of a solar air heater used for drying purposes. *Applied Energy*, 13(2), 83–93.

[2] Ravi, R. K., Kumar, M., Kumar, M., & Saini, R. P. (2022). Experimental investigation on heat transfer and fluid flow characteristics for roughened counter flow solar air collector. *International Journal of Green Energy*, 19(8), 865–878.

[3] Gupta, D., Solanki, S. C., & Saini, J. S. (1997). Thermohydraulic performance of solar air heaters with roughened absorber plates. *Solar Energy*, 61(1), 33–42.

[4] Karwa, R., Solanki, S. C., & Saini, J. S. (1999). Heat transfer coefficient and friction factor correlations for the transitional flow regime in rib-roughened rectangular ducts. *International Journal of Heat and Mass Transfer*, 42(9), 1597–1615.

[5] Ravi, R. K., & Saini, R. P. (2016). A review on different techniques used for performance enhancement of double pass solar air heaters. *Renewable and Sustainable Energy Reviews*, 56, 941–952.

[6] Satcunanathan, S., & Deonarine, S. (1973). A two-pass solar air heater. *Solar Energy*, 15(1), 41–49.

[7] Wijeysundera, N. E., Ah, L. L., & Tjioe, L. E. (1982). Thermal performance study of two-pass solar air heaters. *Solar Energy*, 28(5), 363–370.

[8] Sharma, A., Varun, V., Kumar, P., & Bharadwaj, G. (2013). Heat transfer and friction characteristics of double pass solar air heater having V-shaped roughness on the absorber plate. *Journal of Renewable and Sustainable Energy*, 5, 1–13.

[9] Ravi, R. K., & Saini, R. P. (2016). Experimental investigation on performance of a double pass artificial roughened solar air heater duct having roughness elements of the combination of discrete multi V shaped and staggered ribs. *Energy*, 116, 507–516.

[10] Ravi, R. K., & Saini, R. P. (2018). Nusselt number and friction factor correlations for forced convective type counter flow solar air heater having discrete multi V shaped and staggered rib roughness on both sides of the absorber plate. *Applied Thermal Engineering*, 129, 735–746.

[11] ASHARAE Standard 93-77 Methods of Testing to Determine the Thermal Performance of Solar Collectors, American Society for Heating, Refrigeration, and Air Conditioning Engineering, New York, 1997.

[12] Kline, S. J. (1963). Describing uncertainties in single-sample experiments. *Mechanical Engineering*, 75, 3–8.

[13] Holman, J. P. (2021). *Experimental Methods for Engineers*, 8th edition. Tata McGraw-Hill, New Delhi.

[14] Kumar, A., Saini, R. P., & Saini, J. S. (2012). Experimental investigation on heat transfer and fluid flow characteristics of air flow in a rectangular duct with Multi v-shaped rib with gap roughness on the heated plate. *Solar Energy*, 86(6), 1733–1749.

[15] Hans, V. S., Saini, R. P., & Saini, J. S. (2010). Heat transfer and friction factor correlations for a solar air heater duct roughened artificially with multiple v-ribs. *Solar Energy*, 84(6), 898–911.

[16] *Lewis, M. J. (1975). Optimising the thermohydraulic performance of* rough surfaces. *International Journal of Heat and Mass Transfer*, 18(11), 1243–1248.

Chapter 18

Design and analysis of a solar-based gear pump system

Bhupender Singh, Rajesh Attri,
Bhaskar Nagar, and Arvind Gupta
J.C. Bose University of Science and Technology, YMCA

Ashok Kumar Yadav
Raj Kumar Goel Institute of Technology

18.1 INTRODUCTION

The gear pump was invented by Johannes Kepler in 1600. It is a type of rotary displacement pump in which the pumping chamber is the space between gear teeth and, the meshing action of two gears provides the change in pressure of the pumping chamber to drive the fluid. Gear pumps are highly efficient in high-pressure pump operation. While a piston pump gives high head, vane pump gives high discharge, a gear pump gives a combination of both.

A solar pump mainly consists of the following parts:

  i. Solar panel
 ii. Controller
iii. Motor
 iv. Pump mechanism
  v. Inverter (in case of AC powered pumps)
 vi. Battery

Fluid transportation is one of the many uses for solar energy, which is a plentiful and renewable resource. The durability and straightforward design of gear pumps make them popular in a variety of industries for moving fluids. An opportunity to develop an effective and ecologically friendly pumping system is provided by integrating solar energy with a gear pump system. The design and analysis of such a solar-powered gear pump system are the main topics of this work, which also uses computational fluid dynamics (CFD) simulations to comprehend the fluid dynamics and thermal properties of the system.

A solar-driven pump operates on the principle of 'Photovoltaic Principle' in which PV panels convert sunlight into DC electricity. DC electricity flows

DOI: 10.1201/9781003472629-18

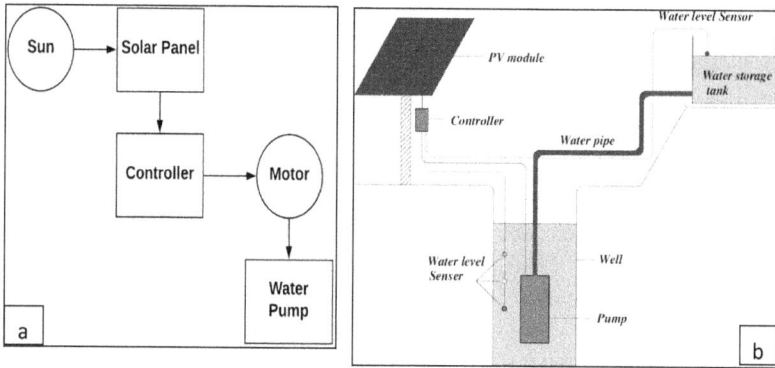

Figure 18.1 (a) Block diagram of a solar-driven pump and (b) working of a solar pump.

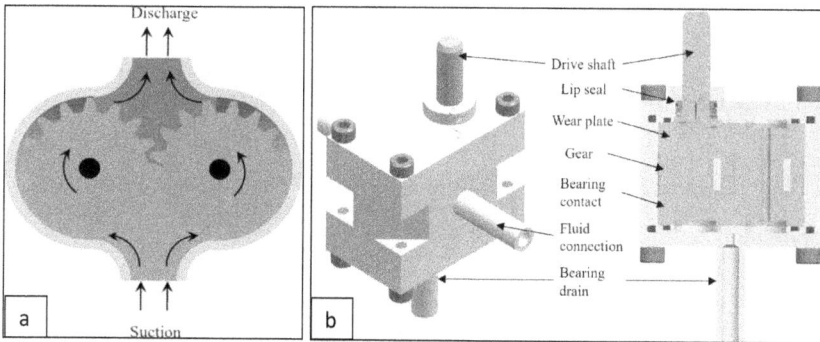

Figure 18.2 (a) External gear pump and (b) configuration of an external gear pump.

through the prime mover of the pump mechanism via controller which regulates the voltage, and hence flow from the pump. The block diagram of a solar-driven pump, as shown in Figures 18.1 and 18.2, shows the working of a solar pump.

18.2 FUNDAMENTALS OF AN EXTERNAL GEAR PUMP

A fluid reservoir, inlet and outlet pipelines, and the actual gear pump are all included in the gear pump assembly. The fluid is moved from the intake to the outlet by the rotating motion of the gear pump, which creates a positive displacement. To ensure seamless energy conversion and fluid conveyance, these components are integrated into the architecture of the solar-powered system.

The external gear pump operates through the action of two meshed external gears. As the gears rotate, the teeth unmesh at the suction side, creating a void and drawing in fluid. The gears then carry the fluid along the wall of the pumping chamber as they continue to rotate. The fluid trapped between the gear teeth is then forced into the discharge region as the gear teeth mesh. The clearances between the gears and the adjacent components create the seal that restricts backflow through the pump. An illustration in Figure 18.2a gives the cross section of an external gear pump with a color gradient to represent fluid pressure within the pump – red denoting high pressure and blue denoting low pressure.

While there is no single external gear pump design, a common configuration is shown in Figure 18.2b, with an exploded rendering shown in Figure 18.3. The core design concept of the pump is a pumping chamber that is formed inside the pump casing between two wear plates. These plates are pressed into the gears and provide axial containment, while the pump casing provides radial confinement of the fluid. Internal sealing is provided by o-ring for static seals and a lip seal for the rotary seal at the drive shaft. The two shafts are supported by journal bearings in the wear plate that counteract the radial load on the gears. There are three fluid connections,

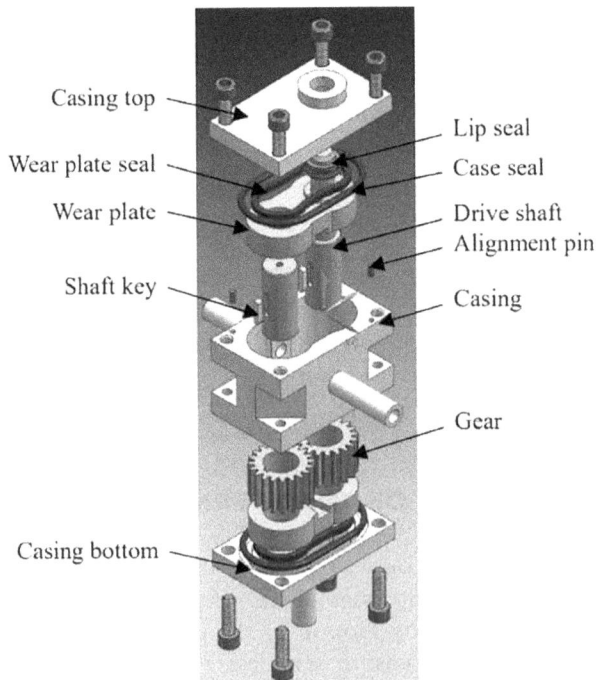

Figure 18.3 Exploded view of an external gear pump.

with two used for the pump inlet/outlet, and the third used as a bearing drain. This allows fluid in the pump to leak through the journal bearings and out through the drain, creating a fluid circuit that continuously lubricates the bearings.

18.3 LITERATURE REVIEW

The investigation of creative techniques to optimize diverse industrial processes has been prompted by the rising demand for sustainable energy solutions. This research focuses on a solar-based gear pump system that combines mechanical pumping with solar energy harvesting to offer a cost-effective and ecologically responsible method of moving fluids. The performance of the system under various situations may be thoroughly examined thanks to the integration of CFD modeling.

The fluid dynamics and heat transfer within the gear pump system are examined using CFD models. Parameters including fluid viscosity, flow rate, temperature distribution, pressure drop, and pump efficiency are taken into account during the simulations. The fluid behavior inside the pump may be seen using the CFD analysis program, and potential areas for improvement can be found.

Thiagarajan et al. [1] improved the ability of external gear machines to lubricate them by inventing a micro-surface linear wedge added to the lateral surfaces of the gear teeth. The author demonstrated how a numerical approach based on CFD can reduce torque loss. Experimentation provided proof of the findings. Utilizing a numerical technique, Vacca et al. [2] investigated the operation of spur external gear units. A mathematical model was used by Borghi et al. [3] to forecast the volumetric efficiency of gear pumps. Experimental data have been used to validate the model. Only by properly engineering the internal fluid dynamics of both lateral plates can the pump performance be improved. If not properly built, these grooves connecting volumes at different pressures can drastically reduce the volumetric efficiency of gear pumps. These grooves designed in the plates can serve a variety of purposes, such as minimizing noise inside the pump [4]. Frosina et al.'s extensive research on the impact of grooves inside of plates is found in their study [5]. Numerous studies recommend utilizing 2D modeling techniques with deforming mesh and volume remeshing to examine gear pumps. Even though 2D modeling techniques produce intriguing results for the study of pumps such external gear pumps, they are unable to predict internal flow behavior as well as 3D models [2–5]. This is dependent on the fact that an external gear pump operates with a flow that is extremely complex due to high pressure and rotational speeds (usually 500–3000 rpm).

Under a range of operational circumstances, such as variations in sun radiation, fluid characteristics, and pump speeds, the efficiency of the solar-based gear pump system is assessed. Key performance metrics are measured and contrasted with conventional pump systems, including energy conversion efficiency, pumping efficiency, and total system efficiency. In terms of energy savings and lower carbon emissions, the analysis shows the benefits of the solar-based strategy.

The search for creative ways to optimize diverse industrial processes has been prompted by the rising demand for sustainable energy solutions [6–19]. To create an effective and ecologically responsible method of transporting fluids, this article focuses on a solar-based gear pump system that integrates solar energy harvesting with mechanical pumping mechanisms. It is possible to analyze the system's performance in detail under various circumstances thanks to the integration of CFD modeling.

Understanding fluid dynamics, heat transfer, and pressure distribution in complicated systems is made possible through CFD simulations. In this study, the behavior of the solar-based gear pump system is modeled using CFD software. The simulation environment defines the system geometry, boundary conditions, and fluid properties.

The Navier-Stokes equations for fluid flow, the energy equation for analyzing heat transfer, and suitable turbulence models are all included in the CFD model to accurately represent flow behavior [20–24]. The simulations shed light on the gear pump system's flow patterns, pressure drop, heat exchange, and temperature distribution.

Solar-powered pumping methods are being investigated in response to the rising demand for sustainable energy sources and effective fluid transportation systems [25–29]. The design and study of a solar-powered gear pump system using CFD modeling is presented in this paper. To enable sustainable fluid transfer in various applications, the suggested system combines solar energy collecting and conversion with a gear pump. The fluid dynamics and thermal behavior of the system are better understood thanks to the CFD simulations, which also help with system optimization for improved performance and efficiency.

The selection of suitable materials, gear profiles, and sealing mechanisms is the main focus of the gear pump design in order to assure effective fluid conveyance. Geometric factors including tooth shape, pressure angle, and clearance are chosen to minimize leakage and improve volumetric effectiveness. The performance of the entire system depends heavily on the mechanical effectiveness and volumetric uniformity of the pump.

Careful study of energy transfer methods, fluid connections, and control techniques is necessary for the integration of the solar collector and gear pump. The heat transfer fluid, which receives the thermal energy from the solar collector, then transfers it to the gear pump, giving it the necessary heat. Control mechanisms are put in place to manage the fluid flow rate and keep both components' working temperatures at their ideal levels.

For the purpose of capturing solar energy and transforming it into thermal energy, the solar collector is the key element. Materials, absorber plate shape, heat transfer fluid, and insulation are all factors that must be taken into account during the design process. Maximizing energy absorption and reducing heat loss requires proper insulation and selective coating. Utilizing simulations, the solar collector design is improved for maximum energy conversion effectiveness [30–38].

To ensure effective fluid conveyance, the gear pump design places a strong emphasis on choosing the right materials, gear profiles, and sealing mechanisms. To decrease leakage and improve volumetric efficiency, geometric characteristics such as tooth profile, pressure angle, and clearance are chosen. The performance of the entire system is greatly influenced by the mechanical effectiveness and volumetric stability of the pump.

The main element for absorbing solar energy and transforming it into thermal energy is the solar collector. The design takes into account factors including insulation, heat transfer fluid, absorber plate geometry, and material selection. To increase energy absorption and reduce heat loss, proper insulation and selective coating are essential. To get the best energy conversion efficiency, simulations are used to optimize the solar collector design.

The design, analysis, and optimization of a solar-powered gear pump system using CFD modeling are all thoroughly studied in this research work. The suggested system seeks to utilize solar energy to effectively run a gear pump, hence promoting the use of sustainable energy sources in a variety of applications. The design of the solar collector, the gear pump, and their integration are all included in the study. After that, a thorough CFD analysis is performed to assess the system's performance under various operating scenarios [39–44]. The outcomes reveal the viability and potential advantages of using solar energy to power gear pumps, demonstrating the efficiency gains made by this hybrid strategy.

18.4 DEVELOPMENT OF SOLID MODEL

The solid model of shell which was completed by Solid Works was saved in.igs extension, then it was imported into ANSYS workbench. The linear elements of SOLID186 were chosen and used the automatic method of ANSYS to mesh grids. The accuracy of Grade 2 was selected to mesh the shell. A typical process of ANSYS structural analysis involves three steps. The first step is to create a finite element model, including the creation or import of geometry model, definition of cell types, and definition of unit physical constants, definition of material properties and the division of finite element mesh. The second step is to apply the loads, including constraints and load. The third step is to solve.

The CFD simulation results provide a thorough insight into the behavior of the solar-based gear pump system. These findings include temperature gradients, heat transfer rates, pressure distribution, velocity profiles, and pressure distribution. Researchers are able to find design changes that improve performance and efficiency by contrasting various circumstances [45].

The solar energy collecting and conversion system and the gear pump assembly are the two primary parts of the solar-powered gear pump system. Photovoltaic (PV) panels, a charge controller, and a battery bank make up the solar system. Solar energy is converted by the PV panels into electrical energy, which the charge controller then transfers to the battery bank for storage. The motor of the gear pump is driven by electrical energy.

Finite element analysis (FEA) is a mathematical representation of a physical system comprising a part/assembly, material properties and boundary conditions. In several situations, product behavior in the real world cannot be approximated by simple hand calculations. A general technique like FEA is a convenient method to represent complex behaviors by accurately capturing physical phenomena using partial differential equations. Meshing is one of the most important steps in performing an accurate simulation using FEA.

18.4.1 Development of model for cover

A mesh is made up of elements that contain nodes (coordinate locations in space) that represent the shape of the geometry. The examination of numerous system parameters is made possible by CFD simulations, which improve performance. Changes in gear design, fluid viscosity, solar panel efficiency, and pipe size can all have an impact on heat transfer and fluid flow, and this impact can be measured. A quantitative foundation for assessing the system's effectiveness, pressure properties, and thermal behavior is provided by the simulations. An FEA solver cannot easily work with irregular shapes, but it is much happier with common shapes like cubes. Meshing is the process of turning irregular shapes into more recognizable volumes called elements. Element size can be taken as 2.56e−002.

Fixed support function of ANSYS software is applied at the holes where nuts are inserted as they apply force opposite to the direction of hydrostatic pressure and should be fixed to avoid any wear and tear [46−49].

From the given Figure 18.4, we can assume that the hydrostatic pressure at center is zero because it is designed in a way that the center is elevated to some extent. Maximum hydrostatic pressure is at the edges of the cover which is 1.6887e−12 Pa as most of the fluid is at high pressure when compressed by the movement of gears. The deformation of the cover becomes smaller and smaller from the center part of the high-pressure area up and down both ends. The maximum deformation occurs at the edges of the high-pressure surface. The maximum deformation is 2.4273e−24 m, which can meet the requirement of the gear pump. The equivalent strain is mainly more at the center and the less at edges. The maximum equivalent elastic strain is 1.06e−21, which can meet the requirement of the gear pump, and the minimum equivalent elastic strain is 2.634e−24.

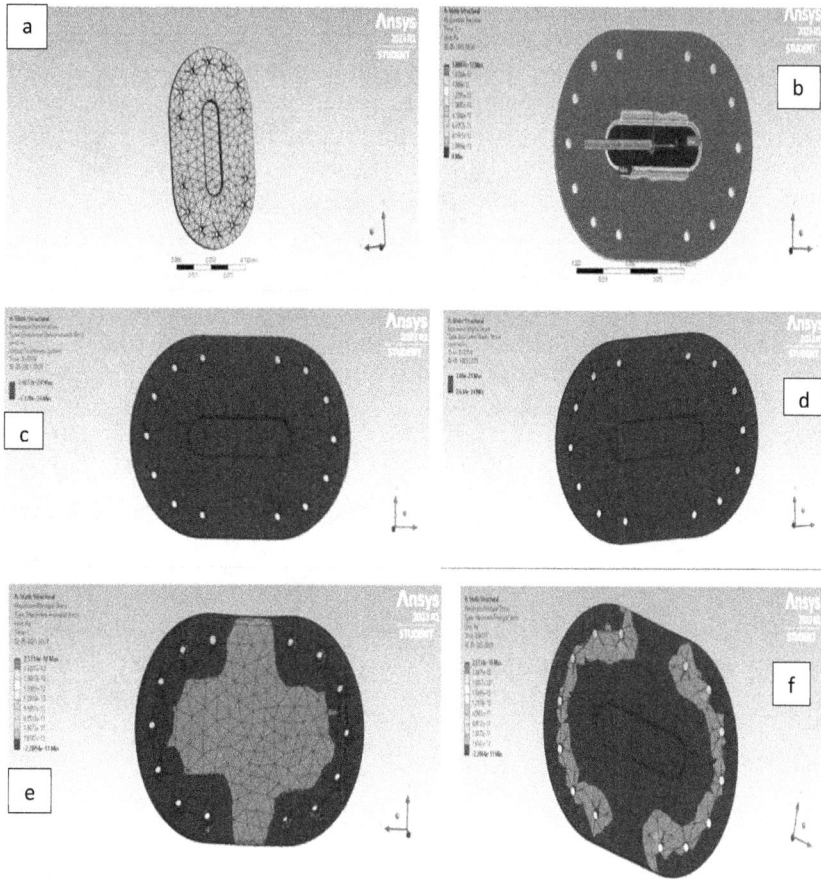

Figure 18.4 (a) Meshing of cover (b) hydrostatic pressure distribution in cover (c) directional deformation in cover (d) equivalent strain in cover (e) stress distribution cover-1 (f) cover-2.

The maximum equivalent stress of the cover is 2.5114e−10 Pa. Stresses are maximum at the ends of the cover and are very less at center and fixed points.

18.4.2 Development of model for shaft

The deformation of the cover becomes smaller and smaller from the center part of the high-pressure area up and down both ends. The maximum deformation occurs at the center of the rotating shaft. The maximum deformation is 5.0806e-11 m, which can meet the requirement of the gear pump.

The equivalent strain is mainly more at the center and at the edges. The maximum equivalent elastic strain is 9.0174e−8, which can meet the requirement of the gear pump and the minimum equivalent elastic strain is 1.875e−9. Figure 18.5a–h represents the same.

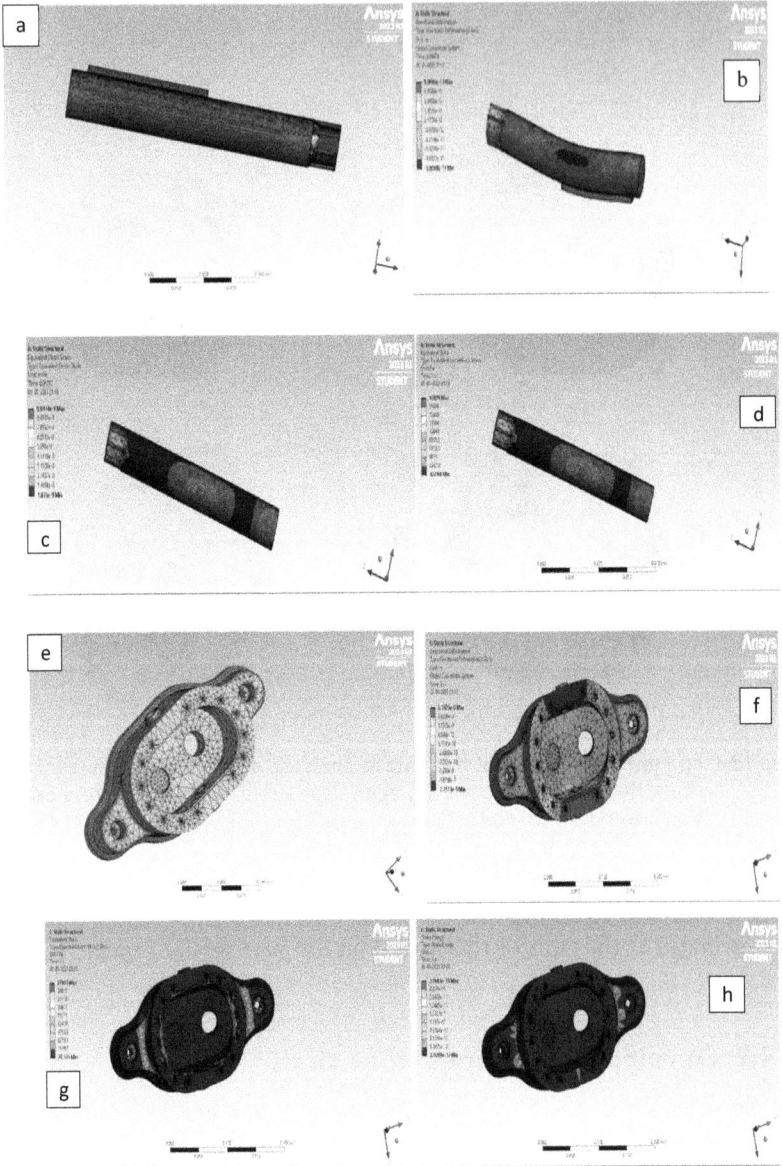

Figure 18.5 (a) Stress distribution in housing (b) strain energy in housing (c) meshing of shaft (d) directional deformation in shaft (e) strain in shaft (f) stress distribution in shaft (g) meshing of housing (h) directional deformation in housing.

The maximum equivalent stress of the cover is 17,829 Pa. Stresses are maximum at the ends of the cover and the center.it is lowest between the edges and the middle part. Strain energy is defined as the energy stored in a body due to deformation. The energy stored is maximum at the ends of the moving shaft with a magnitude of 5.9087e–13 J.

18.4.3 Development of model for housing

The maximum deformation occurs at the edges of the high-pressure surface. The maximum deformation is 2.1025e–9 m, which can meet the requirement of the gear pump. The equivalent strain is mainly minimum at the whole body and maximum at specific points given below. The maximum equivalent elastic strain is 2.903e–7, which can meet the requirement of the gear pump and the minimum equivalent elastic strain is 2.525e–10. The maximum equivalent stress of the cover is 27,915 Pa. Stresses are maximum at the ends of the housing nearer to fixed points. Strain energy is maximum at the ends of the housing with a magnitude of 2.76e–11 m J.

18.4.4 Development of model for gears

Fixed support function of ANSYS software is applied at the center of the spur gears. We will apply rotational velocity to the gears such that the magnitude of the velocity is 1000 rad/s, which is same as that of the motor selected for the design. The driven gear is rotated in clockwise direction and the driven gear is in the anticlockwise direction. The deformation of the cover becomes smaller and smaller from the center part of the high-pressure area up and down both ends. The maximum deformation occurs at the edges of the body of spur gear.

The maximum deformation is 4.15e-7 m, which can meet the requirement of the gear pump. The equivalent strain is mainly more at the center and minimum at the edges of spur gear. The maximum equivalent elastic strain is 0.000931, which can meet the requirement of the gear pump and the minimum equivalent elastic strain is 3.5457e–7. Figure 18.6a–e represents the same. The maximum equivalent stress of the cover is 6.3837e–7 Pa. Stresses are maximum at the center of the driven gear and minimum at the ends of spur gear. Strain energy is defined as the energy stored in a body due to deformation. The energy stored is maximum at the center of the moving gear (driven) with a magnitude of 0.0001893 J. After performing CAE of the solid model under different conditions the designed model is found to be safe and it can work according to the requirements.

18.5 CONCLUSIONS

In brief, the solar-operated pump provides a promising answer for numerous applications requiring green and sustainable pumping operations. With its usage of solar energy, this era gives numerous advantages, such as decreased

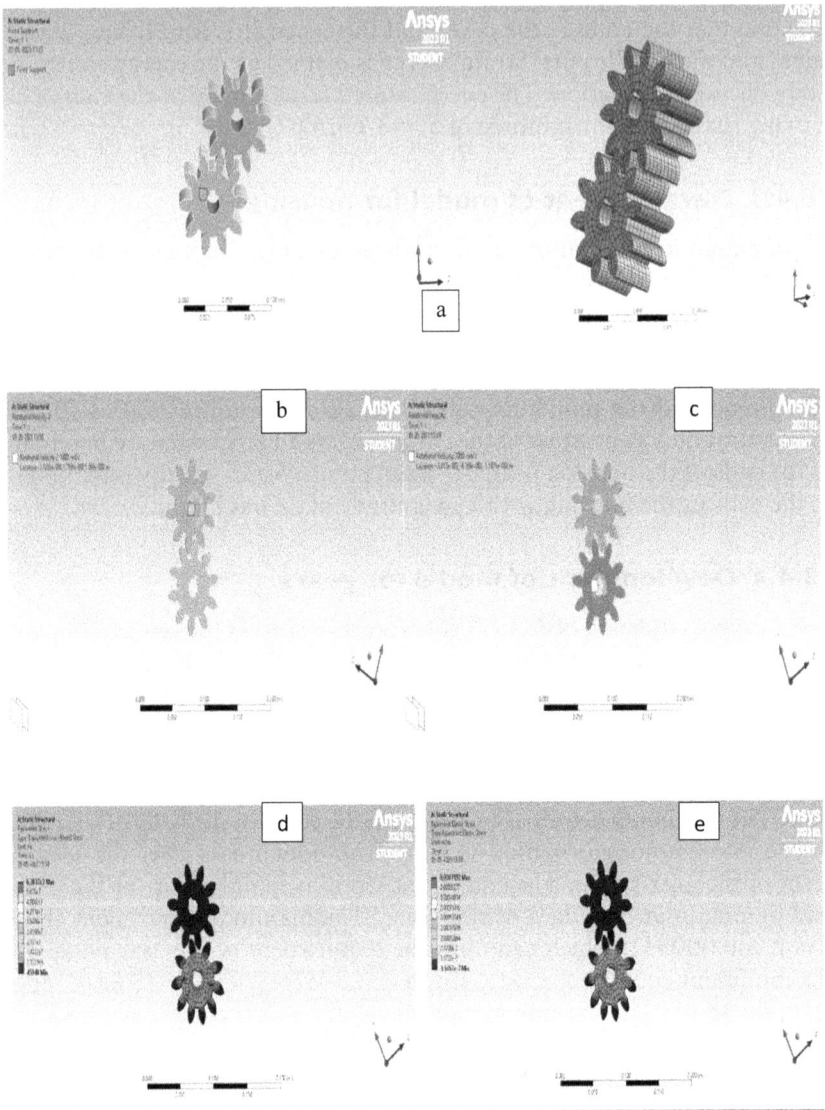

Figure 18.6 (a) Meshing of gears (b) rotational velocity of gear 1 (c) rotational velocity of gear 2 (d) stress distribution (e) equivalent elastic strain.

dependency on fossil fuels, lower operating charges, and environmental friendliness. The aggregate of the pump's simplicity, reliability, and adaptableness with solar energy's renewable nature makes it a possible choice for industries that include agriculture, water management, and renewable energy. Further studies and development on this field can optimize device efficiency, decorate overall performance, and expand its variety of applications. Embracing solar-operated external equipment pumps can contribute to a greener and greater sustainable future even as assembly the ever-growing call for or efficient pumping answers

A viable approach to environmentally friendly fluid transportation is the combination of solar power and a gear pump system. The system's design can be optimized with the use of the CFD modeling approach, which provides useful insights into the fluid dynamics and thermal behavior of the system. As a result of these simulations, effective and dependable solar-based gear pump systems are being developed, advancing the use of renewable energy in fluid transportation. The performance study identifies potential areas for system optimization. Additional advancements in gear design, solar collector effectiveness, and control schemes are a few examples. To evaluate the adaptability and scalability of the solar-based gear pump system, it might also be investigated for use in other sectors and geographical areas.

The design, analysis, and optimization of a solar-powered gear pump system utilizing a CFD modeling approach are all thoroughly studied in this research work. The outcomes demonstrate the potential advantages of combining mechanical pumping systems with solar energy, demonstrating increased sustainability and efficiency. The discovery advances solar-powered industrial processes and sets the path for more investigation and advancement in this area.

REFERENCES

1. Thiagarajan, D., Dhar, S., Vacca, A., Improvement of lubrication performance in external gear machines through micro-surface wedged gears, *Tribol. Trans.* 2017; 60, 337–348.
2. Vacca, A., Guidetti, M., Modelling and experimental validation of external spur gear machines for fluid power applications, *Simul. Model. Pract. Theory* 2011; 19, 2007–2031.
3. Borghi, M., Milani, M., Paltrinieri, F., Zardin, B., Studying the axial balance of external gear pumps. In *Proceedings of the SAE Commercial Vehicle Engineering Congress,* Chicago, IL, 1–3 November, 2005.
4. Castilla, R., Gamez-Montero, P.J., del Campo, D., Raush, G., Garcia-Vilchez, M., Codina, E., Three-dimensional numerical simulation of an external gear pump with decompression slot and meshing contact point, *ASME J. Fluids Eng.* 2015; 137, 041105.
5. Frosina, E., Senatore, A., Buono, D., Pavanetto, M., Olivetti, M., Costin, I., Improving the performance of a two way flow control valve, using a 3D CFD modeling. In *Proceedings of the ASME International Mechanical Engineering Congress and Exposition,* Montreal, QC, Canada, 14–20 November, 2014.
6. Ahmed, A., Yadav, A.K., Singh, A., Agbulut, U., A hybrid RSM-GA-PSO approach on optimization of process intensification of linseed biodiesel synthesis using an ultrasonication reactor: Enhancing fuel properties and engine characteristics with ternary fuel blends, *Energy* 2023. doi:10.1016/j.energy.2023.129077.

7. Dewangan, A., Mallick, A., Yadav, A.K., Islam, S., Agbulut, U., Production of oxy-hydrogen gas and the impact of its usability on CI Engine combustion, performance, and emission behaviors, *Energy* 2023; 278, 127937.

8. Yadav, A.K., Dewangan, A., Mallick, A., Effect of n-butanol and diethyl ether additives on performance and emission characteristics of a diesel engine fuelled with diesel-pongamia biodiesel blends. *J. Energy Eng.* 2018; 144 (6), 04018062.

9. Khan, T.A., Khan, T.A., Yadav, A.K. A hydrodynamic cavitation-assisted system for optimization of biodiesel production from green microalgae oil using a genetic algorithm and response surface methodology approach, *Environ. Sci. Pollut. Res.* 2022; 29, 49465–49477. doi:10.1007/s11356-022-20474-w2022.

10. Ahmed, A., Yadav, A.K., Singh, A., Multi-response optimization to improve the performance and emissions characteristics of A VCR engine fueled with microalgae spirulina (L.): A response surface methodology approach coupled with genetic algorithm, *Environ. Dev. Sustain.* 2023. Doi: 10.1007/s10668-023-04016-z

11. Singh, D., Singh, S., Yadav, A.K., Khan, O., Dewangan, A., From theory to practice: A sustainable solution to water scarcity by using a hybrid solar distiller with a heat exchanger and aluminum oxide nanoparticles, *ACS Omega* 2023; 37 (8), 33543–33553.

12. Dewangan, A., Mallick, A., Yadav, A.K., Ahmad, A., Alqahtani, D., Combined effect of operating parameters and nano particles on performance of a diesel engine: Response surface methodology coupled genetic algorithm approach, *ACS Omega* 2023; 8 (27), 24586–24600.

13. Yadav, A.K., Pal, A., Experimental studies on utilization of *Prunus armeniaca* L. (Wild Apricot) biodiesel as an alternative fuel for CI engine, *Waste Biomass Valor.* 2018; 9 (10), 1961–1969.

14. Yadav, A.K., Khan, M.E., Pal, A., Experimental investigations of performance and emissions characteristics of kusum (*Schleichera oleosa*) biodiesel in a multi-cylinder transportation diesel engine, *Waste Biomass Valor.* 2017; 8 (4), 1331–1341.

15. Yadav, A.K., Khan, M.E., Pal, A., Biodiesel production from Oleander (*Thevetia peruviana*) oil and its performance testing on a diesel engine, *Korean J. Chem. Eng.* 2017; 34 (2), 340–345.

16. Ahmed, A., Yadav, A. K., Singh, A., Process optimization of spirulina microalgae biodiesel synthesis using RSM coupled GA technique: A performance study of a biogas-powered dual-fuel engine, *Int. J. Environ. Sci. Technol.* 2024; 21, 169–188. doi:10.1007/s13762-023-04948-z.

17. Ahmed, A., Yadav, A.K., Singh, A., Application of machine learning and genetic algorithm for prediction and optimization of biodiesel yield from waste cooking oil, *Korean J. Chem. Eng.* 2023; 40, 2941–2956. doi:10.1007/s11814-023-1489-9.

18. Ahmed, A., Yadav, A.K., Singh, A., Enhancement of biogas yield from dual organic waste using hybrid statistical approach and its effects on ternary fuel blend (biodiesel/n-butanol /diesel) powered diesel engine, *Environ. Progr. Sustain. Energy* in press: e14163. doi:10.1002/ep.14163.

19. Ahmed, A., Yadav, A.K., Singh, A., Enhancing waste cooking oil biodiesel yield and characteristics through machine learning, response surface methodology, and genetic algorithm for optimal utilization in CI engines, *Int. J. Green Energy* in press. doi:10.1080/15435075.2023.2253870.

20. Yadav, A.K., Khan, M.E., Pal, A., Ultrasonic assisted production of biodiesel from karabi oil using heterogeneous catalyst, *Biofuels* 2018; 9, 101–112.

21. Yadav, A.K., Vinay, S.B., Optimization of biodiesel production from annona squamosa seeds oil using response surface methodology and its characterization, *Energy Sources, Part A* 2018; 40 (9), 1051–1059.

22. Yadav, A.K., Khan, T.A., Khan, T.A., Kumar, S., Methyl ester of *Gmelina arborea* oil as a substitute for petroleum diesel: An experimental study on its performance and emissions in a diesel engine, *Energy Sources, Part A* 2021; 43 (11), 1307–1314.

23. Khan, O., Yadav, A.K., Khan, M.E., Sharma, D., The ultrasonic assisted optimization of biodiesel production from eucalyptus oil, *Energy Sources, Part A,* 2017; 39, 1556–7036.

24. Dewangan, A., Yadav, A.K., Wax deposition during production of waxy crude oil and its remediation, *Petroleum Sci. Technol.* 2017; 35 (18), 1831–1838.

25. Dewangan, A., Yadav, A.K., Mallick, A., Current scenario of biodiesel development in India: Prospects and challenges, *Energy Sources, Part A* 2018; 40 (20), 2494–2501.

26. Yadav, A.K., Khan, M.E., Pal, A., Kaner biodiesel production through hybrid reactor and its performance testing on a CI engine at different compression ratios, *Egypt. J. Petroleum* 2017; 26, 525–532.

27. Khan, I.A., Singh, S.K., Yadav, A.K., Efficient production of biodiesel from *Cannabis sativa* oil using intensified transesterication (hydrodynamic cavitation) method, *Energy Sources, Part A* 2020; 42 (20), 2461–2470.

28. Dewangan, A., Mallick, A., Yadav, A.K., Comparative study of *Manilkara zapota* and karanja based biodiesel properties and its effect on multi-cylinder diesel engine characteristics, *Energy Sources, Part A* 2022; 44 (2), 5143–5153.

29. Khan, O., Yadav, A.K., Khan, M.E., Characterization of bio-ethanol obtained from eichhornia crassipes plant; its emission and performance analysis on CI engine, *Energy Sources, Part A* 2021; 43 (14), 1793–1803.

30. Ahmed, A., Yadav, A.K., Singh, A., Optimization of cavitation-assisted biodiesel production and fuel properties from *Neochloris oleoabundans* microalgae oil using genetic algorithm and response surface methodology, *J. Process Mech. Eng. E* in press. doi:10.1177/09544089231159832.

31. Yadav, M., Karimi, M.N., Yadav, A.K., Effect of graphene oxide nanoparticles dispersed biodiesel on combustion, performance, and emission characteristics of a diesel engine, *J. Process Mech. Eng. E.* in press. doi:10.1177/09544089231209139.

32. Singh, S., Singh, B., Kumar, S., Yadav, A.K., Temperature-dependent dynamic hysteresis scaling of ferroelectric hysteresis parameters of lead free [(Ba 0.825+xCa0.175-x) (Ti1- x Snx)O3]ceramics, *Ferroelectrics* 2019; 551, 133–142.

33. Dewangan, A., Yadav, A.K., Mallick, A., Optimization of biodiesel production and engine performance from underutilized simarouba oil in compression ignition engine, *Int. J. Oil Gas Coal Technol.* 2020; 25 (3), 357.

34. Ahmed, A., Yadav, A.K., Singh, A., Optimization of biogas yield from anaerobic co-digestion of dual waste for environmental sustainability: ANN, RSM, and GA, approach, *Int. J. Oil Gas Coal Technol.* 2023; 33 (1), 75.

35. Ahmed, A., Yadav, A.K., Singh, A., An environmental impact assessment and optimization study of biodiesel production from microalgae, *Int. J. Global Warming.* Accepted 2023; 31, 294–313.

36. Yadav, A.K., Khan, M.E., Pal, A., A comparative study on ultrasonic cavitation and mechanical stirring method towards efficient production of biodiesel from non-edible oils and performance testing on a C.I. engine, *Int. J. Environ. Waste Manag.* 2016; 18, 349–367.

37. Yadav, A.K., Khan, M.E., Pal, A., Performance and emission characteristics of a stationary diesel engine fuelled by *Schleichera oleosa* oil methyl ester (SOME) produced through hydrodynamic cavitation process, *Egypt. J. Petroleum* 2018; 27, 89–93.

38. Vinay, S.B., Yadav, A.K., Optimization of performance and emission characteristics of CI engine fuelled with mahua oil methyl ester-diesel blend using response surface methodology, *Int. J. Ambient Energy* 2020; 44 (6), 674–685.

39. Yadav, A.K., Khan, M.E., Pal, A., Performance, emission and combustion characteristics of an indica diesel engine operated with yellow oleander (*Thevetia peruviana*) oil biodiesel produced through hydrodynamic cavitation method, *Int. J. Ambient Energy* 2018; 39 (4), 365–371.

40. Yadav, A.K., Pal, A., Ghosh, U., Gupta, S.K., Comparative study of biodiesel production methods from yellow oleander oil and its performance analysis on an agricultural diesel engine, *Int. J. Ambient Energy* 2019; 40 (2), 152–157.

41. Khan, I.A., Singh, S.K., Yadav, A.K., Sharma, D., Enhancement in the performance of a diesel engine fueled with pongamia methyl ester and n-butanol as oxygenated additive, *Int. J. Ambient Energy* 2019; 40 (8), 842–846.

42. Ahmed, A., Yadav, A.K., Singh, A., Biodiesel yield optimization from a third-generation feedstock (microalgae spirulina) using a hybrid statistical approach, *Int. J. Ambient Energy* 2023; 44 (1), 1202–1213.

43. Ahmed, A., Yadav, A.K., Singh, A., Modelling and optimisation of VCR diesel engine parameters fuelled with microalgae methyl-ester: A RSM-coupled Taguchi approach, *Int. J. Ambient Energy* 2023; 44 (1), 2498–2506.

44. Ahmed, A., Yadav, A.K., Biodiesel production from mahua oil: Characterization, optimization, and modeling with a hybrid statistical approach, *Int. J. Ambient Energy* 2023; 44, 2618–2627.

45. Yoon, Y., Park, B.H., Shim, J., Han, Y.O., Hong, B.J., Yun, S.H., Numerical simulation of three-dimensional external gear pump using immersed solid method, *Appl. Therm. Eng.* 2017; 118, 539–550.

46. Khan, T.A., Khan, M.E., Yadav, A.K., Experimental studies on utilization of *Neochloris oleoabundans* microalgae biodiesel as an alternative fuel for diesel engine, *Int. J. Ambient Energy* 2023; 44 (1), 115–123.

47. Singh, D., Yadav, A.K, Kumar, A., Samsher, Energy matrices and life cycle conversion analysis of N-identical hybrid double slope solar distiller unit using Al2O3 nanoparticle, *J. Water Environ. Nanotechnol.* 2023; 8 (3), 267–284.

48. Yadav, A.K., Dewangan, A., Mallick, A., Synthesis and stability study of biodiesel from kachnar seed oil, *J. Energy Eng.* 2018; 144 (5), 04018053.

49. Dewangan, A., Mallick, A., Yadav, A.K., Effect of metal oxide nanoparticles and engine parameters on the performance of a diesel engine: A review, *Mater. Today Proc.* 2020; 21, 1722–1727.

Chapter 19

Solar desalination techniques

Challenges and opportunities

Morteza Khashehchi, Sivasakthivel Thangavel,
Pooyan Rahmanivahid, Milad Heidari,
and Taleb Moazzeni
Global College of Engineering and Technology (GCET) Muscat

Vikas Verma
Tezpur University Assam (Central
University, Government of India)

Ashwani Kumar
Technical Education Department Uttar Pradesh Kanpur
(Under Government of Uttar Pradesh)

19.1 INTRODUCTION

19.1.1 Brief overview of water scarcity as a global challenge

Water scarcity is an escalating global predicament that transcends geographic, economic, and social boundaries, casting a looming shadow over the well-being of both people and the planet (Figure 19.1 shows the effect of water scarcity on the soil) [1–4]. This multifaceted crisis arises from a complex interplay of factors, including population growth, urbanization, climate change, mismanagement of water resources, and unsustainable consumption patterns. As the world's population burgeons, demands for water surge, exerting immense pressure on finite freshwater reserves. Rapid urbanization further exacerbates the strain, concentrating populations in water-stressed areas and often leading to inadequate sanitation and access to clean water for marginalized communities [5]. Climate change, an undeniable reality, alters precipitation patterns, intensifies droughts, and heightens the frequency of extreme weather events, magnifying the unpredictability of water availability.

Agriculture, a cornerstone of human sustenance, is both a major contributor to and a casualty of water scarcity. With mounting food demands, agriculture accounts for a significant portion of global water withdrawals, often depleting water bodies and diminishing aquatic ecosystems.

DOI: 10.1201/9781003472629-19

305

Figure 19.1 Effect of water scarcity on the soil.

Paradoxically, changing climate patterns can disrupt traditional agricultural practices, reducing crop yields and intensifying desertification, further exacerbating the cycle of scarcity. Industries, too, play a pivotal role in this conundrum. Manufacturing processes, energy generation, and industrial activities heavily rely on water, competing with communities and ecosystems for this finite resource. In regions where water scarcity is prevalent, social unrest, migration, and economic destabilization can occur, underscoring the interconnectedness of water security with broader societal and geopolitical stability.

While traditional water sources dwindle, the world's water woes are starkly juxtaposed with the vast expanse of salty seawater that dominates the planet. This is where solar desalination emerges as a beacon of hope, harnessing the immense energy of the Sun to address water scarcity sustainably [6]. By leveraging solar energy to drive desalination processes, this technology presents a viable solution to convert abundant seawater into fresh, potable water. Yet, even in this transformative potential lies a set of intricate challenges and promising opportunities.

In the quest for water security, solar desalination stands as a pioneering avenue (Figure 19.2). However, for the full realization of its potential, it is imperative to comprehend not only its technical intricacies but also the intricate socioeconomic and environmental contexts within which it operates. As this chapter delves into the depths of solar desalination techniques, challenges, and opportunities, it navigates through the waters of innovation and adaptation to chart a course toward a more water-secure world. By intertwining the complexities of water scarcity with the promise of renewable energy-driven solutions, this chapter sets out to uncover the profound impact

Figure 19.2 Solar desalination concept.

of solar desalination on the global stage, emphasizing the need for collaborative efforts, technological advancements, and thoughtful policy interventions to ensure equitable access to clean water for generations to come.

19.1.2 Introduction to solar desalination as a sustainable solution

In a world grappling with the ever-pressing challenges of water scarcity, the notion of harnessing the boundless power of the Sun to transform the very essence of survival presents an awe-inspiring prospect. Enter solar desalination, a technologically innovative and environmentally conscious approach that promises to reshape the landscape of global water security. At its core, solar desalination embodies the marriage of two essential elements [7]:

- The unparalleled potency of solar energy and,
- The dire need for a sustainable solution to address the critical scarcity of fresh water.

In regions characterized by arid climates, dwindling freshwater resources, and a burgeoning demand for water, solar desalination emerges as a beacon of hope, illuminating a pathway toward a future where access to clean and potable water is no longer a luxury but a fundamental right.

The crux of solar desalination lies in its ingenious utilization of solar energy, a resource that remains abundant and untapped in many of the very regions grappling with water shortages. Solar desalination systems capitalize on the Sun's rays to drive a range of processes, from evaporating water to separating salt and impurities, culminating in the production of pure and consumable water. This transformative mechanism not only addresses the dire need for freshwater but also presents a sustainable alternative to the energy-intensive conventional desalination techniques that rely on fossil fuels. By effectively converting solar energy into the mechanical work required for the separation of salt and water molecules, solar desalination mitigates the twin challenges of water scarcity and environmental degradation in one elegant stroke [8].

Moreover, the prospect of solar desalination is closely aligned with the principles of sustainability. The key tenets of sustainability – meeting the needs of the present without compromising the needs of future generations – resonate profoundly with the objectives of solar desalination. By leveraging a renewable energy source, solar desalination circumvents the unsustainable path of overexploiting finite resources, thus mitigating the risk of exacerbating environmental degradation. In addition, solar desalination systems have the potential to be decentralized, empowering local communities with the ability to generate their own water supply and reducing the reliance on long-distance water transportation, which incurs both financial and ecological costs [9].

19.2 IMPORTANCE OF ADDRESSING CHALLENGES AND CAPITALIZING ON OPPORTUNITIES

The importance of addressing challenges and capitalizing on opportunities within the context of solar desalination is a mandate that reverberates far beyond the confines of technology and engineering. It resonates at the core of human existence, intertwining with the very essence of sustenance, progress, and environmental stewardship. As the global community grapples with the escalating specter of water scarcity, exacerbated by factors ranging from climate change and population expansion to urbanization and mismanagement of water resources, the imperative to embrace innovative solutions becomes more pressing than ever. In this crucible of necessity, solar desalination emerges as a beacon of promise, poised to bridge the chasm between dwindling freshwater supplies and burgeoning demand, while simultaneously harnessing the inexhaustible energy of the Sun to fuel our planet's future.

Yet, the trajectory toward widespread implementation of solar desalination is not a smooth one, but rather a dynamic interplay of challenges and prospects that demand our unwavering attention. At the forefront stands the energy conundrum – a formidable challenge that underscores the necessity for continuous advancements in solar collector efficiency, heat transfer optimization, and desalination process innovation. In a world where solar energy availability varies, developing effective energy storage mechanisms becomes paramount to ensure consistent water production. Simultaneously, the cost factor looms large, necessitating concerted efforts to mitigate the financial barriers associated with transitioning to solar desalination. Collaborative research endeavors, technological breakthroughs, and economies of scale must coalesce to render this technology economically feasible, particularly in regions where water scarcity is most acute.

Scaling up solar desalination from experimental endeavors to community-level infrastructure entails a multidimensional tapestry of challenges. Engineering systems that can effectively handle significant water volumes without compromising efficiency and reliability necessitate an intricate understanding of fluid dynamics, material science, and operational intricacies. Moreover, integrating solar desalination into established water supply networks mandates seamless coordination between government bodies, municipalities, and technological innovators. Furthermore, the vital aspect of water quality underscores the indispensability of rigorous monitoring and control mechanisms to ensure that the produced water not only meets but surpasses established potable water standards, mitigating health risks and safeguarding public well-being.

In navigating this intricate labyrinth of challenges, the vista of opportunities unfurls, promising transformative shifts that transcend technological marvels. At the nucleus lies the remarkable synergy between renewable energy and water production, epitomized by the integration of solar desalination with renewable energy grids. This confluence not only exemplifies sustainability in action but also elucidates the potential to optimize energy utilization, curtail environmental footprints, and engineer a harmonious balance between human progress and ecological preservation. The decentralization potential of solar desalination further empowers local communities, particularly those residing in remote and marginalized areas, by affording them control over their water security and obviating the need for extensive water transportation, with its attendant economic and environmental tolls.

Furthermore, the economic and social dimensions intertwine in a complex dance, where the promise of solar desalination can catalyze socioeconomic stability. By alleviating the strain on agriculture and industry, regions grappling with water scarcity can redirect resources toward holistic development, engendering economic prosperity and fortifying the social fabric. In regions where water-related conflicts simmer beneath the surface, the advent of solar desalination can potentially quell tensions by virtue of its capacity to redefine the water landscape, mitigating the factors that often precipitate disputes over access to this finite resource.

19.2.1 Solar desalination techniques: explanation of various solar desalination techniques

The concept of solar desalination techniques unfolds as a tapestry of ingenuity, where the convergence of solar energy and water purification manifests in a symphony of diverse methods. At its core, solar desalination represents a testament to human innovation, harnessing the immense power of the Sun to convert saline and brackish water into life-sustaining freshwater. Within this paradigm, several techniques emerge, each with its unique mechanisms, applications, and implications.

Solar distillation stands as one of the oldest and most intuitive methods, embodying the simplicity of harnessing solar energy to evoke the age-old cycle of evaporation and condensation. Various solar still designs, ranging from basin-type to sloped, and even multi-effect, operate on the principle of heating saline water through solar absorption, inducing evaporation, and subsequently collecting the condensed vapor as freshwater [9]. This decentralized approach holds appeal, particularly in remote areas where access to conventional water sources is limited. However, the efficiency of solar distillation remains a challenge, as optimizing the heating and cooling processes, as well as managing temperature differentials, requires careful engineering.

On a larger scale, solar multistage flash distillation (MSF) capitalizes on the multi-step principle of evaporation and condensation. In MSF, solar energy is harnessed to heat multiple stages, each characterized by flash evaporation of seawater followed by condensation of the vapor. This step-wise process effectively separates water from salt, producing freshwater. MSF systems offer enhanced efficiency and higher production rates compared to solar distillation, but they demand meticulous design to balance energy input, vapor flow, and heat transfer within each stage. Moreover, the challenges of controlling temperature differentials and minimizing heat loss present complex engineering hurdles.

The terrain of solar reverse osmosis (RO) takes us to the heart of membrane-based desalination, where the potency of solar energy powers the intricate dance of separating water molecules from impurities through a semipermeable membrane [10]. Solar energy fuels the pumps that pressurize saline water, enabling its passage through the membrane while retaining the solutes. This process demands fine-tuned control of pressures and flows, ensuring the longevity of the membrane while maintaining efficiency. Solar RO exemplifies the synergy between energy and water purification [10], yet the dependency on consistent solar power availability remains a challenge, necessitating innovations in energy storage to ensure uninterrupted operation.

19.2.2 Solar distillation: types of solar stills and their mechanisms

Solar distillation presents a mechanism that transforms saline or brackish water into fresh, potable water through the orchestration of heat, vapor, and condensation (Figure 19.3). This technique holds profound implications for regions grappling with water scarcity, where traditional freshwater sources are either limited or contaminated. Among the variants of solar distillation, different types of solar stills stand out as the vessels that catalyze this transformative process.

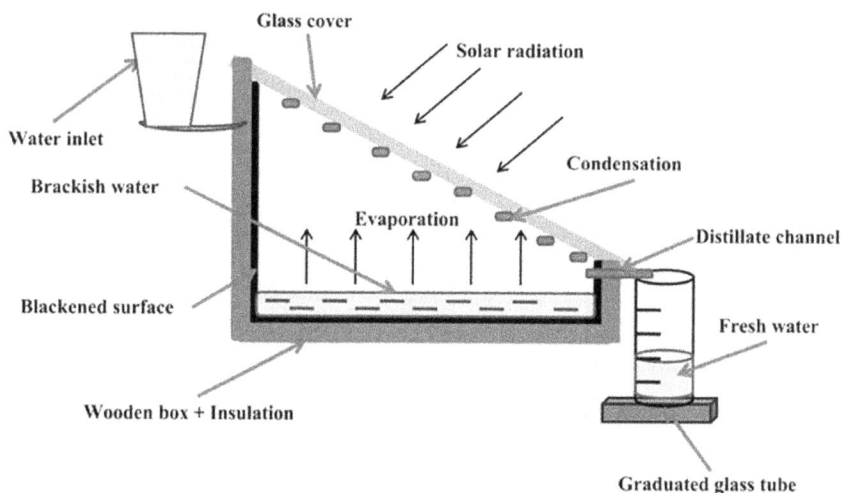

Figure 19.3 Schematic diagram of single slope solar still [11].

The simplest form of solar still is the basin-type solar still, resembling a shallow container with a transparent cover or lid that encapsulates the water to be treated. Figure 19.3 shows a schematic of this technique [11]. As sunlight penetrates the transparent surface, it heats the saline water, initiating the process of evaporation. The water vapor, stripped of impurities, rises and condenses on the cooler inner surface of the lid. Gravity guides the condensed water droplets along the sloped lid to a collection trough, from where they can be collected as freshwater. This rudimentary yet effective design embodies the elegance of solar distillation's core mechanism [11].

As the quest for efficiency advances, the multi-effect solar still emerges as a sophisticated iteration (Figure 19.4). This design features multiple basins stacked in layers, with each basin cascading its vapor to the basin below. This mechanism capitalizes on the residual heat in the vapor stream, utilizing it to accelerate evaporation in subsequent basins. The cumulative effect leads to higher distillation rates, exemplifying the potential of intricately designed systems that harness energy gradients for enhanced performance [12].

The vertical multiple-effect solar still takes the concept of multistage distillation to new heights, literally. Stacked vertically, each basin is equipped with a transparent cover and a separate condensation system. As saline water evaporates and rises through the stack, it encounters cooler surfaces, promoting condensation at each stage. This design optimizes space utilization while delivering high production rates, making it a viable solution for densely populated areas or regions with limited land availability [12].

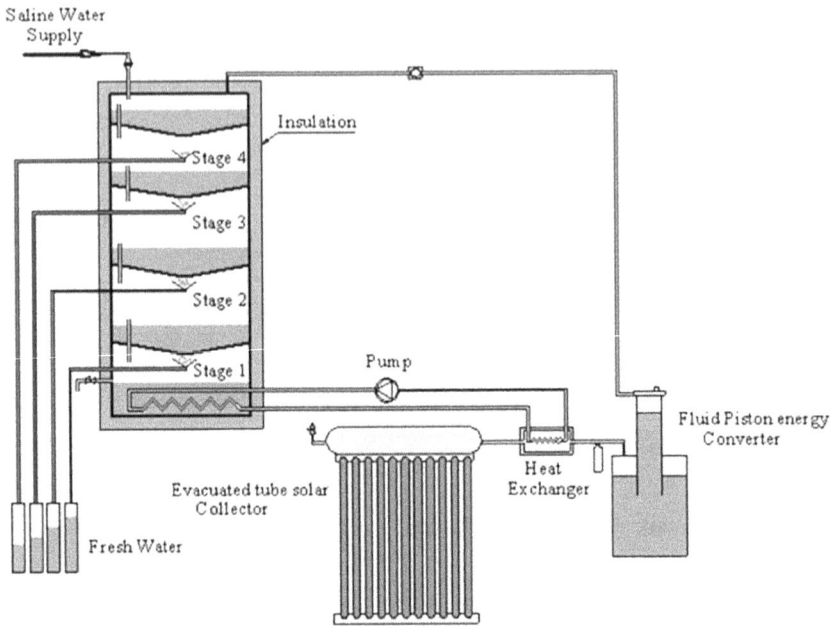

Figure 19.4 Schematic diagram of evacuated multi-effect solar still [12].

Amid these designs, the solar desalination greenhouse stands as a hybrid solution that combines solar distillation with greenhouse technology. In this arrangement, a greenhouse structure encases the solar still, creating a controlled environment that enhances solar absorption and heat retention. The greenhouse effect accelerates evaporation and condensation processes, amplifying freshwater yield while enabling cultivation within the greenhouse space, contributing to food security.

19.2.3 Solar multistage flash distillation: operating principles and efficiency

Solar multistage flash distillation (MSF) emerges as a pinnacle of ingenuity, epitomizing the confluence of solar energy and desalination technology. At its heart lies an intricate dance of evaporation and condensation, choreographed across multiple stages, to unravel the complex web of saline water and fresh water. Operating as a virtuoso performance of physics and engineering, MSF employs the Sun's radiant energy to elevate saline water to its boiling point, inducing a cascade of vaporization and re-condensation that culminates in the separation of impurities from precious freshwater [13] (Figure 19.5).

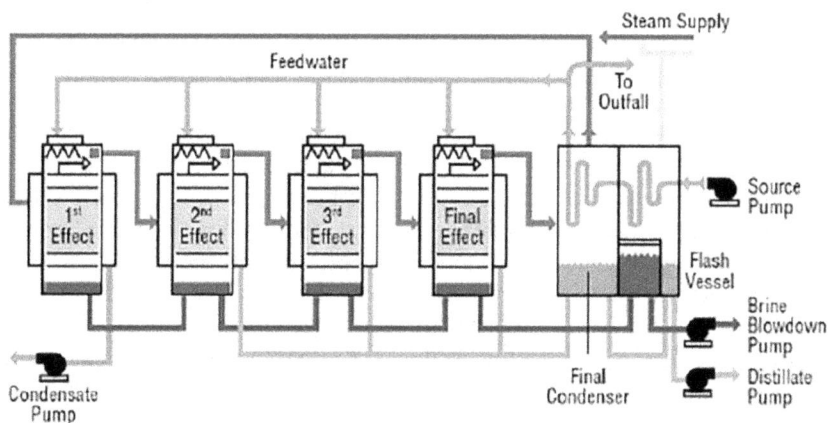

Figure 19.5 Multistage flash process [13].

The operating principles of MSF hinge on the relationship between pressure and boiling points. In a low-pressure environment, the boiling point of water decreases, allowing it to vaporize at lower temperatures. MSF capitalizes on this principle, creating a multi-step process where saline water is subjected to varying pressures and temperatures across a series of interconnected chambers. The initial stage is characterized by the lowest pressure, enabling the water to boil at a relatively low temperature. As the saline water evaporates, it leaves behind salt and other impurities, resulting in vapor rich in freshwater content. This vapor is then directed to the subsequent stage, where it encounters a chamber with higher pressure and subsequently a higher boiling point. The temperature differential between stages fosters vaporization in each stage, followed by re-condensation as the vapor encounters cooler surfaces, eventually yielding purified freshwater [13].

The efficiency of MSF is intricately linked to its intricate engineering. The design of each stage, characterized by the pressure gradient, heat exchange mechanisms, and material selection, determines the overall performance. Achieving optimal pressure differentials, maximizing heat transfer efficiency, and mitigating heat loss are paramount considerations. Engineers often employ strategies such as heat exchangers, baffles, and well-designed insulation to maintain temperature differentials and streamline vapor flow. In addition, advancements in materials, such as improved heat-conductive surfaces and heat-resistant coatings, contribute to refining the efficiency of the system [13].

MSF's efficiency is influenced by the alignment of its operating parameters with local climatic conditions. Solar energy availability, temperature differentials, and ambient conditions play a pivotal role in determining the system's performance. Hence, the design and deployment of MSF systems necessitate a comprehensive analysis of the specific geographical and

climatic context. In addition, the incorporation of energy storage mechanisms, like phase-change materials, can bridge the gap between energy generation and water production, ensuring consistent operation even during
periods of intermittent sunlight.

As MSF systems evolve, the promise of enhanced efficiency is accom-
panied by an array of challenges. The management of pressure differentials and temperature gradients requires sophisticated control mechanisms
to prevent performance degradation or inefficiencies. Furthermore, system
maintenance, scaling, and integration into existing infrastructure pose technical complexities. Moreover, the utilization of solar energy in MSF systems
necessitates innovative approaches to solar collectors, energy storage, and
distribution systems, presenting avenues for research and development [14].

19.2.4 Solar RO: role of solar energy
in powering the process

Solar RO emerges as a paradigm of ingenuity, intertwining the elegance
of physics, chemistry, and engineering to revolutionize desalination
(Figure 19.6). At its nucleus lies the transformative principle of selectively
permeable membranes that, when paired with the boundless energy of the
Sun, render saline water potable [15]. Operating as a symphony of osmotic
pressure, pressure differentials, and molecular dynamics, RO epitomizes
the convergence of science and sustainability. Central to this orchestration
is the pivotal role of solar energy, which not only drives the process but
also heralds a new era of efficiency and environmental responsibility in the
domain of water purification.

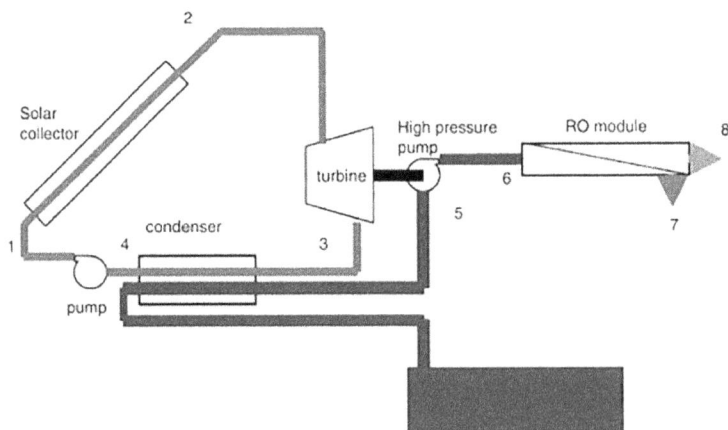

Figure 19.6 Schematic of a solar-powered RO system based on the Rankine cycle [15].

The essence of RO hinges on osmosis – the natural phenomenon where solvent molecules migrate through a semipermeable membrane from an area of lower solute concentration to an area of higher solute concentration. RO, however, subverts this natural flow by applying external pressure to the saline solution, forcing water molecules through the membrane while retaining the majority of solutes. The result is fresh water on one side and a concentrated brine on the other side. It is within this interplay of pressure and molecular behavior that solar energy asserts its transformative influence [15].

The integration of solar energy into RO presents opportunities for enhancing energy efficiency and addressing operational challenges. Traditional RO systems, reliant on grid electricity or fossil fuels, often encounter issues of energy intensity and greenhouse gas emissions. The utilization of solar energy in RO not only ameliorates these concerns but also augments the process's overall efficiency. Furthermore, the intermittency of solar energy can be managed through innovative storage solutions, ensuring continuous operation and alleviating the constraints associated with energy availability.

The design of solar-powered RO systems requires a nuanced approach to optimize energy utilization. Engineers must meticulously calculate the pressure requirements, considering factors such as membrane type, feedwater salinity, and desired freshwater yield. Efficient pump technology, precise pressure control mechanisms, and energy-efficient components are essential to ensure the effective utilization of solar energy. In addition, the geographical location and local solar irradiance dictate the system's performance, necessitating tailored designs that cater to the specific climatic context.

19.2.5 Solar nano-filtration: benefits of using solar energy for pump power

Solar nano-filtration emerges as a frontier of innovation, poised at the intersection of advanced membrane technology and renewable energy (Figure 19.7). At its core lies the principle of selectively permeable membranes, designed to allow the passage of water molecules while blocking larger ions and molecules. As this technique ushers in a new era of precision water purification, the role of solar energy takes center stage, unveiling a cascade of benefits that extend far beyond efficient water treatment. One of the most compelling advantages of employing solar energy in nano-filtration lies in its capacity to power the pumps required for pressurizing the feed water [16].

Traditionally, nano-filtration processes rely on conventional sources of energy, often derived from fossil fuels or grid electricity [16]. The incorporation of solar energy as a primary or supplementary power source revolutionizes this paradigm, introducing a realm of environmental stewardship, energy efficiency, and operational reliability. Solar energy is harnessed to drive the pumps responsible for elevating the pressure of the saline water,

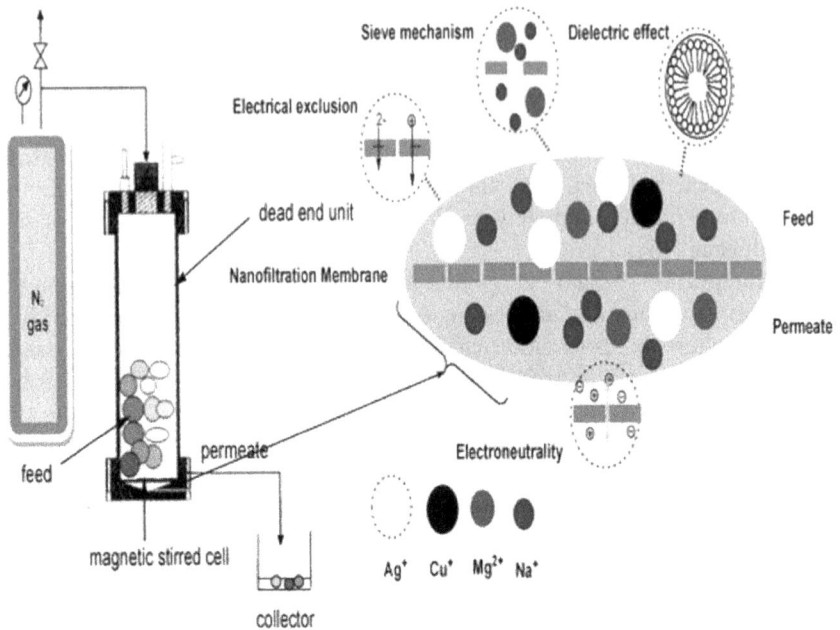

Figure 19.7 Nano-filtration separation mechanism [16].

enabling it to traverse the nano-filtration membrane. This solar-powered pressurization exemplifies a harmonious alignment of resource utilization, as the Sun's boundless energy is harnessed without depleting finite fossil fuel reserves or contributing to carbon emissions.

The benefits of using solar energy for pump power in nano-filtration are multifaceted and transformative. First and foremost, solar-powered pumps alleviate the ecological burden associated with traditional energy sources [16]. Fossil fuels, the stalwarts of conventional energy generation, are notorious for their adverse environmental impacts, from air pollution and greenhouse gas emissions to habitat disruption and resource depletion. By harnessing solar energy, nano-filtration not only circumvents these detrimental consequences but also champions the ethos of sustainability, aligning with the global imperative to transition toward renewable energy sources and mitigate climate change.

Energy efficiency forms another cornerstone of solar-powered nano-filtration. Solar energy, directly converted into mechanical power to drive the pumps, minimizes energy losses associated with energy conversion, transmission, and distribution, which are often encountered in grid-based systems. This direct conversion maximizes the utilization of the available solar energy, enhancing the overall efficiency of the nano-filtration process.

Moreover, the integration of solar energy, coupled with intelligent control systems, enables dynamic adjustment of pump power according to solar irradiance, optimizing energy consumption and adapting to varying operating conditions.

19.3 CHALLENGES IN SOLAR DESALINATION: ENERGY EFFICIENCY AND TECHNOLOGICAL ADVANCES

Energy efficiency stands as a beacon of sustainability in our modern world, an imperative driven by the twin goals of minimizing resource consumption and mitigating environmental impact. At the heart of this endeavor lies an intricate dance between technological innovation and human ingenuity, as humanity strives to harness advanced solutions that propel us toward a future where efficiency and progress coalesce. The convergence of energy efficiency and technological advances is a nexus that holds the potential to reshape industries, economies, and the global environmental trajectory.

Technological leaps have propelled energy efficiency to new heights, unfurling a tapestry of transformative possibilities across sectors. In industries, the marriage of smart sensors, data analytics, and automation has ushered in the era of industrial internet of things (IIoT), where energy consumption is monitored, optimized, and streamlined in real time. From factories to power plants, this synergy has led to the identification of energy waste, the refinement of processes, and the reduction of carbon footprints, thereby fostering a harmonious balance between economic growth and environmental responsibility [17].

The realm of transportation has also experienced a revolution as electric vehicles (EVs) surge to the forefront, underpinned by energy-efficient propulsion systems and advanced battery technologies. EVs exemplify the potential for energy efficiency to transcend sectors, altering not only the way we travel but also how energy is generated, stored, and distributed. Alongside EVs, the advent of autonomous vehicles promises to further optimize travel patterns and energy usage, as these vehicles interact seamlessly with traffic systems, minimizing congestion and fuel consumption.

In buildings, energy efficiency has undergone a renaissance through the incorporation of smart building technologies. From intelligent lighting systems that adjust to occupancy and natural light levels to sophisticated HVAC systems that respond to real-time conditions, buildings have become dynamic ecosystems where energy is allocated and conserved with unprecedented precision. Moreover, the integration of renewable energy sources like solar panels has empowered buildings to generate their own clean energy, mitigating reliance on conventional grids and amplifying the transition to a sustainable energy landscape.

19.3.1 Need for optimizing energy conversion and utilization

The imperative for optimizing energy conversion and utilization resonates with the urgency of our times, where the demands of a rapidly growing global population intersect with the imperative of environmental stewardship. This dual challenge calls for a paradigm shift that transcends mere energy production, delving into the heart of how energy is harnessed, transformed, and ultimately consumed. At the core of this endeavor lies the recognition that our current energy systems are far from optimal, plagued by inefficiencies, wastage, and environmental repercussions. As societies grapple with the consequences of resource depletion and climate change, the need to optimize energy conversion and utilization takes on a profound significance that reverberates across industries, economies, and ecosystems.

The quest for optimized energy conversion begins at the point of generation, where the utilization of energy sources must be maximized to extract the most energy possible. Conventional power plants, whether fueled by coal, natural gas, or nuclear energy, often experience substantial losses during the conversion of primary energy to electricity. Advanced technologies such as combined cycle gas turbines and high-efficiency cogeneration systems exemplify strides toward enhancing energy conversion efficiency [18]. In addition, the rise of renewable energy sources such as solar and wind has underscored the importance of tapping into these sources with utmost efficiency, ensuring that the energy harnessed from sunlight and wind is effectively converted into electricity.

Efficiency in energy utilization spans across sectors, from industries and transportation to residential and commercial spaces. In industries, the optimization of energy use not only minimizes costs but also reduces greenhouse gas emissions. This entails adopting energy-efficient equipment, optimizing production processes, and implementing waste heat recovery systems. In transportation, the shift toward EVs and hybrid technologies aims to curtail energy losses during propulsion, while innovative designs in aircraft, ships, and vehicles enhance fuel efficiency and reduce emissions. In buildings, the integration of smart technologies, efficient lighting, insulation, and HVAC systems ensures that energy is utilized judiciously, reducing both operational costs and environmental impact.

Challenges inevitably accompany the pursuit of optimized energy conversion and utilization. The integration of advanced technologies often requires significant investments and expertise. The transition toward more efficient systems might entail retrofitting existing infrastructure, a process that can be complex and resource-intensive. Moreover, behavioral changes and cultural shifts are often necessary to promote energy-conscious practices on an individual and societal level. Equitable access to energy-efficient technologies also remains a concern, as disparities in economic and technological resources can perpetuate energy inefficiencies in marginalized communities.

19.3.2 Advances in solar collector technology for improved efficiency

Advances in solar collector technology stand as a testament to human ingenuity's capacity to harness nature's most abundant and renewable resource – the Sun – for myriad applications. These innovations exemplify our commitment to maximizing the efficiency of solar energy conversion, unlocking its full potential to address energy needs and mitigate environmental impact. The evolution of solar collectors traverses a dynamic landscape, from fundamental improvements in materials and design to sophisticated integration with storage and control systems. As our understanding of solar energy deepens and technologies mature, solar collectors are at the forefront of transformative change, offering promise across sectors from power generation to heating, cooling, and beyond.

One of the most remarkable strides in solar collector technology is the development of high-efficiency photovoltaic (PV) panels. Traditional solar cells have given way to next-generation PV technologies, such as tandem solar cells and perovskite solar cells, that boast higher conversion efficiencies [19]. Tandem solar cells combine different types of semiconductors, each attuned to specific wavelengths of light, enabling more efficient absorption and conversion of sunlight [20]. Perovskite solar cells exhibit rapid advancements, leveraging innovative materials that deliver impressive performance and ease of manufacturing. Furthermore, bifacial solar panels, which can capture sunlight from both sides, amplify energy production by reflecting sunlight off surrounding surfaces, exemplifying the multifaceted approach to achieving greater efficiency.

Concentrated solar power (CSP) technology has also made notable strides in enhancing solar collector efficiency [21]. CSP systems use mirrors or lenses to focus sunlight onto a small area, creating intense heat that drives a power cycle to generate electricity. Advances in CSP include the development of more precise tracking systems that keep collectors aligned with the Sun, optimizing energy capture throughout the day [21]. In addition, innovative heat storage materials, such as molten salts or phase-change materials, extend the operational window of CSP systems, allowing energy to be stored and dispatched as needed, even after sunset.

In the realm of solar thermal collectors, innovations have focused on maximizing heat absorption and minimizing thermal losses. Vacuum tube collectors, for instance, employ a vacuum layer to insulate the absorber and minimize heat loss through conduction and convection. Evacuated tube collectors have proven highly effective in cold climates, ensuring efficient heat transfer and retention. Similarly, advances in selective coating materials have enabled solar thermal collectors to absorb a broader spectrum of sunlight while minimizing radiative heat losses, optimizing overall performance.

The marriage of nanotechnology with solar collectors has ushered in remarkable improvements in efficiency. Nanostructured materials, with their enhanced surface area and unique optical properties, enable more efficient light absorption and better heat transfer [22]. Nanomaterial-coated surfaces on solar collectors can increase light absorption and reduce reflection, enhancing the overall efficiency of energy capture. Moreover, advances in nanofluids have led to improved heat transfer in solar thermal systems, enhancing their energy conversion efficiency [23].

Incorporating smart technologies has further refined the efficiency of solar collectors. Sensors, actuators, and control systems enable real-time adjustments to collector angles and orientations, ensuring optimal alignment with the Sun's position [24]. This dynamic tracking minimizes shading and maximizes energy capture, especially in locations with variable weather patterns. In addition, the integration of artificial intelligence (AI) and machine learning enhances predictive capabilities, optimizing collector performance based on historical data and real-time conditions [25].

Hybrid solar collector systems offer a synergy of energy generation and thermal application. Photovoltaic-thermal (PV-T) systems, for instance, combine PV panels with heat exchangers, allowing simultaneous electricity generation and thermal energy extraction. These systems utilize excess heat from PV panels to provide hot water or space heating, significantly increasing overall energy conversion efficiency [26].

19.3.3 Innovations in heat transfer systems and membranes

Innovations in heat transfer systems and membranes represent a frontier of scientific and engineering breakthroughs that ripple across diverse industries, from energy to healthcare, manufacturing to environmental conservation. These advances embody humanity's quest for more efficient, sustainable, and precise methods of transferring heat, managing temperature, and controlling material interactions [27]. As the world grapples with resource constraints and environmental challenges, these innovations hold the potential to reshape fundamental processes and unlock new realms of technological possibility.

Heat transfer systems play a pivotal role in various sectors, from power generation to electronics cooling [27]. Advances in heat exchanger designs have led to enhanced thermal efficiency, reduced energy consumption, and increased system reliability. Novel heat exchanger geometries, such as compact and microscale designs, have enabled more efficient heat transfer in confined spaces. In addition, the integration of additive manufacturing techniques allows for intricate, customized heat exchanger structures, optimizing energy utilization in complex applications.

Microfluidic heat transfer systems have gained prominence in biological and pharmaceutical research, offering precise temperature control for various applications. Microscale channels allow for rapid and controlled thermal cycling, vital for processes such as polymerase chain reactions (PCR) and protein analysis. Such innovations accelerate research, minimize reagent consumption, and increase the accuracy of results, paving the way for breakthroughs in medicine, genetics, and biotechnology [28].

Nanofluids, which comprise nanoparticles suspended in a base fluid, have revolutionized heat transfer systems by enhancing thermal conductivity and convective heat transfer rates. These fluids find applications in electronics cooling, solar thermal systems, and industrial processes. Nanofluids offer improved heat dissipation, enabling the design of more compact and efficient cooling solutions, crucial for high-power electronics and advanced machinery [29].

Advancements in membrane technology have transformed industries reliant on precise separation and purification processes [30]. In water treatment, nano-filtration and RO membranes exhibit finer pores, enabling the removal of even smaller contaminants and ions from water sources. These innovations have far-reaching implications for drinking water quality, industrial processes, and environmental preservation.

Membranes also hold promise in energy storage and conversion. Fuel cells, for instance, rely on membranes to facilitate selective ion transport and separate reactants. Innovations in membrane materials and design are crucial for improving fuel cell efficiency, reducing costs, and expanding their applicability across transportation and stationary power systems [30].

In the realm of biotechnology, advancements in biomimetic membranes mimic natural processes, enabling controlled transport of molecules and ions [31]. These membranes are pivotal in drug delivery systems, enabling the targeted and controlled release of pharmaceutical agents. They also find application in artificial organs, enhancing the biocompatibility of medical devices.

Energy-efficient buildings benefit from innovations in thermal insulating materials and coatings [32]. Aerogels, for instance, offer exceptional insulation properties while being lightweight and environmentally friendly. Phase-change materials, which absorb and release thermal energy during phase transitions, optimize temperature regulation, reducing the need for energy-intensive heating and cooling systems.

19.4 OPPORTUNITIES IN SOLAR DESALINATION

The field of solar desalination offers a range of promising opportunities that align with the growing need for sustainable water supply solutions. As water scarcity becomes an increasingly pressing global issue, the integration of solar energy into desalination processes presents numerous benefits that address challenges related to water availability, environmental conservation, and energy efficiency. These opportunities span technological advancements, economic growth, environmental preservation, and improved access to clean water resources, making solar desalination a compelling area of exploration and development.

19.4.1 Case studies and real-world applications

Case studies and real-world applications provide tangible evidence of the effectiveness and potential of solar desalination in addressing water scarcity challenges. These practical examples showcase how solar desalination technologies are being implemented, yielding positive outcomes and inspiring further adoption across diverse contexts.

19.4.1.1 Examples of successful solar desalination projects

Showcasing examples of successful solar desalination projects provides concrete evidence of the feasibility and impact of this innovative approach to addressing water scarcity challenges:

The Al Khafji Solar Desalination Plant in Saudi Arabia stands as a pioneering example of large-scale solar desalination. By employing multi-effect distillation technology powered by solar thermal energy, the plant produces a substantial 60,000 cubic meters of fresh water daily. This integration of solar power significantly reduces energy consumption and greenhouse gas emissions, setting a precedent for sustainable water solutions [33].

Australia's solar desalination initiatives demonstrate the adaptability of the technology. In Whyalla, a solar desalination plant utilizes solar energy to desalinate brackish groundwater, offering a local community a reliable source of sustainable water. Similarly, the Solar Desalination Plant in Victoria showcases how solar energy can power RO technology, ensuring water security for the region [33].

The Solar Water Enterprise project in Tunisia exemplifies a community-driven approach to solar desalination. By integrating solar desalination with local water distribution networks, the project provides clean water to rural communities. This locally focused initiative highlights the direct positive impact solar desalination can have on enhancing water security in underserved areas [34].

Portable solar desalination units deployed in disaster-stricken regions and remote locations underscore the technology's versatility. These units provide immediate relief and access to clean water during emergencies, demonstrating the potential for solar desalination to address urgent water needs in challenging circumstances.

Research initiatives, such as the Solar Desalination Challenge by the United States Department of Energy, showcase ongoing efforts to drive innovation in the field. Collaborative teams work on developing affordable solar desalination systems that cater to agricultural, municipal, and industrial water demands. These projects embody the spirit of innovation and collaboration required to advance solar desalination technology [35].

19.5 HIGHLIGHT THE IMPACT OF SOLAR DESALINATION ON COMMUNITIES AND REGIONS

Highlighting the impact of solar desalination on communities and regions underscores the profound and multifaceted benefits that this technology brings to areas grappling with water scarcity:

Solar desalination directly addresses water security concerns by providing a reliable and sustainable source of fresh water. Communities and regions that previously relied on unpredictable or dwindling water sources now have access to a consistent supply, fostering resilience against water shortages.

The improved access to clean and safe drinking water positively impacts public health. Solar desalination eliminates the risk of waterborne diseases, promoting better hygiene practices and enhancing the overall well-being of communities.

In regions where water availability is limited, solar desalination supports agricultural activities. The availability of a consistent water source enables communities to engage in irrigation, leading to increased agricultural productivity, food security, and economic growth.

Reliable water supply from solar desalination spurs industrial development. Industries that require substantial water resources, such as manufacturing and energy production, can operate without straining local water supplies, contributing to economic stability and job creation.

Solar desalination systems are often designed to withstand extreme weather conditions, making them an ideal choice in areas prone to climate-related uncertainties. These systems provide a consistent water supply even during droughts or disruptions to conventional water sources.

Community engagement is a hallmark of solar desalination projects. Involving local communities in their development and management fosters a sense of ownership and empowerment. This community engagement strengthens social cohesion and allows communities to actively participate in addressing their water challenges.

Solar desalination plays a crucial role in disaster response and recovery. In emergency situations, where conventional water infrastructure may be compromised, solar desalination can provide communities with an independent and immediate source of clean water.

19.6 LESSONS LEARNED FROM THESE CASE STUDIES AND THEIR IMPLICATIONS FOR WIDER ADOPTION

Analyzing lessons learned from these case studies and their implications for wider adoption reveals valuable insights into the potential of solar desalination as a solution to water scarcity:

Technological Viability
Local Context Matters
Community Engagement
Collaboration and Partnerships
Economic Viability
Scalability and Adaptability
Policy and Regulatory Support
Technology Transfer
Climate Resilience
Community Ownership

In summary, lessons learned from successful solar desalination case studies emphasize the technical viability, adaptability, and socioeconomic dimensions of the technology. These lessons provide valuable guidance for wider adoption, highlighting the need for context-specific solutions, collaboration, community engagement, and supportive policy frameworks. As the global water crisis intensifies, these insights pave the way for the broader integration of solar desalination as a transformative tool for addressing water scarcity challenges.

19.7 CONCLUSIONS

19.7.1 Summarize the key points discussed in the chapter

In this chapter, we delved into the realm of solar desalination techniques, exploring the challenges and opportunities they present in the context of addressing water scarcity. We began by acknowledging the global challenge of water scarcity, characterized by diminishing freshwater resources and a growing population. The introduction of solar desalination as a sustainable solution was highlighted, focusing on its potential to harness solar energy for converting seawater into fresh water. We emphasized the importance of addressing challenges and capitalizing on opportunities to enhance the viability of solar desalination.

The chapter expanded to explain various solar desalination techniques, starting with solar distillation. Different types of solar stills and their mechanisms were discussed, along with their advantages and limitations. Moving on, we explored more advanced techniques like solar multistage flash distillation (MSF) and solar RO, detailing their operating principles and the role of solar energy in powering these processes. In addition, we highlighted the benefits of using solar energy for pump power in solar nano-filtration, showcasing its potential to reduce energy consumption.

Energy efficiency and technological advances emerged as crucial themes. The need for optimizing energy conversion and utilization was explored, focusing on advancements in solar collector technology and heat transfer systems. Innovations in these areas were identified as key drivers for improving the overall efficiency of solar desalination systems.

We examined cost implications and discussed initial investment challenges along with strategies for cost reduction through research and innovation. Scaling up and integration were addressed, emphasizing engineering challenges during the transition to larger-scale applications and integration with existing infrastructure for effective water supply.

Water quality and sustainability were highlighted as essential considerations. We discussed ensuring consistent production of high-quality water and managing brine disposal to minimize environmental impact. Furthermore, the symbiotic relationship between energy and water production was explored, underlining the potential for renewable energy synergy and the role of solar desalination in promoting water security and resilience.

Case studies were presented to showcase successful solar desalination projects, each demonstrating unique applications and lessons learned. These examples underscored the impact of solar desalination on communities and regions, influencing public health, agricultural sustainability, industrial growth, and more. The lessons derived from these case studies informed implications for wider adoption, including the importance of local context, community engagement, collaborations, and economic feasibility.

19.7.2 Emphasize the role of solar desalination in a sustainable water future

Emphasizing the role of solar desalination in a sustainable water future underscores its significance as a transformative solution that addresses pressing water scarcity challenges. Solar desalination's integration of renewable energy and advanced technologies positions it as a cornerstone of sustainable water supply systems, with profound implications for both environmental preservation and human well-being. By harnessing the power of the Sun to drive desalination processes, solar desalination offers a range of benefits that collectively contribute to shaping a more secure and resilient water future.

At the heart of its contribution is the capacity to provide a consistent source of fresh water, irrespective of geographical constraints or climate variability. Solar desalination's ability to function independently of traditional water sources ensures reliable access to clean water, particularly in arid and drought-prone regions where conventional sources are insufficient. This resilience is further strengthened by its low carbon footprint, as solar energy replaces fossil fuels, reducing greenhouse gas emissions and environmental impact.

Solar desalination enhances environmental stewardship through its efficient use of resources. Its reliance on renewable energy minimizes reliance on finite resources while simultaneously reducing energy consumption, compared to conventional desalination methods. This synergy between energy and water production exemplifies a holistic approach to sustainability, optimizing resource utilization and minimizing negative ecological footprints.

19.7.3 Encourage continued research, innovation, and collaboration in the field

Encouraging continued research, innovation, and collaboration in the field of solar desalination is essential to unlocking its full potential and ensuring a sustainable water future. As the world faces escalating water scarcity challenges, a proactive approach that embraces cutting-edge technologies and cross-disciplinary partnerships becomes imperative.

Research drives the advancement of solar desalination technology, refining efficiency, scalability, and cost-effectiveness. Continued exploration of novel materials, membranes, and system designs can lead to breakthroughs that revolutionize the efficiency and affordability of these systems. Research efforts should also focus on addressing challenges such as brine disposal and environmental impact, paving the way for more environmentally friendly solutions.

Innovation is the driving force behind transformative change. By fostering a culture of innovation, we can explore new concepts, experiment with novel approaches, and find creative solutions to complex challenges. Advances in nanotechnology, materials science, and automation hold the potential to significantly enhance solar desalination efficiency and overall performance. Encouraging innovation requires an open-minded approach that welcomes diverse perspectives and promotes experimentation.

Collaboration amplifies the impact of research and innovation. Bringing together experts from various fields, including engineering, chemistry, materials science, and renewable energy, fosters a synergy that accelerates progress. Partnerships between academia, industry, governments, and local communities create a holistic ecosystem for knowledge exchange, technology transfer, and practical implementation.

REFERENCES

[1] UN Water. "Water Scarcity." Available at: https://www.unwater.org/water-facts/scarcity/.

[2] Global Water Forum. "Water Scarcity." Available at: https://www.globalwaterforum.org/water-scarcity/.

[3] United Nations Sustainable Development Goals (SDGs). Available at: https://sdgs.un.org/goals/goal6.

[4] Intergovernmental Panel on Climate Change (IPCC). "Climate Change and Land." Available at: https://www.ipcc.ch/srccl/.

[5] N.P. Tseole, T. Mindu, C. Kalinda, M.J. Chimbari, Barriers and facilitators to water, sanitation and hygiene (WaSH) practices in Southern Africa: A scoping review, *PLoS One*, 17(8), 2022, e0271726.

[6] S. Gorjian, B. Ghobadian, Solar desalination: A sustainable solution to water crisis in Iran, *Renewable and Sustainable Energy Reviews*, 48, 2015, 571–584.

[7] E. Deniz, *Solar-Powered Desalination. Desalination Updates.* InTech, 2015. doi:10.5772/60436.

[8] International Renewable Energy Agency, Water Desalination Using Renewable Energy Technology Brief, IEA-ETSAP and IRENA(c) Technology Brief I12, Copyright © IEA-ETSAP and IRENA 2012.

[9] Al-Mezeini, Saif Salim Saif, Mohd Asim Siddiqui, Mohammad Shariq, Talal M. Althagafi, Inas A. Ahmed, Mohammad Asif, Sultan J. Alsufyani, Saud A. Algarni, Niyaz Ahamed M.B., Khalda M. A. Elamin, and et al. Design and experimental studies on a single slope solar still for water desalination, *Water* 15(4), 2023, 704.

[10] H.R. Lotfy, J. Staš, H. Roubík, Renewable energy powered membrane desalination – Review of recent development, *Environmental Science and Pollution Research International*, 29(31), 2022, 46552–46568. doi:10.1007/s11356-022-20480-y. Erratum in: *Environmental Science and Pollution Research International*, 2022 May 21.

[11] A. Kumar, S. Vyas, D.N. Nkwetta, Experimental study of single slope solar still coupled with parabolic trough collector, *Materials Science for Energy Technologies*, 3, 2020, 700–708.

[12] G. Gopi, M. Premalatha, G. Arthanareeswaran, Transient mathematical modelling and investigation of radiation and design parameters on the performance of multi-effect solar still integrated with evacuated tube collector, *Energy Conversion and Management: X*, 14, 2022, 100210.

[13] K. Yadav, M. Gudjonsdottir, G. Axelsson, M. Ómarsdóttir, A. Sircar, M. Shah, A. Yadav, Geothermal-solar integrated multistage flash distillation cogeneration system: A cleaner and sustainable solution, *Desalination*, 566, 2023, 116897.

[14] A. Al-Othman, M. Tawalbeh, M. El Haj Assad, T. Alkayyali, A. Eisa, Novel multi-stage flash (MSF) desalination plant driven by parabolic trough collectors and a solar pond: A simulation study in UAE, *Desalination*, 443, 2018, 237–244.

[15] R. Kumar, A.K. Shukla, M. Sharma, G. Nandan, Thermodynamic investigation of water generating system through HDH desalination and RO powered by organic Rankine cycle, *Materials Today: Proceedings*, 46(Part 11), 2021, 5256–5261.

[16] M.A. Farahat, H.E.S. Fath, I.I. El-Sharkawy, S. Ookawara, M. Ahmed, Energy/exergy analysis of solar driven mechanical vapor compression desalination system with nano-filtration pretreatment, *Desalination*, 509, 2021, 115078.

[17] T.A. Siddique, N.K. Dutta, N. Roy Choudhury, Nanofiltration for arsenic removal: Challenges, recent developments, and perspectives, *Nanomaterials (Basel)*, 10(7), 2020, 1323. doi:10.3390/nano10071323. PMID: 32640523; PMCID: PMC7407220.

[18] M. Jaszczur, M. Dudek, Z. Kolenda, Thermodynamic analysis of advanced gas turbine combined cycle integration with a high-temperature nuclear reactor and cogeneration unit, *Energies*, 13, 2020, 400.

[19] S.A. Olaleru, J.K. Kirui, D. Wamwangi, K.T. Roro, B. Mwakikunga, Perovskite solar cells: The new epoch in photovoltaics, *Solar Energy*, 196, 2020, 295–309.

[20] E. Akoto, V. Isahi, V. Odari, C. Maghanga, F. Nyongesa, Monolith Cs1-xRbxSnI3 perovskite - silicon 2T tandem solar cell using SCAPS-1D, *Results in Optics*, 12, 2023, 100470.

[21] Z. Kee, Y. Wang, J. Pye, A. Rahbari, Small-scale concentrated solar power system with thermal energy storage: System-level modelling and techno-economic optimisation, *Energy Conversion and Management*, 294, 2023, 117551.

[22] S. Gagliardi, F. Rondino, C. Paoletti, M. Falconieri, On the morphology of nanostructured TiO_2 for energy applications: The shape of the ubiquitous nanomaterial. Nanomaterials, 12, 2022, 2608.

[23] X. Li, G. Zeng, X. Lei, The stability, optical properties and solar-thermal conversion performance of SiC-MWCNTs hybrid nanofluids for the direct absorption solar collector (DASC) application, *Solar Energy Materials and Solar Cells*, 206, 2020, 110323.

[24] Á. de la Puente-Gil, M. de Simón-Martín, A. González-Martínez, A.-M. Diez-Suárez, J.-J. Blanes-Peiró, The Internet of Things for the intelligent management of the heating of a swimming pool by means of smart sensors, *Sensors*, 23, 2023, 2533.

[25] M. Fayyazi, P. Sardar, S.I. Thomas, R. Daghigh, A. Jamali, T. Esch, H. Kemper, R. Langari, H. Khayyam, Artificial intelligence/machine learning in energy management systems, control, and optimization of hydrogen fuel cell vehicles, *Sustainability*, 15, 2023, 5249.

[26] M. Herrando, K. Wang, G. Huang, T. Otanicar, O.B. Mousa, R.A. Agathokleous, Y. Ding, S. Kalogirou, N. Ekins-Daukes, R.A. Taylor, C.N. Markides, A review of solar hybrid photovoltaic-thermal (PV-T) collectors and systems, *Progress in Energy and Combustion Science*, 97, 2023, 101072.

[27] M. Momeni, A.Fartaj, Numerical thermal performance analysis of a PCM-to-air and liquid heat exchanger implementing latent heat thermal energy storage, *Journal of Energy Storage*, 58, 2023, 106363.

[28] A. Naghdloo, E. Ghazimirsaeed, A. Shamloo, Numerical simulation of mixing and heat transfer in an integrated centrifugal microfluidic system for nested-PCR amplification and gene detection, *Sensors and Actuators B: Chemical*, 283, 2019, 831–841.

[29] V.M. Avargani, S. Zendehboudi, M.A. Zamani, Performance evaluation of various nano heat transfer fluids in charging/discharging processes of an indirect solar air heating system, *Energy*, 274, 2023, 127281.

[30] K. Qadeer, A. Al-Hinai, L.F. Chuah, N.R. Sial, A.H. Al-Muhtaseb, R. Al-Abri, M.A. Qyyum, M. Lee, Methanol production and purification via membrane-based technology: Recent advancements, challenges and the way forward, *Chemosphere*, 335, 2023, 139007.

[31] Y.J. Lim, G.S. Lai, Y. Zhao, Y. Ma, J. Torres, R. Wang, A scalable method to fabricate high-performance biomimetic membranes for seawater desalination: Incorporating pillar arene water nanochannels into the polyamide selective layer, *Journal of Membrane Science*, 661, 2022, 120957.

[32] N. Ijjada, R.R. Nayaka, Review on properties of some thermal insulating materials providing more comfort in the building, *Materials Today: Proceedings*, 58(Part 4), 2022, 1354–1359.

[33] World Health Organization & ExpandNet, *Nine Steps for Developing a Scaling-up Strategy*. World Health Organization, Geneva, Switzerland, 2010. https://apps.who.int/iris/handle/10665/44432.

[34] Zhengxuan Liu, Ying Sun, Chaojie Xing, Jia Liu, Yingdong He, Yuekuan Zhou, and Guoqiang Zhang, Artificial intelligence powered large-scale renewable integrations in multi-energy systems for carbon neutrality transition: Challenges and future perspectives, *Energy and AI*, 10, 2022, 100195.

Chapter 20

Atmospheric water generation with different grades of molecular sieve

Manoj Kumar
National Institute of Technology, Srinagar, J&K

Avadhesh Yadav
National Institute of Solar Energy Gurugram, Haryana

Ravi Kant Ravi
Dr. B.R.Ambedkar National Institute of Technology, Punjab

Ved Prakash Sharma
National Institute of Technology, Srinagar, J&K

Venkateswara R. Kode
Christian Brothers University, USA

20.1 INTRODUCTION

Summer in India has become a real challenge for everyone. With this season, water becomes the most precious commodity which can be compared with the importance of gold to a family. India's population is 142.8 crore in 2023, which is 18% of the total population of the world. To cater this much of the population, there is a requirement for huge resources of water. A per the reports of the Indian government think tank, India is facing extreme water stress currently, and in coming years, this situation is going to be more critical. Much research has been done to solve this alarming situation of water, Kumar and Yadav (2015a) conducted experiments on the solar-based experimental setup and confirmed the design parameters in the atmospheric conditions of NIT Kurukshetra, Haryana. Kumar and Yadav (2015b) conducted experiments with $Cacl_2$ and sawdust as composite desiccant material and generated 180 ml/kg/day of water in October 2014. Kumar and Yadav (2016a) conducted experiments with composite desiccant material of $Cacl_2$ and floral foam. The result shows that this composite desiccant material can generate water at the rate of 0.35 ml/cm³/day. The only problem was that this water was confirming to the acidic side of the pH scale. Kumar and Yadav

 DOI: 10.1201/9781003472629-20

(2016 b) performed experiments for the comparison of the water generation capacity of silica gel, activated alumina, and molecular sieve. It is concluded that silica gel performs better as compared to other desiccant materials. Kumar and Yadav (2017), conducted experiments with the help of $Cacl_2$, vermiculite, and sawdust as composite desiccant material to produce water from atmospheric air. It has been found that the proposed composite desiccant material can generate 195 ml/kg/day of water within 8–9 hours. Kumar and Yadav (2019) conducted experiments for producing water by using vermiculite/saw wood/$Cacl_2$. It is noted that the experiments conducted with the proposed innovative composite material can generate 130 ml/kg/day while confirming the freshwater parameters as designated by WHO. Srivastava and Yadav (2018) performed experiments to increase the water generation rate of desiccant materials with the help of a Scheffler reflector. Ibrahim Samir et al. (2023) conducted experiments with desiccant material (silica gel) and saw duct. It has been found that silica gel has more capacity to generate water as compared to sawdust. Karmakar et al. (2020) performed experiments in southern Indian climatic conditions with silica gel. It has been found that desiccant material can produce 98 ml/kg/day of water. Kumar et al. (2022) noted that molecular sieves are useful when there is a requirement to dry a space or package without compromising its properties.

Many researchers have made significant contributions to water generation technology. But none of the researchers has compared the three different grades of molecular sieve i.e., 4A, 5A, and 13 X for the water generation capacity which is considered to be the most prominent desiccant material for adsorption of moisture. The manuscript has been written to present the experimental results conducted at NIT Kurukshetra, Haryana, India.

20.2 EXPERIMENTAL SETUP

The experimental setup consists of SGDBS made up of fiber-reinforced plastic (FRP) material, as shown in the figure. The reason for using FRP is its durability in atmospheric conditions.

The SGDBS has the following components:

1. **FRP Container:** The three containers of size 0.6 m×0.6 m×0.3 m with side window size 0.3 m×0.3 m have been used because of their durability in a wide range of seasonal conditions in northern India, as shown in Figures 20.1 and 20.2. On the front side of the container, a slope has been provided for the smooth collection of water after condensation.
2. **Glazing:** A glazing of 3 mm glass thickness has been taken as per the design recommendation of Kumar and Yadav (2015a). This glass creates a greenhouse-like effect inside the container, which collects the Sun rays having shorter wavelengths and traps the longer wavelength.

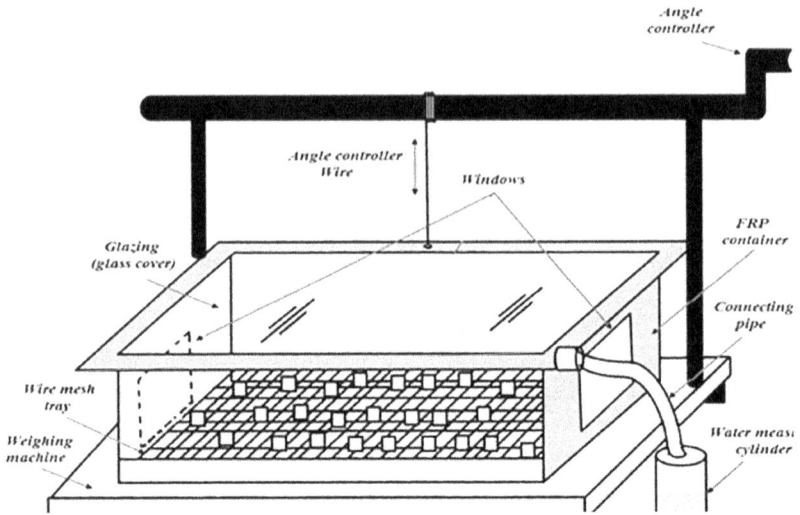

Figure 20.1 Schematic diagram for solar glass desiccant-based system (SGDBS).

3. **Wire Mesh Tray:** A plastic tray having a wire mesh size of 3 mm×3 mm to hold the molecular sieves of different sizes has been placed inside the SGDBS as shown in Figures 20.1 and 20.2.
4. **Water Measuring Cylinder:** For the collection of water, a measuring container with an outside measuring marking, has been used. This is used to collect the condensed water from the collecting tray.
5. **Desiccant Material:** Molecular sieve of size 4A, 5A, and 13x has been used as desiccant material. 1 kg quantity of molecular sieve has been used for carrying out the experiments. The desiccant material has been purchased from SORBEAD INDIA Pvt. Ltd., INDIA. The material properties of the desiccant material with all necessary parameters are shown in Table 20.1.

20.3 MEASURING INSTRUMENTS AND DEVICES

During the days of the experiments, the following instruments and devices have been used for the collection of data:

1. RTD PT100 thermocouples with a resolution of 0.1, have been used to measure the temperature at different locations of the experimental setup.
2. A sling Psychrometer has been used for the measurement of wet bulb temperature and dry bulb temperature of atmospheric air.

(a)

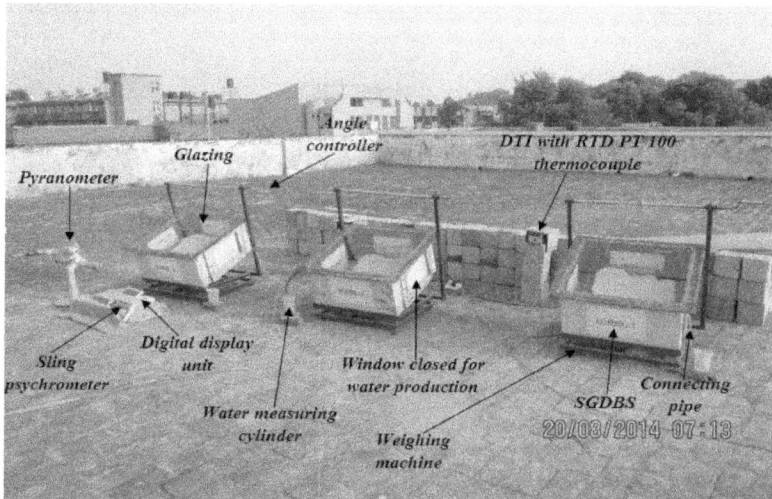

(b)

Figure 20.2 Experimental setup (a) adsorption process and (b) regeneration process.

3. A Pyranometer has been used for the measurement of solar intensity.
4. A make of "THOMSON Electronic weighing scale" with a resolution of 0.0001 kg, has been used for the measurement of adsorption and regeneration rate.

The experimental work has been carried out in August. The process of regeneration is carried out in day time whereas the adsorption is carried out in the nighttime.

Table 20.1 Properties of molecular sieves

Product	Molecular sieve 4A (beads) mm	Molecular sieve 5A (beads) mm	Molecular sieve 13 X (beads) mm
Size	1.6–2.7	1.6–2.7	1.6–2.7
Pore diameter	4 Å	5 Å	10 Å
Loss on attrition (% wt)	0.1	0.24	0.1
Bulk density (g/mL)	0.735	0.677	0.674
Adsorption capacity for CO_2 (% wt)	18.54	18.54	19.50
Crushing strength (N)	30	31	45
Adsorption capacity (%)	14.97% @ 60% RH	21.74% @ 80% RH	32% @ 100% RH
Heat of adsorption	1100 kcal/kg of water adsorbed	1100 kcal/kg of water adsorbed	1100 kcal/kg of water adsorbed

20.4 SYSTEM OPERATION

For the start of the experiment, the molecular sieves are kept on the tray (wire mesh). The adsorption process starts when the side windows of SGDBS are opened at 1900 hours. This process continues till the saturation of the moisture inside the solid desiccant material. On the next day at 0600 hours, the side windows are closed and the regeneration process starts after the rise of the sun. The solar radiations of shorter wavelength pass through the glass and enter the SGDBS. After falling on the surface of the desiccant material it raises its temperature. It gets converted to a longer wavelength, which remains trapped inside the SGDBS and creates a greenhouse-like effect. Due to this effect, desiccant material releases moisture in the form of vapors. This moisture starts to condense on the inner side of the glass and forms a faction of very small drops. These drops slide along the slope of the glass and reach the water-collecting tray and finally go to the water-collecting container through the connecting pipe. The collected water quantity was monitored after every 30 minutes on the experimental day.

20.4.1 Analysis of experimental data

The adsorption rate for the desiccant material is given by [14]

$$GA = m_{ws} \frac{dw}{dt} \qquad (20.1)$$

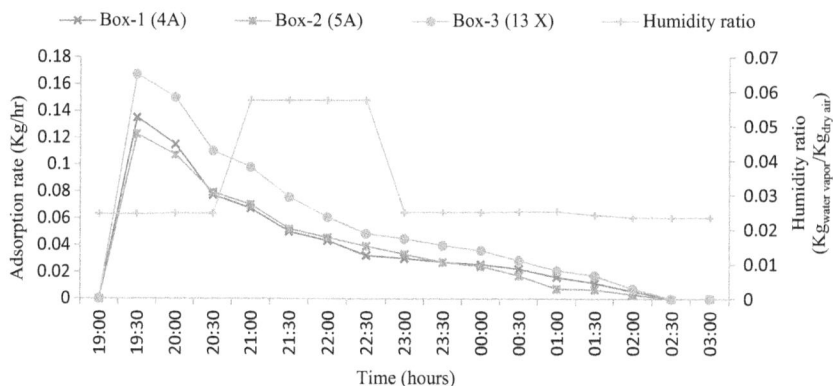

Figure 20.3 Variation of adsorption rate and humidity ratio with time for desiccant material.

20.5 RESULTS AND DISCUSSION

Case 1: Adsorption Rate of Molecular Sieves

In this case, the aim was to observe the adsorption rate for different grades of molecular sieves i.e., 4A, 5A, and 13X. Three different boxes with molecular sieves of different grades have been taken, which are notated as box 1 (4A), box 2 (5A), and box 3 (13X) in Figure 20.3.

The adsorption process started at 19:00 hours. At the start of the experiment, the maximum adsorption rate was 0.132328 kg/h for box 1, 0.124888 kg/hour for box 2, and 0.168020 kg/h for box 3. The reason for the adsorption rate is the emptiness of pores of molecular sieves which decreases with the progression of time. In box 3 high adsorption rate is because of its high pore diameter and slightly low bulk density as that of box 1 and box 2. All three boxes reached the saturation of moisture at 0300 hours.

Case 2: Temperature Distribution

a. **Variation in Solar Intensity and Desiccant Material Temperature with Time**

Figure 20.4 shows that molecular sieves 4A, 5A, and 13X gained maximum temperatures of 87.4°C, 87.6°C, and 86.8°C respectively. The desiccant material temperature in the three boxes is more or less the same. This is because the heat of adsorption of material is the same for given molecular sieves. Also, the design parameter of all three boxes (which are responsible for heating the material by solar radiation) is the same.

b. **Variation in Solar Intensity and Inner Space Temperature with Time**

The maximum temperature accumulated in box 3 as compared to the other two boxes. This is because due to the higher regeneration

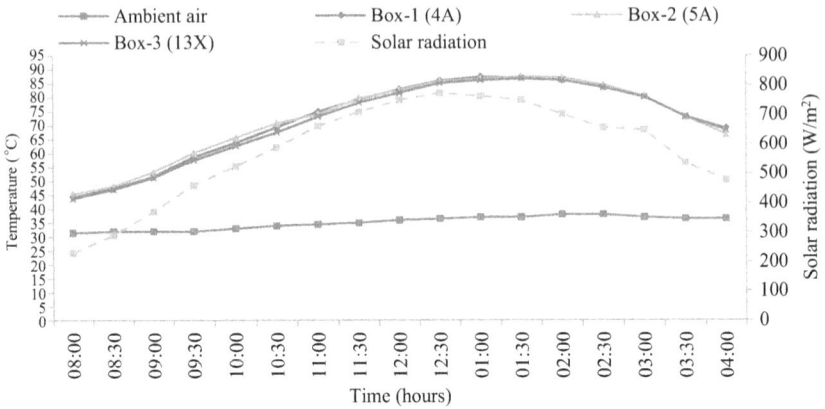

Figure 20.4 Variation in solar intensity and temperature with time.

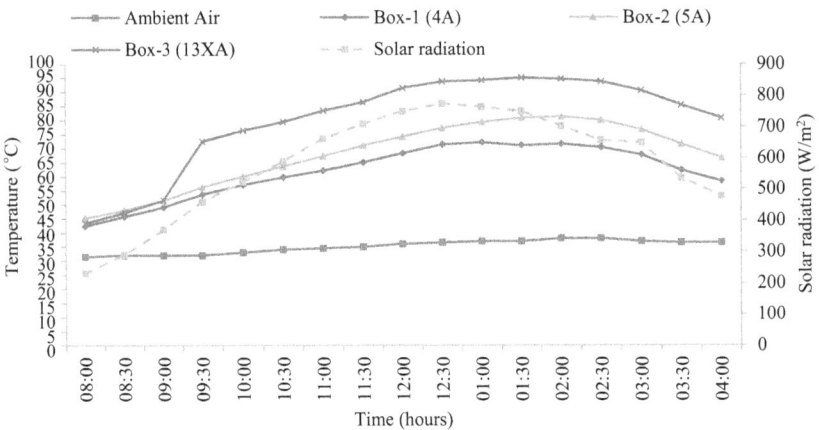

Figure 20.5 Variation in solar radiation and internal space temperature with time.

rate, the accumulation of vapor is greater which reduces the loss of heat to the surrounding by radiation. The maximum temperature inside box 3, box 2, and box 1 is 95.1°C, 81.3°C, and 71.6°C, respectively, as shown in Figure 20.5.

c. **Variation in Solar Intensity and Inner Glass Temperature with Time**

Figure 20.6 depicts the variation in solar intensity and inner glass temperature with time. It has been observed that box 3 has a higher inner glass temperature i.e., 74.4°C at 1300 hours as that of box 1 and box 2. This is because of the release of heat by desiccant material as it is exposed to the sun. In the case of box 2 and box 3, the variation of inner glass temperature is almost the same which is 66.7°C.

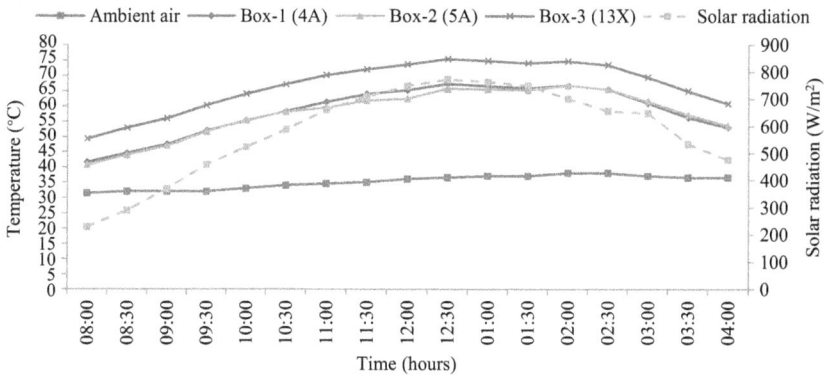

Figure 20.6 Variation of solar radiation and inner glass temperature with time.

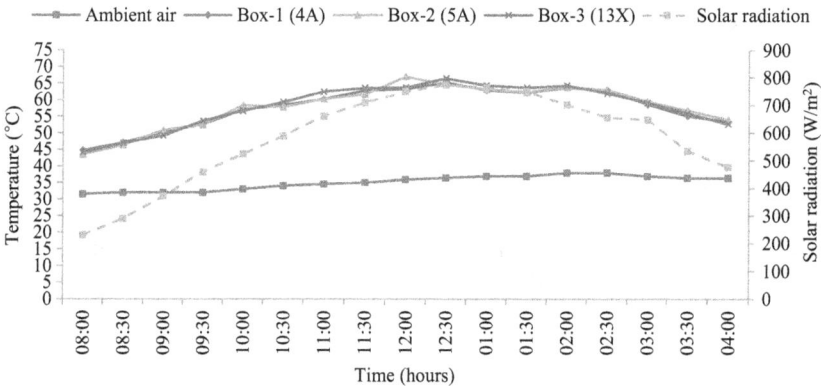

Figure 20.7 Variation of solar radiation and outer glass temperature with time.

d. Variation in Solar Intensity and Outer Glass Temperature with Time
Figure 20.7 depicts that the outer glass temperature of all three boxes is almost the same because of surface heating phenomena. All three boxes gained maximum temperatures of 66.4°C, 64.5°C, and 65.1°C for box 3, box 2, and box 1, respectively.

Case 3: Water Generated

The water generation rate of box 3 is higher at all times during the days of the experiment. This is because the absorption rate is higher for desiccant material in box 3. The maximum water generated by box 3, box 2, and box 1 is 40, 30, and 25 ml/kg/day, respectively, as shown in Figure 20.8.

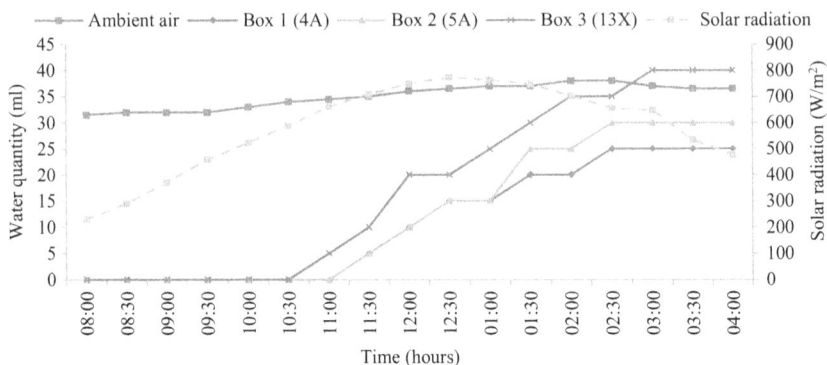

Figure 20.8 Variation in solar intensity and water generation capacity by three boxes with time.

20.6 EXPERIMENTAL ERROR

The experimental error gives the difference between true and measured values. This is very common and to some extent, it can not be avoided and thus can't be seen as a mistake. However, by understanding the different ways and types of experimental error, its impact on the measured/recorded results can be reduced. Since this manuscript records the experimental results and thus it is necessary to calculate the experimental error for measured and recorded data.

Percentage error in measuring:

1. temperatures $= 0.3\%$
2. Intensity $= 2\%$
3. water quantity $= 0.025\%$
4. Total error for experimental setup $=$ (Total number of thermocouples % error in measuring temperature) + (No. of pyranometer × Percentage error in measuring intensity) + (No. of measuring container × Percentage error in measuring water quantity) $= 4 \times 0.3 + 1 \times 2 + 1 \times 0.025 = 3.225\%$
5. weighing machine $= 0.1\%$

20.7 CONCLUSIONS

The following conclusions can be drawn for the experimental setup that uses using molecular sieve as desiccant material for the generation of water from atmospheric air:

1. The experimental results show that molecular sieves of different grades can be used for atmospheric water generation.
2. The result for the maximum quantity of water generation shows that molecular sieves of grade 13X can generate 1.33 times water as compared to that of molecular sieves of grade 5A and 1.6 times molecular sieves of grade 4A.
3. The water adsorption capacity of the molecular sieve is very high, but for complete regeneration, a high temperature is required, which is not achieved by SGDBS, as it is suitable for the regeneration of medium-temperature desiccant materials.

NOMENCLATURE

GA   adsorption rate (kg/hour)
mws weight of desiccant on a wet basis (kg)
w   moisture content in desiccant ($kg_{water\ vapor}/kg_{desiccant}$)
Y   annual yield

REFERENCES

Ibrahim Samir, A., Al-Samari, A., Banu Arshi, P.S., and Mahbubul, L.M., "The contribution of adsorbent materials (silica gel and sawdust) in removing water vapor: A case study in Iraq", *Diyala Journal of Engineering Sciences*, Vol. 16, no. 1, pp. 103–114, 2023.

Karmakar, A., Mileo, P. G. M., Bok, I., Peh, S. B., Zhang, J., Yuan, H., Maurin, G., and Zhao, D., "Thermo-responsive MOF/polymer composites for temperature-mediated water capture and release", *Angewandte Chemie International Edition*, Vol. 59, no. 27, pp. 11003–11009, 2020.

Kumar, M., Mehla, N., Srivastava, S., and Ravi, K.R., "Water generation from atmospheric air by using desiccant material- nature-based solution – A review", *World Journal of Engineering*, Vol. 20, no. 5, pp. 956–973, 2022.

Kumar, M. and Yadav, A., "Experimental investigation of design parameters of solar glass desiccant box type system for water production from atmospheric air", *Journal of Renewable and Sustainable Energy*, Vol. 7, p. 033122, 2015a.

Kumar, M. and Yadav, A., "Experimental investigation of solar-powered water production from atmospheric air by using desiccant material 'CaCl₂/saw wood'", *Desalination*, Vol. 367, pp. 216–222, 2015b.

Kumar, M. and Yadav, A., " Solar-driven technology for freshwater production from atmospheric air by using the composite desiccant material 'CaCl₂/floral foam'", *Environment Development Sustainability*, Vol. 18, pp. 1151–1165, 2016a.

Kumar, M. and Yadav, A., "Comparative study of solar-powered water production from atmospheric air using different desiccant materials", *International Journal of Sustainable Engineering*, Vol. 9, pp. 390–400, 2016b.

Kumar, M. and Yadav, A., "Composite desiccant material 'CaCl₂/vermiculite/saw wood': A new material for freshwater production from atmospheric air", *Applied Water Science*, Vol. 7, pp. 2103–2111, 2017.

Kumar, M. and Yadav, A., "Water generation from atmospheric air by using different composite desiccant material", *International Journal of Ambient Energy*, Vol. 40, pp. 343–349, 2019.

Srivastava, S. and Yadav, A., " Water generation from atmospheric air by using composite desiccant material through fixed concentrating solar thermal power", *Solar Energy*, Vol. 169, pp. 302–315, 2018.

Index

For Product Safety Concerns and Information please contact our EU
representative GPSR@taylorandfrancis.com
Taylor & Francis Verlag GmbH, Kaufingerstraße 24, 80331 München, Germany

www.ingramcontent.com/pod-product-compliance
Lightning Source LLC
Chambersburg PA
CBHW060757220326
41598CB00022B/2471

9 7 8 1 0 3 2 7 5 3 3 7 9